Translational Aspects of Cardiovascular Biology: From Bench to Bedside

Translational Aspects of Cardiovascular Biology: From Bench to Bedside

Editor

Gaetano Santulli

MDPI • Basel • Beijing • Wuhan • Barcelona • Belgrade • Manchester • Tokyo • Cluj • Tianjin

Editor
Gaetano Santulli
Medicine (Division of Cardiology)
Albert Einstein College of Medicine
New York City
United States

Editorial Office
MDPI
St. Alban-Anlage 66
4052 Basel, Switzerland

This is a reprint of articles from the Special Issue published online in the open access journal *Biology* (ISSN 2079-7737) (available at: www.mdpi.com/journal/biology/special_issues/Cardiovascular_Biology).

For citation purposes, cite each article independently as indicated on the article page online and as indicated below:

LastName, A.A.; LastName, B.B.; LastName, C.C. Article Title. *Journal Name* **Year**, *Volume Number*, Page Range.

ISBN 978-3-0365-7575-9 (Hbk)
ISBN 978-3-0365-7574-2 (PDF)

Cover image courtesy of Gaetano Santulli

Contents

About the Editor

Gaetano Santulli

A physician-scientist, Prof. Dr. Gaetano Santulli, MD, PhD completed his postdoctoral training at Columbia University Medical Center, supported by an American Heart Association (AHA) fellowship and a Scientist Development Grant. His expertise comprises both clinical and basic research topics, including hypertension, diabetes, heart failure, arrhythmias, and vascular disorders. He published more than 200 peer-reviewed papers on topics spanning from calcium to microRNA and mitochondrial pathophysiology.

He is the Editor-in-Chief of *Frontiers in Cardiovascular Endocrinology* and serves on the editorial boards of several medical journals, including *European Journal of Preventive Cardiology, Cardiovascular Diabetology, Hypertension, BMC Medicine, Cardiovascular Research, Atherosclerosis, ESC Heart Failure, Nutrients, Molecular Medicine, European Heart Journal Cardiovascular Pharmacotherapy, Journal of Molecular and Cellular Cardiology, Cell Signalling, Non-coding RNA, Cellular and Molecular Life Sciences, Scientific Reports*, and *PLoS One*. His NIH-funded laboratory is currently studying the molecular mechanisms linking alterations in intracellular calcium fluxes to cardiometabollic disease.

Preface to "Translational Aspects of Cardiovascular Biology: From Bench to Bedside"

Cardiovascular disease remains the primary cause of death worldwide, and the search for novel mechanisms and therapeutics is desperately needed. Therefore, basic and translational studies in the cardiovascular field represent the best strategy to identify novel therapeutic targets and improve the quality of life of patients with cardiovascular disorders. On these grounds, this book has the main objective of gathering peer-reviewed scientific papers with a robust basic research setting and/or translational potential, exploring novel molecular and cellular mechanisms underlying cardiovascular disorders, as well as pre-clinical and clinical investigations testing new hypotheses that can advance the biomedical field.

Gaetano Santulli

Editor

biology

MDPI

Editorial

Translational Aspects of Cardiovascular Biology: From Bench to Bedside

Gaetano Santulli [1,2,3]

1 Division of Cardiology, Department of Medicine, Wilf Family Cardiovascular Research Institute, Fleischer Institute for Diabetes and Metabolism (*FIDAM*), Einstein Institute for Neuroimmunology and Inflammation (*INI*), Albert Einstein College of Medicine, New York, NY 10461, USA; gsantulli001@gmail.com
2 Department of Advanced Biomedical Sciences, International Translational Research and Medical Education (*ITME*) Consortium, "*Federico II*" University, 80131 Naples, Italy
3 Department of Molecular Pharmacology, Einstein-Mount Sinai Diabetes Research Center (*ES-DRC*), Einstein Institute for Aging Research, Albert Einstein College of Medicine, New York, NY 10461, USA

Cardiovascular disease is the leading cause of death worldwide, and the search for novel mechanisms and therapeutics is desperately needed. Therefore, basic and translational studies in the cardiovascular field represent the best strategy to identify novel therapeutic targets, as well as to improve the quality of life of patients with cardiovascular disorders (Figure 1). This Editorial introduces the Special Issue "Translational Aspects of Cardiovascular Biology: From Bench to Bedside", published in the Journal *Biology*. This Special Issue gather peer-reviewed scientific papers, which deal with cardiovascular medicine, from basic science to pre-clinical and clinical investigations, highlighting the crucial importance of translational medicine in testing groundbreaking hypotheses to advance the biomedical field.

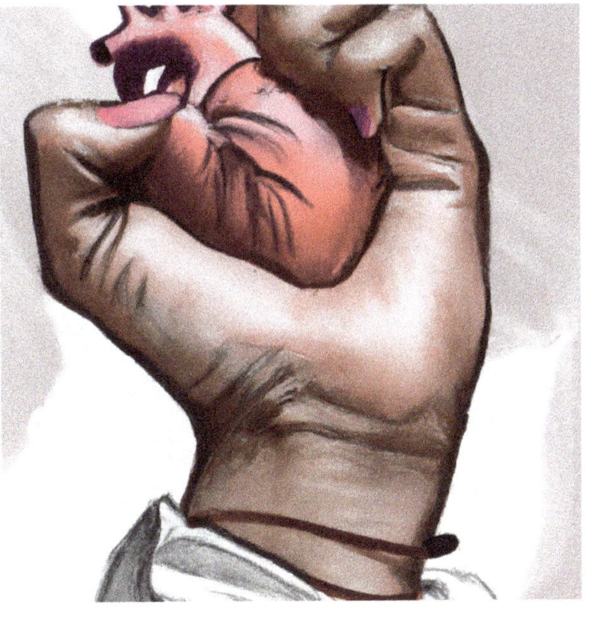

Figure 1. Artistic representation of the importance of translational medicine in the challenging quest for new therapies to cure cardiovascular disease.

Citation: Santulli, G. Translational Aspects of Cardiovascular Biology: From Bench to Bedside. *Biology* **2023**, *12*, 658. https://doi.org/10.3390/biology12050658

Received: 21 April 2023
Accepted: 23 April 2023
Published: 27 April 2023

The Special Issue opens with three basic research papers, which investigate endothelial cells. The first article shows that a specific non-coding RNA, microRNA-4432, is able to specifically target the gene encoding for fibroblast growth factor binding protein 1 (FGFBP1) in human brain microvascular endothelial cells, and this paper demonstrates that this microRNA significantly reduces endothelial oxidative stress, a well-established feature of hypertension [1]. The second study provides a demonstration of the extracellular metabolism of nicotinamide adenine dinucleotide (NAD+) and nicotinamide mononucleotide (NMN), multifunctional metabolites involved in a number of cellular processes, and which is vastly different in its expression in the vascular endothelium obtained from different species and locations [2]. The third paper, instead, deals with the functions and intracellular mechanisms of endothelial cells in human liver grafts [3].

Subsequently, a couple of in vitro investigations provide new insights on how several anticancer drugs, including Paclitaxel, Carboplatin, Doxorubicin, and Cyclophosphamide, can alter the biophysical characteristics of red blood cells [4]. These studies also examine why inhibiting isocitrate dehydrogenase 1 (IDH1) may help prevent foam cell formation by reducing oxidized low-density lipoprotein-induced ferroptosis in macrophages [5]. The following two papers harness two different animal models, mouse and swine, respectively, to show that dextran sodium sulfate-induced chronic colitis has no significant impact on femoral artery endothelial function or ischemic limb recovery [6]. Additionally, after myocardial infarction, neuregulin (NRG-1β) exhibits pro-myogenic and anti-cachectic actions in respiratory muscles [7]. In two elegant reviews, Bouhamida and colleagues explained the main effects of the interplay between hypoxia signaling on mitochondrial dysfunction and inflammation in cardiovascular diseases and cancer [8], while Nie and collaborators summarized the therapeutic potential of long non-coding RNAs (lncRNAs) in cardiac fibrosis [9].

The last section of the Special Issue includes a series of clinical studies, investigating tissue and serum biomarkers in degenerative aortic stenosis [10], identifying reliable biomarkers in obesity-related atrial fibrillation [11], establishing retinol-binding protein-4 as a predictor of insulin resistance in patients with coronary artery disease and type 2 diabetes mellitus [12], and verifying the correlation between carbonic anhydrase isozymes and the evolution of myocardial infarction in diabetic patients [13].

A double-blind, randomized, placebo-controlled clinical trial concludes the Special Issue, showing that vaccination with the pneumococcal vaccine *Prevenar-13* does not result in IgM against oxidized low-density lipoproteins, in contrast with previous findings in rodents [14].

Acknowledgments: We thank all the contributors to this Special Issue.

Conflicts of Interest: The author declares no conflict of interest.

References

1. Avvisato, R.; Mone, P.; Jankauskas, S.S.; Varzideh, F.; Kansakar, U.; Gambardella, J.; De Luca, A.; Matarese, A.; Santulli, G. miR-4432 Targets FGFBP1 in Human Endothelial Cells. *Biology* **2023**, *12*, 459. [CrossRef] [PubMed]
2. Jablonska, P.; Mierzejewska, P.; Tomczyk, M.; Koszalka, P.; Franczak, M.; Kawecka, A.; Kutryb-Zajac, B.; Braczko, A.; Smolenski, R.T.; Slominska, E.M. Differences in Extracellular NAD(+) and NMN Metabolism on the Surface of Vascular Endothelial Cells. *Biology* **2022**, *11*, 675. [CrossRef]
3. Vasuri, F.; Germinario, G.; Ciavarella, C.; Carroli, M.; Motta, I.; Valente, S.; Cescon, M.; D'Errico, A.; Pasquinelli, G.; Ravaioli, M. Trophism and Homeostasis of Liver Sinusoidal Endothelial Graft Cells during Preservation, with and without Hypothermic Oxygenated Perfusion. *Biology* **2022**, *11*, 1329. [CrossRef]
4. Skverchinskaya, E.; Levdarovich, N.; Ivanov, A.; Mindukshev, I.; Bukatin, A. Anticancer Drugs Paclitaxel, Carboplatin, Doxorubicin, and Cyclophosphamide Alter the Biophysical Characteristics of Red Blood Cells, In Vitro. *Biology* **2023**, *12*, 230. [CrossRef] [PubMed]
5. Li, B.; Wang, C.; Lu, P.; Ji, Y.; Wang, X.; Liu, C.; Lu, X.; Xu, X.; Wang, X. IDH1 Promotes Foam Cell Formation by Aggravating Macrophage Ferroptosis. *Biology* **2022**, *11*, 1392. [CrossRef] [PubMed]

6. Wu, H.; Zhu, Q.; Liu, X.; Hao, H.; Sun, Z.; Wang, M.; Hill, M.A.; Xu, C.; Liu, Z. Recovery of Ischemic Limb and Femoral Artery Endothelial Function Are Preserved in Mice with Dextran Sodium Sulfate-Induced Chronic Colitis. *Biology* **2022**, *11*, 1169. [CrossRef] [PubMed]
7. Galindo, C.L.; Nguyen, V.T.; Hill, B.; Easterday, E.; Cleator, J.H.; Sawyer, D.B. Neuregulin (NRG-1beta) Is Pro-Myogenic and Anti-Cachectic in Respiratory Muscles of Post-Myocardial Infarcted Swine. *Biology* **2022**, *11*, 682. [CrossRef] [PubMed]
8. Bouhamida, E.; Morciano, G.; Perrone, M.; Kahsay, A.E.; Della Sala, M.; Wieckowski, M.R.; Fiorica, F.; Pinton, P.; Giorgi, C.; Patergnani, S. The Interplay of Hypoxia Signaling on Mitochondrial Dysfunction and Inflammation in Cardiovascular Diseases and Cancer: From Molecular Mechanisms to Therapeutic Approaches. *Biology* **2022**, *11*, 300. [CrossRef]
9. Nie, X.; Fan, J.; Wang, D.W. The Function and Therapeutic Potential of lncRNAs in Cardiac Fibrosis. *Biology* **2023**, *12*, 154. [CrossRef]
10. Kapelouzou, A.; Geronikolou, S.; Lidoriki, I.; Kontogiannis, C.; Kaklamanis, L.; Tsourelis, L.; Cokkinos, D.V. Tissue and Serum Biomarkers in Degenerative Aortic Stenosis-Insights into Pathogenesis, Prevention and Therapy. *Biology* **2023**, *12*, 347. [CrossRef] [PubMed]
11. Xie, Z.; Liu, C.; Lu, X.; Chen, Z.; Zhang, N.; Wang, X.; Li, X.; Li, Y. Identification and Verification of Biomarkers and Immune Infiltration in Obesity-Related Atrial Fibrillation. *Biology* **2023**, *12*, 121. [CrossRef]
12. Perumalsamy, S.; Ahmad, W.A.W.; Huri, H.Z. Retinol-Binding Protein-4-A Predictor of Insulin Resistance and the Severity of Coronary Artery Disease in Type 2 Diabetes Patients with Coronary Artery Disease. *Biology* **2021**, *10*, 858. [CrossRef] [PubMed]
13. Magheru, S.; Magheru, C.; Maghiar, F.; Sachelarie, L.; Marc, F.; Moldovan, C.M.; Romila, L.; Hoza, A.; Farcas, D.M.; Gradinaru, I.; et al. Correlation between Carbonic Anhydrase Isozymes and the Evolution of Myocardial Infarction in Diabetic Patients. *Biology* **2022**, *11*, 1189. [CrossRef]
14. Grievink, H.W.; Gal, P.; Ozsvar Kozma, M.; Klaassen, E.S.; Kuiper, J.; Burggraaf, J.; Binder, C.J.; Moerland, M. The Effect of a 13-Valent Conjugate Pneumococcal Vaccine on Circulating Antibodies against Oxidized LDL and Phosphorylcholine in Man, a Randomized Placebo-Controlled Clinical Trial. *Biology* **2020**, *9*, 345. [CrossRef]

biology

Communication

miR-4432 Targets FGFBP1 in Human Endothelial Cells

Roberta Avvisato [1,2,3,†], Pasquale Mone [1,2,†], Stanislovas S. Jankauskas [1,2], Fahimeh Varzideh [1,2], Urna Kansakar [1,2], Jessica Gambardella [1,2,3], Antonio De Luca [4], Alessandro Matarese [5] and Gaetano Santulli [1,2,3,6,7,8,*]

1 Division of Cardiology, Department of Medicine, Albert Einstein College of Medicine, New York, NY 10461, USA
2 Wilf Family Cardiovascular Research Institute, Albert Einstein College of Medicine, New York, NY 10461, USA
3 Department of Advanced Biomedical Sciences, *"Federico II"* University, 80131 Naples, Italy
4 Department of Mental and Physical Health and Preventive Medicine, University of Campania *"Luigi Vanvitelli"*, 81100 Caserta, Italy
5 *"Antonio Cardarelli"* Hospital, 80100 Naples, Italy
6 Department of Molecular Pharmacology, Albert Einstein College of Medicine, New York, NY 10461, USA
7 Fleischer Institute for Diabetes and Metabolism (*FIDAM*), New York, NY 10461, USA
8 Einstein-Mount Sinai Diabetes Research Center (*ES-DRC*), Albert Einstein College of Medicine, New York, NY 10461, USA
* Correspondence: gaetano.santulli@unina.it
† These authors contributed equally to this work.

Simple Summary: The inner layer of blood vessels is formed by endothelial cells. When these cells do not work properly, several issues ensue in the human body. One of these issues is elevated blood pressure, also known as hypertension, which is an established risk factor for ischemic heart disease, stroke, chronic kidney disease, and dementia. However, the exact mechanisms linking dysfunctional endothelium and hypertension are not fully defined. In this work, we discovered that a small nucleic acid (miR-4432) is able to target and inhibit a specific gene (fibroblast growth factor binding protein 1, *FGFBP1*) in human brain microvascular endothelial cells, and we demonstrate for the first time that this miR-4432 significantly reduces endothelial oxidative stress, a well-established feature of hypertension. Taken together, our findings provide unprecedented mechanistic insights and open the field to new studies aimed at ameliorating endothelial dysfunction by harnessing miR-4432-based strategies.

Abstract: MicroRNAs (miRs) are small non-coding RNAs that modulate the expression of several target genes. Fibroblast growth factor binding protein 1 (FGFBP1) has been associated with endothelial dysfunction at the level of the blood–brain barrier (BBB). However, the underlying mechanisms are mostly unknown and there are no studies investigating the relationship between miRs and FGFBP1. Thus, the overarching aim of the present study was to identify and validate which miR can specifically target FGFBP1 in human brain microvascular endothelial cells, which represent the best in vitro model of the BBB. We were able to identify and validate miR-4432 as a fundamental modulator of FGFBP1 and we demonstrated that miR-4432 significantly reduces mitochondrial oxidative stress, a well-established pathophysiological hallmark of hypertension.

Keywords: blood–brain barrier; blood pressure; cerebrovascular disease; endothelial dysfunction; hBMECs; hypertension; HUVEC; microRNA; miRNA; miR-4432-3p

Citation: Avvisato, R.; Mone, P.; Jankauskas, S.S.; Varzideh, F.; Kansakar, U.; Gambardella, J.; De Luca, A.; Matarese, A.; Santulli, G. miR-4432 Targets FGFBP1 in Human Endothelial Cells. *Biology* **2023**, *12*, 459. https://doi.org/10.3390/biology12030459

Academic Editor: Matthias S. Leisegang

Received: 14 February 2023
Revised: 9 March 2023
Accepted: 13 March 2023
Published: 16 March 2023

1. Introduction

Hypertension is a leading risk factor for ischemic heart disease, stroke, chronic kidney disease, and dementia [1]. It is a multifactorial disease involving interactions among genetic, environmental, demographic, vascular, and neuroendocrine factors [2,3]. Endothelial dysfunction is an established hallmark of hypertension [4–6]; however, the exact molecular mechanisms linking dysfunctional endothelial cells (ECs) and high blood pressure are not fully understood. Several genome-wide association studies (GWAS) have identified a number of genes associated with hypertension [7,8], but only a few of these genes have been functionally validated. In 2019, the International Consortium of Antihypertensive Pharmacogenomics Studies (ICAPS) recognized fibroblast growth factor binding protein 1 (FGFBP1) as one of the genes involved in the regulation of blood pressure [9]. FGFBP1 is a key promoter of the development of the blood–brain barrier (BBB) [10], an aspect that is especially relevant considering that ECs are a major component of the BBB [11], which is crucial for maintaining neuronal and glial function [12]. Specifically, FGFBP1 has been implied in refining and maintaining barrier characteristics in the mature BBB endothelium [13].

MicroRNAs (miRs) are a relatively well conserved group of small (~21 nucleotides) non-coding RNAs that modulate the expression of their target genes: miRNAs can bind the 3′ untranslated region (UTR) of specific genes, thereby inhibiting their expression. Thus, miRNAs have been involved in numerous pathological and physiological processes [14,15]. Others and ourselves have, in the last decades, identified a variety of miRs involved in the regulation of endothelial function [16,17].

Since FGFBP1 has been previously linked to the modulation of the BBB, and precisely to endothelial function, the central scope of the present study is to detect which miR can target FGFBP1 in hBMECs (human brain microvascular endothelial cells).

2. Results

2.1. miR-4432 Targets FGFBP1 in a Conservative Manner

We applied bioinformatic analyses and functional experiments which led, for the first time to our knowledge, to the identification of hsa-miR-4432-3p (miR-4432) as a crucial modulator of FGFBP1 transcription, in a manner that is highly conserved across different species, including primates such as chimpanzee (*Pan troglodytes*), orangutan (*Pongo abelii*), macaque (*Macaca mulatta*), and gorilla (*Gorilla gorilla*), although it is not detected in mouse (*Mus musculus*) and rat (*Rattus norvegicus*), as shown in Figure 1.

SEQUENCE (5′-3′)	*SPECIES*
AGC GA- --- -AG A-- -G- --- -UC UUU GAG CU ---	HUMAN
AGC GA- --- -AG A-- -G- --- -UC UUU GAG CU ---	CHIMP
AGC GA- --- -AG A-- -G- --- -UC UUU GAG CU ---	GORILLA
AGC GA- --- -AG A-- -G- --- -UC UUU GAG CU ---	ORANGUTAN
AGC GA- --- -AG A-- -G- --- -UC UUU GAG CU ---	MACAQUE
AGC GA- --- -AG A-- -G- --- -UC UUU GAG CU ---	BABOON
AGC GA- --- -AG A-- -G- --- -UC UUU GAG CU ---	HORSE
AGC GA- --- -AG A-- -G- --- -UC UUU GAG CU ---	DOG
AGU GA- --- -GG UC- --- --- -UC UUU GAG CU ---	PANDA
AG- --- --- --- --- --- --- --- --- --- -U ---	MOUSE
AG- --- --- --- --- --- --- --- --- --- -U ---	RAT
AAU GG- --- -AG AUC UG- --- -UC CUC CAG GU ---	GUINEA PIG
AGU AA- --- -GU AC- --- --- -UC UUG GAG CU ---	PIG
AGG GA- --- -GG GCC UG- --- -UC UUU GAG CU ---	ELEPHANT
AGU AA- --- -UC CCC CUG CCC CUC UUU GAG CU ---	MACAW
CGU GA- --- -GG GCU GG- --- -UC GUU GAG -U ---	ARMADILLO
AAC AA- --- -AG UCU UC- --- -UC UUU GAG CU ---	PLATYPUS
AGC CAA GUC CUU UCU UUG UUU CUC UUU GAG CU ---	ALLIGATOR
AGU AA- --- -GG AC- --- --- -UC UUU GAG CU ---	DOLPHIN
AGU AA- --- -GG AC- --- --- -UC UUU GAG CU ---	WHALE
250 ... ...260.. ... ..3′	

Figure 1. Identification of miR-4432 as a specific modulator of FGFBP1; the complementary nucleotides between the target region of FGFBP1 3′-UTR and hsa-miR-4432-3p, highlighted in yellow, are conserved across a number of different species, including primates.

Furthermore, we designed a mutant construct of FGFBP1 3′-UTR ("FGFBP1 MUT") that harbors nucleotide substitutions at the level of the miR-4432 binding sites of FGFBP1 3′-UTR, as illustrated in Figure 2.

```
5' AGCGAAGAGUCUUU FGFBP1 3'UTR
          |||||||
3' GAACGUCUCAGAAA miR-4432

5' AGCGAACACUGUUU FGFBP1 MUT
```

Figure 2. The designed mutant construct of FGFBP1 3′-UTR (FGFBP1 MUT) that harbors nucleotide substitutions at the level of the miR-4432 binding sites (indicated in red) of FGFBP1 3′-UTR, proving that miR-4432 specifically targets the 3′UTR of FGFBP1.

2.2. *miR-4432 Regulates FGFBP1 Transcription in Endothelial Cells*

We first verified that miR-4432 is actually expressed in two different types of endothelial cells, namely hBMECs, which remain the best in vitro model of the BBB [18], and human umbilical vascular endothelial cells (HUVECs), and that its expression is regulated by miR-4432 mimic and miR-4432 inhibitor, as shown in Figure 3.

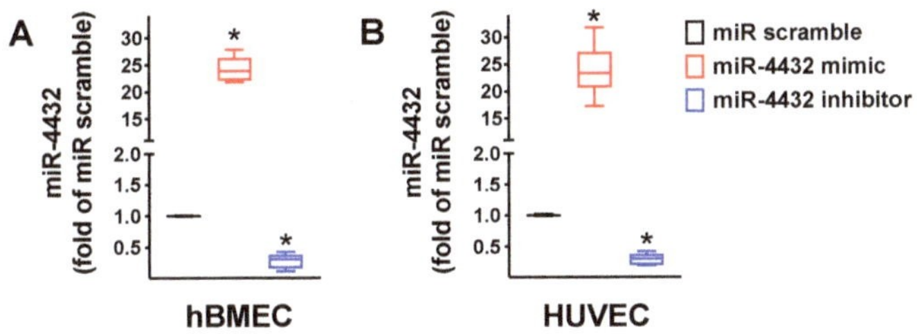

Figure 3. RT-qPCR showing that miR-4432 is expressed in both hBMECs (**A**) and HUVECs (**B**). All the assays were carried out in quadruplicate; the graphs indicate the median and the 5th to 95th percentiles; *: $p < 0.01$ vs. miR-scramble.

Then, we performed a series of experiments in hBMECs to test whether miR-4432 is a regulator of FGFBP1 transcription. Through luciferase assays, we determined that FGFBP1 is a target of miR-4432 (Figure 4); these findings were also endorsed in HUVECs (Supplementary Figure S1).

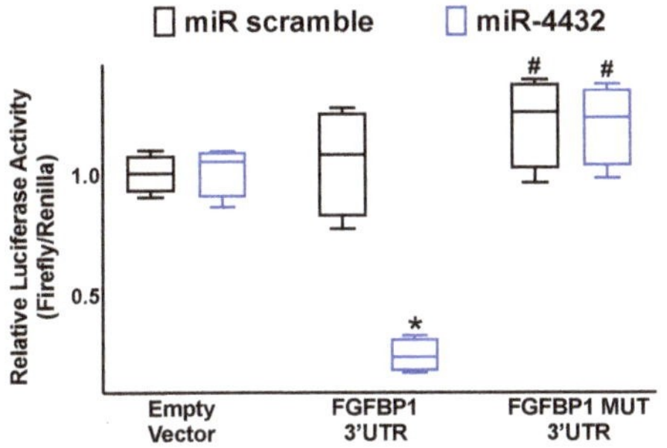

Figure 4. miR-4432 targets FGFBP1. Luciferase activity was quantified in hBMECs forty-eight hours after the transfection, utilizing the vector without FGFBP1 3′-UTR ("Empty Vector"), the vector that included the WT FGFBP1 3′-UTR ("FGFBP1 3′-UTR"), and the vector that included the mutated form of the FGFBP1 3′-UTR ("FGFBP1 MUT 3′UTR"); a miR-scramble (non-targeting miR) was used as an additional control. All the assays were carried out in quadruplicate; the graphs indicate the median and the 5th to 95th percentiles; *: $p < 0.01$ vs. miR-scramble; #: $p < 0.05$ vs. FGFBP1 3′UTR.

2.3. FGFBP1 Expression Is Controlled by miR-4432

As depicted in Figure 5, we experimentally proved that miR-4432 significantly diminishes the mRNA expression of FGFBP1 in hBMECs.

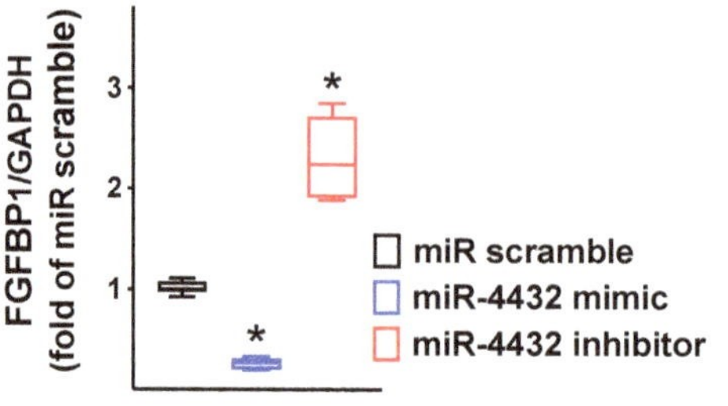

Figure 5. In hBMECs, FGFBP1 transcription was diminished by miR-4432 and augmented by miR-4432 inhibitor. FGFBP1 mRNA was quantified via RT-qPCR in hBMECs that had been transfected for forty-eight hours with the miRs indicated in the figure; values were normalized to GAPDH (glyceraldehyde-3-phosphate-dehydrogenase). All the assays were carried out at least in triplicate; the graph shows the medians and the 5th to 95th percentiles; *: $p < 0.01$ vs. miR-scramble. Sequences of the primers that have been used for the RT-qPCR are shown in Table 1.

These findings were then confirmed by immunoblot at the protein level (Figure 6), as well.

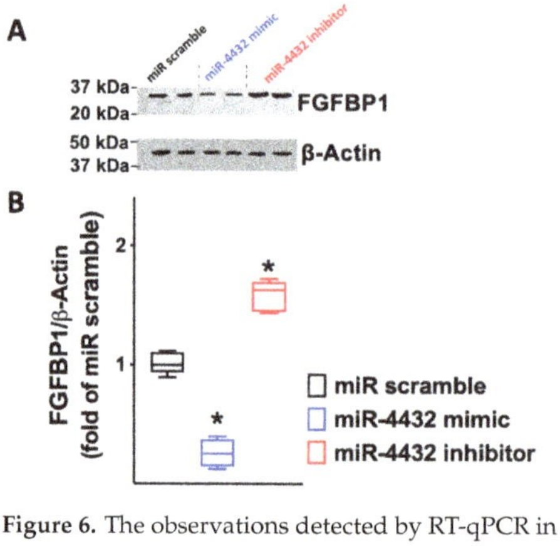

Figure 6. The observations detected by RT-qPCR in terms of mRNA were upheld by Western blots, as shown in the representative blots, showing two biological replicates per condition (**A**) and their quantification (**B**). All assays were carried out at least in triplicate; the graph represents the medians and the 5th to 95th percentiles; *: $p < 0.01$ vs. miR-scramble.

2.4. miR-4432 Regulates Mitochondrial Oxidative Stress in Human ECs

The next logical step was to gain more insights into the physiological and disease-related consequences of the interaction between miR-4432 and FGFBP1. The generation of mitochondrial reactive oxygen species (ROS) induced by the known vasoconstrictor angiotensin II (Ang II) in ECs [19] has been mechanistically implied in the pathogenesis of hypertension [20–22] and previous investigations have evidenced that the upregulation of FGFBP1 can increase oxidative stress signaling, leading to pro-hypertensive effects [23].

On these grounds, we quantified, by MitoSOX, the ROS production induced by Ang II in hBMECs transfected with miR-4432 mimic, miR-4432 inhibitor, or, as control, miR-scramble. Strikingly, we observed that mitochondrial oxidative stress was significantly reduced by miR-4432 mimic and increased by miR-4432 inhibitor (Figure 7).

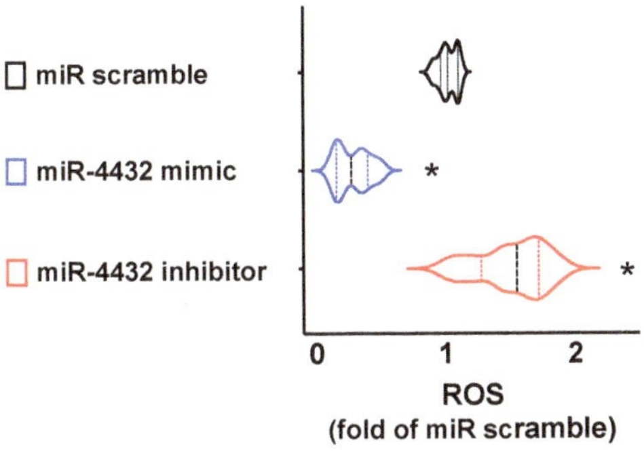

Figure 7. In hBMECs, the production of mitochondrial ROS (reactive oxygen species) was significantly diminished by miR-4432 mimic and increased by miR-4432 inhibitor. Mitochondrial ROS generation induced by Ang II (200 nMol, 4 h) was quantified using MitoSOX Red in hBMECs that had been transfected for forty-eight hours with the miRs indicated in the figure. All the assays were carried out at least in triplicate; the violin plots show the median (dashed line) and the quartiles (dotted lines); *: $p < 0.01$ vs. miR-scramble.

To mechanistically prove the functional role of FGFBP1, we repeated the ROS quantification after the knock-down of FGFBP1, showing that in the absence of FGFBP1 there is no significant effect of miR-4432 (Figure 8).

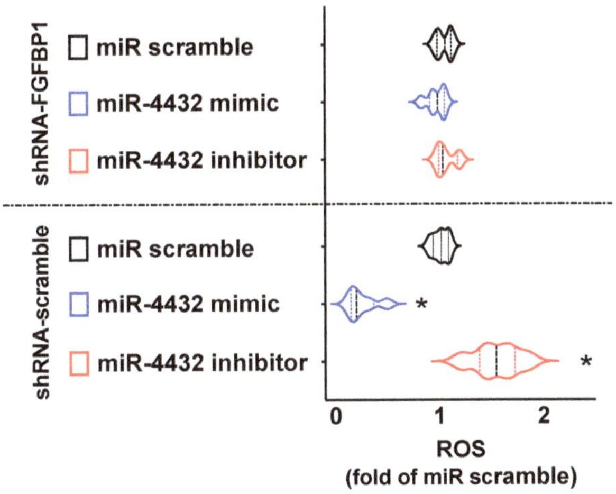

Figure 8. Mitochondrial ROS generation induced by Ang II in hBMECs was not affected when measured after having silenced FGFBP1 (top), whereas it was significantly blunted by miR-4432 mimic and increased by miR-4432 inhibitor when the cells had been treated with a shRNA scramble. All the assays were carried out at least in triplicate; the violin plots show the median (dashed line) and the quartiles (dotted lines); *: $p < 0.01$ vs. miR-scramble.

3. Discussion

The experimental observation herein reported indicates that miR-4432 targets FGFBP1 in human ECs, representing a novel potential strategy against numerous diseases characterized by endothelial dysfunction, including hypertension [24–29].

Consistent with our results, hypertensive patients have been shown to have approximately 1.5- and 1.4-fold higher expression of FGFBP1 mRNA and protein compared to normotensive subjects [30], further corroborating the crucial role of FGFBP1 in the pathophysiology of hypertension.

A genetic polymorphism in the human FGFBP1 gene has been associated with a higher gene expression and an increased risk of familial hypertension [30]. Preclinical studies in spontaneously hypertensive rats substantiated a contribution of the FGFBP1 genomic locus to hypertension and to glomerular damage [31]. In addition, the induction of FGFBP1 in a transgenic mouse model resulted in sustained hypertension and increased vascular sensitivity to the vasoconstrictor angiotensin II (Ang II) via ROS and MAP kinase pathway signaling [23,32]. Taken together, these pieces of evidence indicate that FGFBP1 can finely control steady-state blood pressure, most likely by regulating vascular sensitivity to endogenous Ang II.

Another study explored the indirect relationship between FGFBP1 and miRs in human umbilical vein ECs, showing that miR-146a promotes angiogenesis by increasing FGFBP1 expression via targeting CREB3L1 (Cyclic AMP Responsive-Element-Binding Protein-3-Like 1) [33]. In agreement with these data, FGFBP1 has been shown to be significantly upregulated in the hemolytic uremic syndrome associated with human immunodeficiency virus (HIV-HUS), which is characterized by endothelial damage and microcystic tubular dilation [34,35]; furthermore, the inhibition of FGFBP1 was shown to be beneficial in preventing brain vessel damage triggered by acute kidney injury [32].

Intriguingly, FGFBP1 is also expressed in keratinocytes, infiltrating mononuclear cells, and Kaposi's Sarcoma spindle cells [36,37]; its activation during the process of wound healing in the skin can induce angiogenic lesions that closely resemble Kaposi's Sarcoma [36]. Equally importantly, FGFBP1 can promote hepatocellular carcinoma metastasis [38], and patients with pancreatic cancer who express higher FGFBP1 levels have been shown to have a worse prognosis [39].

So, FGFBP1 is generally considered an indicator of early stages of pancreatic and colorectal adenocarcinoma [40], and as a biomarker it is very useful in predicting bacillus Calmette–Guérin response in bladder cancer [41]. It has been shown to be significantly upregulated in early dysplastic lesions of the human colon as well as in primary and metastatic colorectal cancers, whereas its knock-down led to anti-proliferative effects [42–44]. Therefore, its targeting using miR-based approaches could also lead to novel strategies in oncology.

Last but not least, the FGF signaling pathway has been shown to be intimately involved in the regulation of the vascular tone, with important roles in a number of homeostatic processes including blood pressure regulation, inflammation, shock, and ischemia-reperfusion, as well as injury/repair situations involving the vasculature, nervous system and dermal wound healing [45,46], and it also affects vascular morphogenesis of pre-endothelial cells of the embryo [47]. One of the main limitations of our study is having performed just in vitro assays; however, the FGFBP1 targeting by miR-4432 was confirmed in two different cell types (i.e., hBMECs and HUVECs). Additional studies are necessary to confirm the effects of miR-4432 in the pathobiology of hypertension and other cardiovascular and cerebrovascular disorders.

In summary, we established that FGFBP1 is expressed in ECs and that miR-4432 finely controls its expression levels both at the mRNA and protein level.

4. Methods

4.1. Cells and Other Reagents

hBMECs were purchased from Neuromics (Catalog code number: HEC02; Minneapolis, MN, USA). HUVECs were purchased from ThermoFisher Scientific (Catalog code number: C0035C; Waltham, MA, USA). Cells were cultured at early passages (3–7) under standard conditions (37 °C, 5% CO_2), as previously described [48]. In some assays, the cells were transfected with *pcDNA3.1-FGFBP1* plasmids obtained from GenScript (Piscataway, NJ, USA). All other reagents were obtained from Merck (Darmstadt, Germany).

4.2. Identification of miR-4432 as a Modulator of FGFBP1

To ascertain which miRs could specifically target the 3′-UTR of FGFBP1, we harnessed Target Scan Human 8.0, as reported previously [48]. The effects of miR-4432 on FGFBP1 gene transcription were assessed in hBMECs cells through a luciferase-reporter that contained the 3′-UTR of the predicted miR interaction site, in both the WT and mutated forms. The mutant of FGFBP1 3′-UTR (FGFBP1-MUT, see Figures 1 and 2), which contained substituted nucleotides in the region of the predicted miR-4432 binding-site of FGFBP1 3′-UTR, was designed via the NEBase Changer and Q5-site-directed mutagenesis kit (New England-Biolabs, Ipswich, MA, USA) as previously reported [48].

Using Lipofectamine-RNAiMAX (Thermo Fisher Scientific), hBMECs were transfected (66% transfection efficiency) with 0.05 μg of the 3′-UTR plasmid as well as miR-4432 mimic (a chemically synthesized double-stranded RNA that mimics endogenous miR-4432, MedChemExpress, Monmouth Junction, NJ, USA) or miR-4432 inhibitor (a steric blocking oligonucleotide that hybridizes with mature miR-4432 and inhibits its function, IDT, Coralville, IA, USA), or a negative control (non-targeting scramble, IDT), reaching a final concentration of 50 nMol/L [48]. Utilizing the Luciferase-Reporter Assay System (Promega, Madison, WI, USA), we quantified Firefly-and-Renilla luciferase activities forty-eight hours after the transfection, as previously described [48]. In some experiments, endothelial cells were transfected with shRNA-FGFBP1 or shRNA-scramble (Origene, Rockville, MD, USA), following the manufacturer's instructions. TaqMan microRNA Assays (Thermo Fisher Scientific) were used to quantify mature miR-4432 using U18 as endogenous control, as described in the literature [16]. FGFBP1 expression was assessed via RT-qPCR as previously reported [48], normalizing to glyceraldehyde 3-phosphate dehydrogenase (GAPDH). The sequences of oligonucleotide primers (Merck, Darmstadt, Germany) are shown in Table 1.

Table 1. Primer sequences used for RT-qPCR assays.

Genes	F/R	Sequence (5′-to-3′)	bp
FGFBP1	*Forward*	GG AGG AGC TGT GAG TAA CGT	113
	Reverse	TG TCA GGT AGA GTG CAA GGG	
GAPDH	*Forward*	GG CTC CCT TGG GTA TAT GGT	94
	Reverse	TT GAT TTT GGA GGG ATC TCG	

FGFBP1 stands for fibroblast growth factor binding protein 1; GAPDH stands for glyceraldehyde-3-phosphate-dehydrogenase; bp indicates base pairs.

4.3. Immunoblotting

Immunoblotting assays were performed as previously described and validated by our group [16,49]; the intensity of the bands was quantified using FIJI ("Fiji Is Just Image J") software. The antibody for FGFBP1 was purchased from ThermoFisher Scientific (Catalog code number: PA5-77220); the antibody for β-Actin was purchased from abcam (Cambridge, MA, USA; Catalog code number: ab8229).

4.4. Mitochondrial ROS

Mitochondrial ROS generation was assessed using MitoSOX Red (catalog code number: #M36008; Thermo Fisher Scientific) in hBMECs cells treated with Ang II (400 nMol for 4 h), as previously described [50].

4.5. Statistical Analysis

All data were expressed as means ± standard error of the means (SEMs). The statistical analyses were carried out using GraphPad 9 (Dotmatics, San Diego, CA, USA). Statistical significance, set at $p < 0.05$, was tested using the non-parametric Mann–Whitney U test or a two-way ANOVA followed by Bonferroni multiple comparison test, as appropriate.

5. Conclusions

Taken together, our results indicate for the first time, to the best of our knowledge, that miR-4432 specifically targets the 3′UTR of FGFBP1, thereby representing a novel potential strategy against hypertension, cerebrovascular disease, and other disorders characterized by endothelial dysfunction.

Supplementary Materials: The following supporting information can be downloaded at: https://www.mdpi.com/article/10.3390/biology12030459/s1, Figure S1: Validation of FGFBP1 targeting by miR-4432 in HUVECs.

Author Contributions: Conceptualization, P.M., J.G. and G.S.; methodology, R.A., P.M., S.S.J., F.V., U.K. and J.G.; formal analysis, J.G., A.D.L., A.M. and G.S.; investigation, R.A., P.M., J.G. and A.M.; writing—original draft preparation, R.A., P.M., F.V., J.G., A.D.L. and A.M.; writing—review and editing, G.S. All authors have read and agreed to the published version of the manuscript.

Funding: The Santulli's Research Laboratory is supported in part by the National Institutes of Health (NIH); specifically, by the National Institute of Diabetes and Digestive and Kidney Diseases (NIDDK; R01 DK123259, R01 DK033823), the National Heart, Lung, and Blood Institute (NHLBI; T32 HL144456, R01 HL164772, R01 HL159062, R01 HL146691), the National Center for Advancing Translational Sciences (NCATS; UL1 TR002556-06), to G.S.; by the Diabetes Action Research and Education Foundation (to G.S.); and by the Monique Weill Caulier and Irma T. Hirschl Trusts (to G.S.). S.S.J. is supported in part by a postdoctoral fellowship of the American Heart Association (POST836407 AHA). F.V. is supported in part by a postdoctoral fellowship of the American Heart Association (POST995561 AHA). U.K. is supported in part by a postdoctoral fellowship of the American Heart Association (POST1026190 AHA). J.G. is supported in part by a postdoctoral fellowship of the American Heart Association (POST35211151 AHA).

Institutional Review Board Statement: Not applicable.

Informed Consent Statement: Not applicable.

Data Availability Statement: All the data supporting the reported results are contained within this article and its supplementary material.

Conflicts of Interest: The authors declare no conflict of interest.

References

1. Zhou, B.; Perel, P.; Mensah, G.A.; Ezzati, M. Global epidemiology, health burden and effective interventions for elevated blood pressure and hypertension. *Nat. Rev. Cardiol.* **2021**, *18*, 785–802. [CrossRef]
2. Bruno, R.M.; Masi, S.; Taddei, M.; Taddei, S.; Virdis, A. Essential Hypertension and Functional Microvascular Ageing. *High Blood Press. Cardiovasc. Prev.* **2018**, *25*, 35–40. [CrossRef]
3. Saxena, T.; Ali, A.O.; Saxena, M. Pathophysiology of essential hypertension: An update. *Expert Rev. Cardiovasc. Ther.* **2018**, *16*, 879–887. [CrossRef]
4. Vanhoutte, P.M.; Shimokawa, H.; Feletou, M.; Tang, E.H. Endothelial dysfunction and vascular disease—A 30th anniversary update. *Acta Physiol.* **2017**, *219*, 22–96. [CrossRef]
5. Daiber, A.; Steven, S.; Weber, A.; Shuvaev, V.V.; Muzykantov, V.R.; Laher, I.; Li, H.; Lamas, S.; Munzel, T. Targeting vascular (endothelial) dysfunction. *Br. J. Pharmacol.* **2017**, *174*, 1591–1619. [CrossRef] [PubMed]
6. Ruppert, V.; Maisch, B. Genetics of human hypertension. *Herz* **2003**, *28*, 655–662. [CrossRef]
7. Kolifarhood, G.; Sabour, S.; Akbarzadeh, M.; Sedaghati-Khayat, B.; Guity, K.; Rasekhi Dehkordi, S.; Amiri Roudbar, M.; Hadaegh, F.; Azizi, F.; Daneshpour, M.S. Genome-wide association study on blood pressure traits in the Iranian population suggests ZBED9 as a new locus for hypertension. *Sci. Rep.* **2021**, *11*, 11699. [CrossRef]
8. Mompeo, O.; Freidin, M.B.; Gibson, R.; Hysi, P.G.; Christofidou, P.; Segal, E.; Valdes, A.M.; Spector, T.D.; Menni, C.; Mangino, M. Genome-Wide Association Analysis of Over 170,000 Individuals from the UK Biobank Identifies Seven Loci Associated with Dietary Approaches to Stop Hypertension (DASH) Diet. *Nutrients* **2022**, *14*, 4431. [CrossRef] [PubMed]
9. Singh, S.; Warren, H.R.; Hiltunen, T.P.; McDonough, C.W.; El Rouby, N.; Salvi, E.; Wang, Z.; Garofalidou, T.; Fyhrquist, F.; Kontula, K.K.; et al. Genome-Wide Meta-Analysis of Blood Pressure Response to beta(1)-Blockers: Results From ICAPS (International Consortium of Antihypertensive Pharmacogenomics Studies). *J. Am. Heart Assoc.* **2019**, *8*, e013115. [CrossRef]
10. Cottarelli, A.; Corada, M.; Beznoussenko, G.V.; Mironov, A.A.; Globisch, M.A.; Biswas, S.; Huang, H.; Dimberg, A.; Magnusson, P.U.; Agalliu, D.; et al. Fgfbp1 promotes blood-brain barrier development by regulating collagen IV deposition and maintaining Wnt/beta-catenin signaling. *Development* **2020**, *147*, dev185140. [CrossRef]

11. Propson, N.E.; Roy, E.R.; Litvinchuk, A.; Kohl, J.; Zheng, H. Endothelial C3a receptor mediates vascular inflammation and blood-brain barrier permeability during aging. *J. Clin. Investig.* **2021**, *131*, e140966. [CrossRef]
12. Kadry, H.; Noorani, B.; Cucullo, L. A blood-brain barrier overview on structure, function, impairment, and biomarkers of integrity. *Fluids Barriers CNS* **2020**, *17*, 69. [CrossRef] [PubMed]
13. Ben-Zvi, A.; Liebner, S. Developmental regulation of barrier- and non-barrier blood vessels in the CNS. *J. Intern. Med.* **2022**, *292*, 31–46. [CrossRef]
14. Stavast, C.J.; Erkeland, S.J. The Non-Canonical Aspects of MicroRNAs: Many Roads to Gene Regulation. *Cells* **2019**, *8*, 1465. [CrossRef] [PubMed]
15. Ho, P.T.B.; Clark, I.M.; Le, L.T.T. MicroRNA-Based Diagnosis and Therapy. *Int. J. Mol. Sci.* **2022**, *23*, 7167. [CrossRef]
16. Santulli, G.; Wronska, A.; Uryu, K.; Diacovo, T.G.; Gao, M.; Marx, S.O.; Kitajewski, J.; Chilton, J.M.; Akat, K.M.; Tuschl, T.; et al. A selective microRNA-based strategy inhibits restenosis while preserving endothelial function. *J. Clin. Investig.* **2014**, *124*, 4102–4114. [CrossRef] [PubMed]
17. Rafehi, H.; El-Osta, A. HDAC Inhibition in Vascular Endothelial Cells Regulates the Expression of ncRNAs. *Noncoding RNA* **2016**, *2*, 4. [CrossRef] [PubMed]
18. Naik, P.; Cucullo, L. In vitro blood-brain barrier models: Current and perspective technologies. *J. Pharm. Sci.* **2012**, *101*, 1337–1354. [CrossRef]
19. Zhu, N.; Zhang, D.; Chen, S.; Liu, X.; Lin, L.; Huang, X.; Guo, Z.; Liu, J.; Wang, Y.; Yuan, W.; et al. Endothelial enriched microRNAs regulate angiotensin II-induced endothelial inflammation and migration. *Atherosclerosis* **2011**, *215*, 286–293. [CrossRef]
20. Taniyama, Y.; Griendling, K.K. Reactive oxygen species in the vasculature: Molecular and cellular mechanisms. *Hypertension* **2003**, *42*, 1075–1081. [CrossRef]
21. De Giusti, V.C.; Caldiz, C.I.; Ennis, I.L.; Perez, N.G.; Cingolani, H.E.; Aiello, E.A. Mitochondrial reactive oxygen species (ROS) as signaling molecules of intracellular pathways triggered by the cardiac renin-angiotensin II-aldosterone system (RAAS). *Front. Physiol.* **2013**, *4*, 126. [CrossRef] [PubMed]
22. Welch, W.J. Angiotensin II-dependent superoxide: Effects on hypertension and vascular dysfunction. *Hypertension* **2008**, *52*, 51–56. [CrossRef] [PubMed]
23. Tassi, E.; Lai, E.Y.; Li, L.; Solis, G.; Chen, Y.; Kietzman, W.E.; Ray, P.E.; Riegel, A.T.; Welch, W.J.; Wilcox, C.S.; et al. Blood Pressure Control by a Secreted FGFBP1 (Fibroblast Growth Factor-Binding Protein). *Hypertension* **2018**, *71*, 160–167. [CrossRef]
24. Piotti, A.; Novelli, D.; Meessen, J.; Ferlicca, D.; Coppolecchia, S.; Marino, A.; Salati, G.; Savioli, M.; Grasselli, G.; Bellani, G.; et al. Endothelial damage in septic shock patients as evidenced by circulating syndecan-1, sphingosine-1-phosphate and soluble VE-cadherin: A substudy of ALBIOS. *Crit. Care* **2021**, *25*, 113. [CrossRef] [PubMed]
25. Mauricio, M.D.; Aldasoro, M.; Ortega, J.; Vila, J.M. Endothelial dysfunction in morbid obesity. *Curr. Pharm. Des.* **2013**, *19*, 5718–5729. [CrossRef]
26. Bordy, R.; Totoson, P.; Prati, C.; Marie, C.; Wendling, D.; Demougeot, C. Microvascular endothelial dysfunction in rheumatoid arthritis. *Nat. Rev. Rheumatol.* **2018**, *14*, 404–420. [CrossRef]
27. Bijl, M. Endothelial activation, endothelial dysfunction and premature atherosclerosis in systemic autoimmune diseases. *Neth. J. Med.* **2003**, *61*, 273–277.
28. Budhiraja, R.; Parthasarathy, S.; Quan, S.F. Endothelial dysfunction in obstructive sleep apnea. *J. Clin. Sleep Med.* **2007**, 3, 409–415. [CrossRef]
29. Kim, J.A.; Montagnani, M.; Koh, K.K.; Quon, M.J. Reciprocal relationships between insulin resistance and endothelial dysfunction: Molecular and pathophysiological mechanisms. *Circulation* **2006**, *113*, 1888–1904. [CrossRef]
30. Tomaszewski, M.; Charchar, F.J.; Nelson, C.P.; Barnes, T.; Denniff, M.; Kaiser, M.; Debiec, R.; Christofidou, P.; Rafelt, S.; van der Harst, P.; et al. Pathway analysis shows association between FGFBP1 and hypertension. *J. Am. Soc. Nephrol.* **2011**, *22*, 947–955. [CrossRef]
31. Braun, M.C.; Herring, S.M.; Gokul, N.; Monita, M.; Bell, R.; Hicks, M.J.; Wenderfer, S.E.; Doris, P.A. Hypertensive renal disease: Susceptibility and resistance in inbred hypertensive rat lines. *J. Hypertens* **2013**, *31*, 2050–2059. [CrossRef] [PubMed]
32. Zhao, L.; Cao, X.; Li, L.; Wang, X.; Wang, Q.; Jiang, S.; Tang, C.; Zhou, S.; Xu, N.; Cui, Y.; et al. Acute Kidney Injury Sensitizes the Brain Vasculature to Ang II (Angiotensin II) Constriction via FGFBP1 (Fibroblast Growth Factor Binding Protein 1). *Hypertension* **2020**, *76*, 1924–1934. [CrossRef] [PubMed]
33. Zhu, H.Y.; Bai, W.D.; Liu, J.Q.; Zheng, Z.; Guan, H.; Zhou, Q.; Su, L.L.; Xie, S.T.; Wang, Y.C.; Li, J.; et al. Up-regulation of FGFBP1 signaling contributes to miR-146a-induced angiogenesis in human umbilical vein endothelial cells. *Sci. Rep.* **2016**, *6*, 25272. [CrossRef] [PubMed]
34. Ray, P.E.; Tassi, E.; Liu, X.H.; Wellstein, A. Role of fibroblast growth factor-binding protein in the pathogenesis of HIV-associated hemolytic uremic syndrome. *Am. J. Physiol. Regul. Integr. Comp. Physiol.* **2006**, *290*, R105–R113. [CrossRef]
35. Liu, X.H.; Aigner, A.; Wellstein, A.; Ray, P.E. Up-regulation of a fibroblast growth factor binding protein in children with renal diseases. *Kidney Int.* **2001**, *59*, 1717–1728. [CrossRef] [PubMed]
36. Ray, P.E.; Al-Attar, A.; Liu, X.H.; Das, J.R.; Tassi, E.; Wellstein, A. Expression of a Secreted Fibroblast Growth Factor Binding Protein-1 (FGFBP1) in Angioproliferative Kaposi Sarcoma. *J. AIDS Clin. Res.* **2014**, *5*, 309. [CrossRef] [PubMed]
37. Ueyama, T.; Sakuma, M.; Nakatsuji, M.; Uebi, T.; Hamada, T.; Aiba, A.; Saito, N. Rac-Dependent Signaling from Keratinocytes Promotes Differentiation of Intradermal White Adipocytes. *J. Investig. Dermatol.* **2020**, *140*, 75–84.e6. [CrossRef] [PubMed]

38. Huang, W.; Chen, Z.; Shang, X.; Tian, D.; Wang, D.; Wu, K.; Fan, D.; Xia, L. Sox12, a direct target of FoxQ1, promotes hepatocellular carcinoma metastasis through up-regulating Twist1 and FGFBP1. *Hepatology* **2015**, *61*, 1920–1933. [CrossRef]
39. Zhang, Z.; Liu, M.; Hu, Q.; Xu, W.; Liu, W.; Sun, Q.; Ye, Z.; Fan, G.; Xu, X.; Yu, X.; et al. FGFBP1, a downstream target of the FBW7/c-Myc axis, promotes cell proliferation and migration in pancreatic cancer. *Am. J. Cancer Res.* **2019**, *9*, 2650–2664.
40. Tassi, E.; Wellstein, A. The angiogenic switch molecule, secreted FGF-binding protein, an indicator of early stages of pancreatic and colorectal adenocarcinoma. *Semin Oncol.* **2006**, *33*, S50–S56. [CrossRef]
41. Li, F.; Zhang, H.; Wang, Y.; Yao, Z.; Xie, K.; Mo, Q.; Fan, Q.; Hou, L.; Deng, F.; Tan, W. FGFBP1 as a potential biomarker predicting bacillus Calmette-Guerin response in bladder cancer. *Front. Immunol.* **2022**, *13*, 954836. [CrossRef] [PubMed]
42. Ray, R.; Cabal-Manzano, R.; Moser, A.R.; Waldman, T.; Zipper, L.M.; Aigner, A.; Byers, S.W.; Riegel, A.T.; Wellstein, A. Up-regulation of fibroblast growth factor-binding protein, by beta-catenin during colon carcinogenesis. *Cancer Res.* **2003**, *63*, 8085–8089.
43. Schulze, D.; Plohmann, P.; Hobel, S.; Aigner, A. Anti-tumor effects of fibroblast growth factor-binding protein (FGF-BP) knockdown in colon carcinoma. *Mol. Cancer* **2011**, *10*, 144. [CrossRef]
44. Czubayko, F.; Liaudet-Coopman, E.D.; Aigner, A.; Tuveson, A.T.; Berchem, G.J.; Wellstein, A. A secreted FGF-binding protein can serve as the angiogenic switch in human cancer. *Nat. Med.* **1997**, *3*, 1137–1140. [CrossRef]
45. Zhou, M.; Sutliff, R.L.; Paul, R.J.; Lorenz, J.N.; Hoying, J.B.; Haudenschild, C.C.; Yin, M.; Coffin, J.D.; Kong, L.; Kranias, E.G.; et al. Fibroblast growth factor 2 control of vascular tone. *Nat. Med.* **1998**, *4*, 201–207. [CrossRef] [PubMed]
46. Folkman, J.; Klagsbrun, M. Vascular physiology. A family of angiogenic peptides. *Nature* **1987**, *329*, 671–672. [CrossRef] [PubMed]
47. Doetschman, T.; Shull, M.; Kier, A.; Coffin, J.D. Embryonic stem cell model systems for vascular morphogenesis and cardiac disorders. *Hypertension* **1993**, *22*, 618–629. [CrossRef] [PubMed]
48. Mone, P.; Gambardella, J.; Wang, X.; Jankauskas, S.S.; Matarese, A.; Santulli, G. miR-24 Targets the Transmembrane Glycoprotein Neuropilin-1 in Human Brain Microvascular Endothelial Cells. *Noncoding RNA* **2021**, *7*, 9. [CrossRef] [PubMed]
49. Gambardella, J.; Jankauskas, S.S.; Kansakar, U.; Varzideh, F.; Avvisato, R.; Prevete, N.; Sidoli, S.; Mone, P.; Wang, X.; Lombardi, A.; et al. Ketone bodies rescue mitochondrial dysfunction via epigenetic remodeling. *JACC Basic Transl. Sci.* 2023, *in press*.
50. Mone, P.; Varzideh, F.; Jankauskas, S.S.; Pansini, A.; Lombardi, A.; Frullone, S.; Santulli, G. SGLT2 Inhibition via Empagliflozin Improves Endothelial Function and Reduces Mitochondrial Oxidative Stress: Insights From Frail Hypertensive and Diabetic Patients. *Hypertension* **2022**, *79*, 1633–1643. [CrossRef]

biology

MDPI

Article

Differences in Extracellular NAD$^+$ and NMN Metabolism on the Surface of Vascular Endothelial Cells

Patrycja Jablonska [1], Paulina Mierzejewska [1], Marta Tomczyk [1], Patrycja Koszalka [2], Marika Franczak [1], Ada Kawecka [1], Barbara Kutryb-Zajac [1], Alicja Braczko [1], Ryszard T. Smolenski [1] and Ewa M. Slominska [1,*]

[1] Department of Biochemistry, Medical University of Gdansk, 80-211 Gdansk, Poland; patrycja.jablonska@gumed.edu.pl (P.J.); paulina.mierzejewska@gumed.edu.pl (P.M.); marta.tomczyk@gumed.edu.pl (M.T.); marika.franczak@gumed.edu.pl (M.F.); kaweckaada@gumed.edu.pl (A.K.); b.kutryb-zajac@gumed.edu.pl (B.K.-Z.); alicja.braczko@gumed.edu.pl (A.B.); ryszard.smolenski@gumed.edu.pl (R.T.S.)

[2] Institute of Medical Biotechnology and Experimental Oncology, Intercollegiate Faculty of Biotechnology of University of Gdansk and Medical University of Gdansk, Medical University of Gdansk, 80-211 Gdansk, Poland; patrycja.koszalka@gumed.edu.pl

* Correspondence: eslom@gumed.edu.pl

Simple Summary: Nicotinamide adenine dinucleotide (NAD$^+$) is a multifunctional metabolite involved in many key cellular processes. Outside the cell, NAD$^+$ or its metabolites are important signaling molecules, related especially to calcium homeostasis, which controls the functioning of the heart. The cleavage of NAD$^+$ or its precursor, nicotinamide mononucleotide (NMN), produces derivatives entering the cell to rebuild the intracellular NAD$^+$ pool, which is important for cells with high energy turnover. Abnormalities in NAD$^+$ and NMN metabolism can lead to cell aging and the development of cardiovascular diseases. In this study, we demonstrated that the extracellular metabolism of NAD$^+$ and NMN is vastly different in the vascular endothelium obtained from different species and locations. This may have implications for strategies to modulate the NAD$^+$ system and may cause difficulties for comparing the results of different reports.

Abstract: The disruption of the metabolism of extracellular NAD$^+$ and NMN may affect related signaling cascades and pathologies, such as cardiovascular or respiratory system diseases. We aimed to study NAD$^+$ and NMN hydrolysis on surface endothelial cells of diverse origins and with genetically modified nucleotide catabolism pathways. We tested lung endothelial cells isolated from C57BL/6 J wild-type (WT) and C57BL/6 J CD73 knockout (CD73 KO) mice, the transfected porcine iliac artery endothelial cell line (PIEC) with the human E5NT gene for CD73 (PIEC CD73), and a mock-transfected control (PIEC MOCK), as well as HMEC-1 and H5V cells. Substrate conversion into the product was followed by high-performance liquid chromatography (HPLC). We showed profound differences in extracellular NAD$^+$ and NMN metabolism related to the vessel origin, species diversity, and type of culture. We also confirmed the involvement of CD38 and CD73 in NAD$^+$ and NMN cleavage.

Keywords: extracellular NAD$^+$ metabolism; extracellular NMN metabolism; vascular endothelial cells; CD38; CD73

Citation: Jablonska, P.; Mierzejewska, P.; Tomczyk, M.; Koszalka, P.; Franczak, M.; Kawecka, A.; Kutryb-Zajac, B.; Braczko, A.; Smolenski, R.T.; Slominska, E.M. Differences in Extracellular NAD$^+$ and NMN Metabolism on the Surface of Vascular Endothelial Cells. *Biology* **2022**, *11*, 675. https://doi.org/10.3390/biology11050675

Academic Editor: Gaetano Santulli

Received: 7 March 2022
Accepted: 25 April 2022
Published: 27 April 2022

1. Introduction

The primary metabolic function of nicotinamide adenine dinucleotide (NAD$^+$) is related to redox reactions [1], but, in addition, NAD$^+$ may also serve as a substrate in various signaling processes. The main enzyme classes responsible for the catabolism of NAD$^+$ in the cell are sirtuins, poly(ADP-ribose) polymerases (PARPs), and cyclic ADP-ribose synthases/NAD$^+$-glycohydrolases. Enzymes that participate in NAD$^+$-dependent signaling pathways are involved in controlling the necessary balance of NAD$^+$ concentration as well as cell cycle progression, transcriptional regulation, and DNA repair, and have, therefore, been identified as promising targets in many diseases [2–4].

Maintaining a constant level of NAD^+ in a cell is essential for its proper functioning. Disturbed balance in the processes of NAD^+ synthesis and degradation may be the source of the development of many pathologies. The main approach to regulating NAD^+ levels in cells is through the efficient recycling of endogenous precursors. Nucleotide derivatives can support the maintenance of intracellular NAD^+ pools by passing into the cell from the extracellular space [5]. Extracellular NAD^+ is degraded to pyridine and purine metabolites by different types of surface-located enzymes, which are widely expressed on the plasma membrane of various cells and tissues [6]. The enzymes that take part in the degradation of extracellular NAD^+ are mainly CD38-NAD^+-glycohydrolase, which hydrolyzes NAD^+ to nicotinamide (Nam) and ADP-ribose (ADPR) and nicotinamide mononucleotide (NMN) to Nam, and CD73-ecto-5′-nucleotidase, which is responsible for the hydrolysis of NAD^+ to NMN and AMP, then NMN to nicotinamide riboside (NR). Enzymes of minor significance, but that should also be mentioned, are ecto-nucleotide pyrophosphatase/phosphodiesterase (eNPP1, CD203a) and alkaline phosphatase (ALP). All of these enzymes are known to exist on the surface of endothelial cells. However, the importance of extracellular NAD^+ metabolism has not yet been thoroughly investigated [7,8].

The endothelium is a single layer of physiologically active cells that line the luminal surface of the entire vascular system [9]. It takes part in the secretion and active transport of chemical substances. The most important are prostacyclin (dilates blood vessels and inhibits platelet aggregation), nitric oxide (dilates vessels and inhibits platelet aggregation), von Willebrand factor (stimulates platelet aggregation), and thrombomodulin (inhibits blood clotting). The vascular endothelium takes part in vasoconstriction and vasodilation, and thus influences the regulation of blood pressure and tissue blood supply. It is involved in blood clotting (hemostasis and fibrinolysis), processes of atherosclerosis and angiogenesis, as well as inflammatory reactions. Endothelial dysfunction is a known hallmark of many vascular diseases [10,11]. Extracellular enzymes located on endothelial cells are an integral part of metabolism, which enable the homeostatic integration and control of the vascular inflammatory and immune responses of cells at the site of injury. The continuous development of therapeutic strategies targeting ectonucleotidases shows promise in the treatment of vascular thrombosis, impaired inflammation, and abnormal immune reactivity [12,13].

In this work, we measured the hydrolysis of extracellular NAD^+ and NMN on different types of endothelial cells in terms of vascular origin, species diversity, and type of culture (immortalized or primary cell line). We have shown some significant differences that will help to understand the heterogeneity of NAD^+-dependent metabolism and may constitute potential therapeutic strategies in the future.

2. Materials and Methods

2.1. Reagents

Adenosine (CAS 58-61-7), ADPR (adenosine 5′-diphosphoribose, CAS 68414-18-6), AOPCP (adenosine-5′-alpha,beta-methylene diphosphate, CAS 3768-14-7), deamino-NAD^+ (nicotinamide hypoxanthine dinucleotide sodium salt, CAS 104809-38-3), nicotinamide (CAS 98-92-0), NAD^+ (nicotinamide adenine dinucleotide, CAS 53-84-9), NMN (nicotinamide mononucleotide, CAS 1094-61-7), Dulbecco's modified Eagle's medium (DMEM), fetal bovine serum (FBS), Hanks Balanced Salt Solution (HBSS), L-glutamine, phosphate-buffered saline (PBS), penicillin/streptomycin, RPMI1640 medium, trypsin-EDTA solution, and endothelial cell growth supplement (ECGS) were obtained from Sigma-Aldrich (Poznan, Poland).

2.2. Animals Maintenance and Murine Lung Endothelial Cells Isolation

All experiments on mice were conducted following a Guide for the Care and Use of Laboratory Animals published by the European Parliament, Directive 2010/63/EU, and were performed with the approval of the Local Ethical Committee for Animal Experimentation in Bydgoszcz (27/2016). C57BL/6 J CD73 knockout (CD73 KO) mice were obtained from Heinrich-Heine-Universität in Düsseldorf, Germany [13]. The isolation of murine lung endothelial cells (LECs) from C57BL/6 J wild-type (WT) ($n = 6$) and CD73 KO ($n = 5$) mice was described previously [14]. Briefly, mice were fed a standard chow diet until 8 weeks and were then anesthetized with a mixture of ketamine/xylazine followed by opening the chest. Murine lungs were harvested, minced on a Petri dish, incubated with collagenase A (2.5 mg/mL solution in 0.1% BSA in HEPES), and then filtered through a 70-µm strainer. The cells were washed with PBS, resuspended in DMEM with D-valine, 10% FBS, ECGS (15 mg/500 mL), 2 mM L-glutamine, and 1% penicillin-streptomycin (v/v), and plated into a 25 cm^2 tissue-culture flask. After reaching the specified density, cells were sorted using mouse CD31 MicroBeads on the MACS column (Miltenyi Biotec, 130-097-418) according to the manufacturer's protocol. The sorted cells were further cultured in an endothelial cell medium. For the experiments, cells between two to five passages were plated at 12–16 × 10^4 cells/well in 24-well plates. The cells were maintained in a 5% CO_2 humidified atmosphere at 37 °C.

2.3. Cell Culture Conditions of Other Endothelial Cell Types

The endothelial cell lines human dermal microvascular endothelial cells (HMEC-1), murine immortalized heart endothelial cells (H5V), and transfected porcine iliac artery endothelial cells (PIEC) with human E5NT gene (PIEC CD73), as well as mock-transfected control (PIEC MOCK), were used in this study. The PIECs were kindly provided by Marco de Giorgi, who transfected them using F2A (the 2A sequence of Foot and Mouth Disease Virus) technology and Lipofectamine 2000 (Invitrogen, Waltham, MA, USA), and the transfection efficiency was evaluated by FACS analyses [8]. HMEC-1 were grown in MDCB 131 medium supplemented with 10% FBS, ECGS (15 mg/500 mL), hydrocortisone (1 µg/mL), 2 mM L-glutamine, and 1% penicillin-streptomycin (v/v). H5V cells were cultured in DMEM with 4.5 g/L glucose medium supplemented with 10% FBS, 2 mM L-glutamine, 1 mM sodium pyruvate, and 1% penicillin/streptomycin (v/v). PIEC MOCK and PIEC CD73 were cultured in RPMI 1640 medium supplemented with 10% FBS, 2 mM L-glutamine, and 1% penicillin–streptomycin (v/v). Cells were maintained in a 5% CO_2 humidified atmosphere at 37 °C. Cells were cultivated in 25 cm^2 and 75 cm^2 tissue-culture flasks. For the experiments, the cells were plated at 4–6 × 10^4 cells/well in 24-well plates. The medium was changed the next day, when the cells had adhered.

2.4. Determination of Cell-Surface NAD^+ and NMN-Degrading Activities in Cell Cultures

After reaching 90–100% confluence in 24-well culture plates, LECs WT, LECs CD73 KO, HMEC-1, H5V, PIEC MOCK, and PIEC CD73 were rinsed with PBS, and 1 mL HBSS was added to each well. Then, to measure the total NAD^+ and NMN hydrolysis activity, NAD^+ or NMN (in a final concentration of 50 µM [15]) was added to the buffer and after 0, 30, 60, and 120 min of incubation at 37 °C, samples were collected and analyzed with the HPLC method using the LC system (Agilent Technologies 1100 series, Santa Clara, CA, USA) as described earlier [16]. The sample peaks were integrated and quantified using a ChemStation (Agilent Technologies, Santa Clara, CA, USA) chromatography data system (Figure S2, Supplementary Materials).

The cell residue was dissolved in 0.5 mol/L NaOH, and the protein concentration was measured with the Bradford method according to the manufacturer's protocol. The results of the cell-surface NAD^+ and NMN degrading activities were expressed as the sum of products increased over time (nmol/min/g of protein).

2.5. Immunofluorescence Analysis

The distribution of CD73 and CD38 in H5V, HMEC-1, PIEC CD73, and PIEC-MOCK was detected by immunofluorescent staining in a 96-well optical bottom plate at a density 1 × 104 cells/well in a total volume of 200 µL cell culture medium. After 24 h, the cells were rinsed 3 times with PBS and fixed in ice-cold methanol for 5 min, and then rinsed with PBS. To reduce non-specific antibody binding, the cells were incubated with PAD solution (1% Bovine Serum Albumin and 10% normal goat serum solution in PBS). Thereafter, the cells were incubated with rabbit polyclonal anti-CD73 (1:100, Novus, Centennial, CO, USA) and mouse monoclonal anti-CD38 (1:100, Novus, Centennial, CO, USA) primary antibodies diluted in PBS for 1 h at room temperature. Next, the cells were incubated with goat anti-rabbit antibodies labeled with Alexa Fluor 488 (1:600, JacksonImmuno, West Grove, PA, USA) and goat anti-mouse antibodies labeled with goat anti-mouse Alexa Fluor 594 (1:600, JacksonImmuno, West Grove, PA, USA) for 30 min at room temperature. Negative controls were obtained by incubation without primary antibodies (data not shown). The cell nuclei were stained with 4′,6-diamidino-2-phenylindole (DAPI) (1:1000, Thermo Fisher Scientific, Waltham, MA, USA) for 5 min. The stained cells were imaged and analyzed as described previously [17] with an AxioCam MRc5 camera and an AxioObserved.D1 inverted fluorescent microscope (Carl Zeiss AG, Jena, Germany), and analyzed using Zen image processing software (version 3.3). The total CD73 and CD38-positive areas in cells were measured and the mean fluorescence intensity was calculated.

2.6. Determination of Particular Ecto-Enzymes Engaged in the Extracellular NAD^+ and NMN Catabolism on the Surface of the Endothelial Cells

To determine the specific cell-surface ecto-enzymes engaged in the metabolism of NAD^+ and NMN, 1 mL of HBSS with inhibitors for CD38 and CD73 was added: a competitive inhibitor of CD38 (150 μM deamino-NAD^+, Figure S1, Supplementary Materials) and CD73 inhibitor (50 μM AOPCP [18]), respectively. NAD^+ and NMN were added as substrates at a final concentration of 50 μM, respectively. After 0, 30, 60, and 120 min, 50 μL of the sample was collected, centrifuged (14,000× g/15 min/4 °C), and analyzed by HPLC. The degradation rates were shown as nmol/min/g of protein.

2.7. Statistical Analysis

Values were presented as the mean ± SEM. Statistical analysis was performed using an unpaired Student's t-test and one- or two-way ANOVA followed by Tukey's or Bonferroni's posttest (GraphPad software, San Diego, CA, USA). A p-value of 0.05 was considered to indicate a significant difference.

3. Results

3.1. NAD^+ and NMN Hydrolysis Are Different for Various Endothelial Cell Types

The comparison of the total NAD^+ and NMN hydrolysis on endothelial cell lines showed that the metabolism differed depending on the species and organ origin. The most active cells were transfected PIEC with a gene for CD73 (Figure 1a,b). The PIEC CD73 had the strongest signal for CD73 in immunofluorescent staining (Figure 1e), confirming the successful transfection of the NT5E gene compared to the PIEC MOCK control (Figure 1f). H5V, on the other hand, had the lowest signal for CD73 (Figure 1d), which translated into lower total NMN hydrolysis (Figure 1b). CD38 was also present in the cell types tested; however, despite its presence on HMEC-1 cells, no NAD^+ hydrolysis was observed in the analyzed incubation time (Figure 1a).

In contrast to NAD^+ hydrolysis, total NMN hydrolysis was observed on the surface of HMEC-1 after 2 h of incubation (Figure 1b). This may be due to the significant presence of CD73 on these cells (Figure 1g). Small amounts of NMN were also metabolized on H5V and PIEC MOCK cells. Figure 1c shows a diagram of NAD^+ and NMN degradation in the extracellular space. CD73 is an abundant endothelial enzyme responsible for the hydrolysis of NAD^+ to NMN and AMP, and then to NR and Ado, respectively. CD38, in turn, is involved in the metabolism of NAD^+ and NMN, leading to the production of Nam.

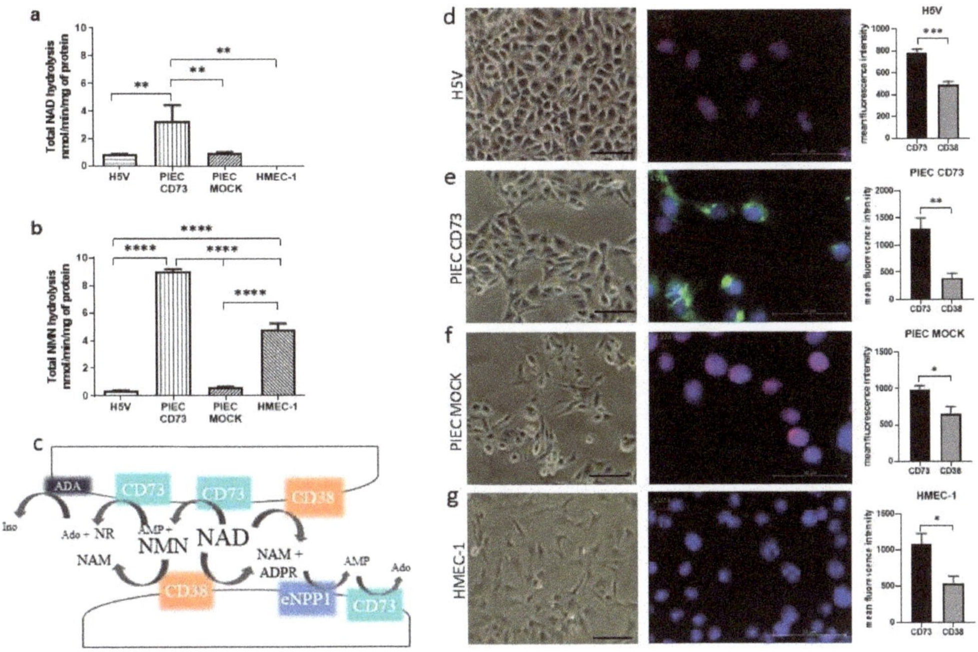

Figure 1. Nicotinamide adenine dinucleotide (NAD^+) and nicotinamide mononucleotide (NMN) hydrolysis are different for various endothelial cell types. Comparison of total NAD^+ (**a**) and NMN (**b**) hydrolysis on endothelial cell lines. Results are presented as the mean ± SEM; n = 3–6; ** $p < 0.01$ **** $p < 0.0001$ with one-way ANOVA followed by Tukey's multiple comparisons test. Diagram of NAD^+ and NMN metabolism in the extracellular space (**c**). Immunofluorescence staining and quantification for CD73 (green signal) and CD38 (red signal) on cell lines. Cell nuclei were counterstained with DAPI (blue signal) for H5V (**d**), PIEC CD73 (**e**), PIEC MOCK (**f**), and HMEC-1 (**g**). Results are presented as the mean ± SEM; n = 5; * $p < 0.05$, ** $p < 0.01$, *** $p < 0.001$ by student unpaired t-test.

3.2. PIEC Cells Mainly Produce NMN, AMP, and NR

In NAD^+ hydrolysis on the surface of PIEC MOCK cells (Figure 2a), the main products are NMN and AMP. In the case of the transfected PIEC CD73, the hydrolysis was enhanced and was characterized by a more intensive transformation of NAD^+. In addition to NMN and AMP, and other products of their metabolism (such as NR, Ado, and Ino), Nam was also created. This indirectly confirms the successful transfection of NT5E, the gene for CD73 that is responsible for the hydrolysis of NAD^+ to NMN and NMN to NR, but also leads to the generation of AMP from NAD^+ and further Ado. For NMN hydrolysis, the only product was NR (Figure 2b). When comparing PIEC MOCK and PIEC CD73 cells, there was about a 10-fold increase in the concentration of the nascent NR.

3.3. Nam and ADPR Are the Main Products of NAD^+ and NMN Metabolism on the Surface of LECs

LEC cells were the most metabolically active in NAD^+ and NMN hydrolysis (Figure 3a,b). The main products that were formed on the surface of the endothelium were Nam and ADPR, followed by Ado in the case of NAD^+ hydrolysis (Figure 3c), and Nam and NR in the case of NMN hydrolysis (Figure 3d). In NAD^+ hydrolysis, cells isolated from CD73 KO mice showed lower conversion to Ado and reduced Nam formation (Figure 3c). In NMN hydrolysis, less NR was formed on LECs CD73 KO cells compared to the control (Figure 2b). This could confirm the involvement of CD73 in the formation of these products on the surface of the lung endothelium.

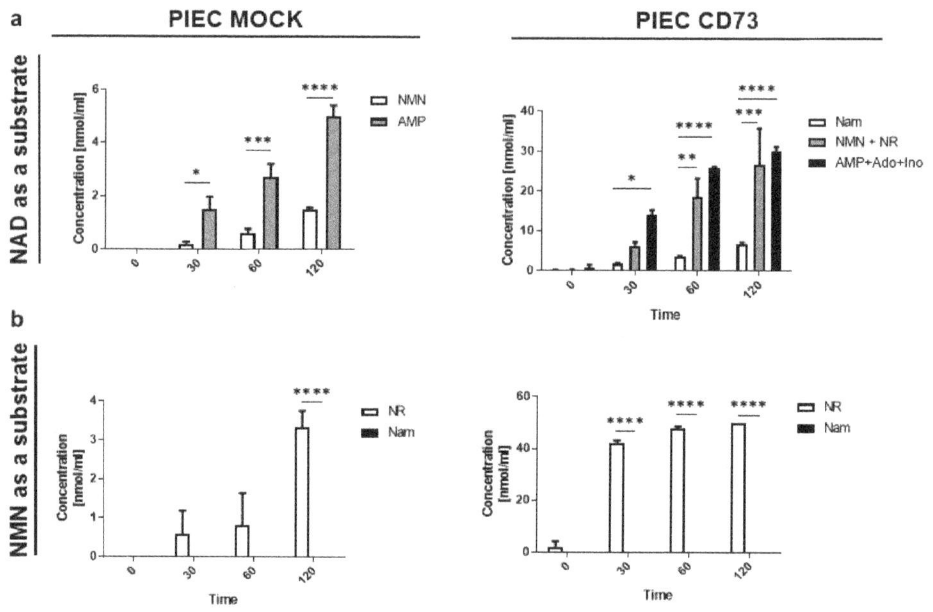

Figure 2. Transfected pig endothelial cells (PIEC) with the human CD73 gene show enhanced NAD$^+$ and NMN metabolism. Concentration of products after incubation with 50 μM NAD$^+$ (**a**) and 50 μM NMN (**b**) on the PIEC MOCK (control) and PIEC CD73. Results are presented as the mean ± SEM; n = 3–6 with two-way ANOVA followed by Tukey's posttest. * $p < 0.05$, ** $p < 0.01$, *** $p < 0.001$, **** $p < 0.0001$.

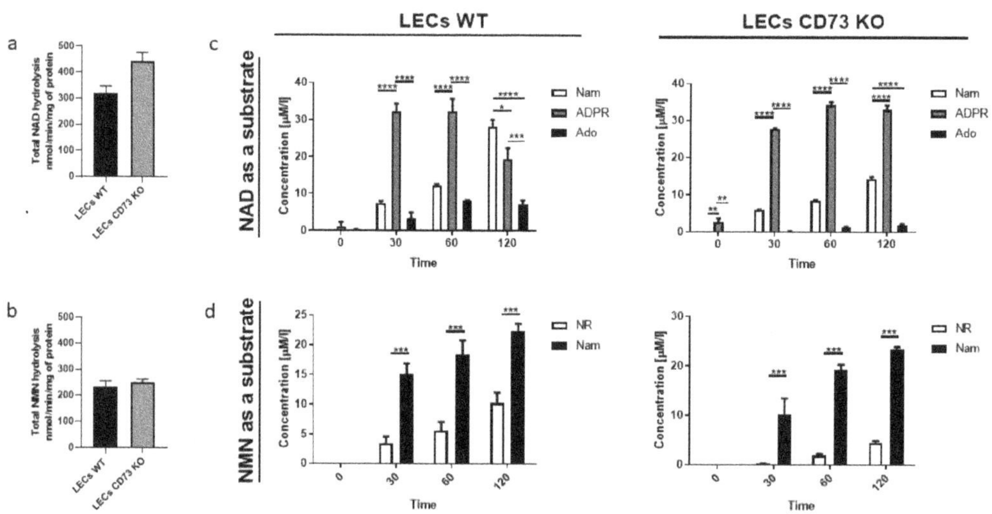

Figure 3. NAD$^+$ and NMN metabolism are different on LECs isolated from WT mice and CD73 KO mice. Total NAD (**a**) and NMN (**b**) hydrolysis on LECs WT and LECs CD73 KO. Concentration of products after incubation with 50 μM NAD$^+$ (**c**) and 50 μM NMN (**d**) on the LECs WT and LECs CD73 KO. Results are presented as the mean ± SEM; n = 3–6. * $p < 0.05$, ** $p < 0.01$, *** $p < 0.001$, **** $p < 0.0001$ with two-way ANOVA followed by Bonferroni's posttest.

3.4. LEC and PIECs Are Characterized by Different NAD^+ and NMN Metabolism

When we compared the substrate consumption of NAD^+ and NMN on LECs cells, as well as PIEC, in an experiment with inhibitors of the main enzymes involved in the metabolism of extracellular NAD^+ and its derivative NMN, we observed that, after the inhibition of CD38 activity on the surface of LECs, the NAD^+ concentration significantly increased on both LECs WT and LECs CD73 KO (Figure 4a). The knockout of CD73 caused less NAD^+ consumption, suggesting that CD73 may also be involved in its metabolism. Due to the lack of sufficient cells for the experiment, we did not perform CD38 inhibition for NMN hydrolysis. Nevertheless, we can assume that the formation of Nam and NR that we observed in Figure 2b is due to both CD38 and CD73 activity, since CD73 knockout affects NMN consumption (Figure 4b).

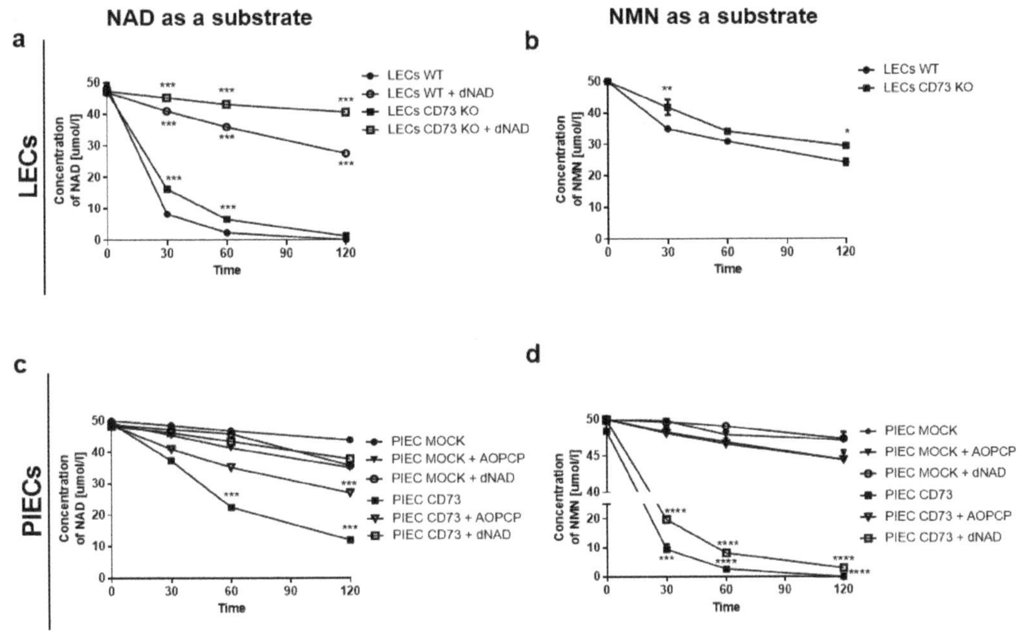

Figure 4. LEC and PIECs characterized by different NAD^+ and NMN consumption—CD38 is mainly present on the LECs, while CD73 is primarily characteristic of PIEC. Mean comparison of the changes in the NAD^+ (**a**,**c**) and NMN (**b**,**d**) concentration over time on LECs WT and CD73 KO with an inhibitor of CD38–dNAD^+ and CD73–AOPCP, and on PIEC MOCK and PIEC CD73 with these inhibitors. $n = 3–6$; * $p < 0.05$, ** $p < 0.001$, *** $p < 0.001$, **** $p < 0.0001$ vs. LECs WT or PIEC MOCK with a two-way ANOVA followed by Bonferroni's posttest.

A characteristic feature of PIEC CD73 cells is the increased metabolic activity of both NAD^+ and NMN (Figure 4c,d) due to the insertion of the additional human CD73 gene. The inhibition of CD73 (by AOPCP) did not affect the NAD^+ hydrolytic activity on the surface of PIEC MOCK cells, while it significantly reduced NAD consumption in PIEC CD73 cells. Moreover, the inhibition of CD38 by dNAD had no such effect (Figure 4c).

The NMN intake on PIEC CD73 cells was mainly due to the activity of CD73, as shown in Figure 4d. The inhibition of CD38 did not reduce NMN consumption on the PIEC CD73; the cells behaved as though they were untreated, suggesting that this enzyme was only slightly involved in NMN metabolism. In contrast, on PIEC MOCK, no significant changes in NMN consumption were observed following the inhibition of CD38 and CD73.

4. Discussion

In this study, we compared the metabolism of NAD+ and NMN in various vascular endothelial cells and confirmed that CD38 and CD73 are the main enzymes responsible for their hydrolysis. In addition, our studies showed that endothelial cells are characterized by different metabolisms of NAD+ and NMN. In the case of NAD+ metabolism on HMEC-1 cells, we demonstrated its absence within the studied time frame as compared to NMN, which may indicate a diverse affinity of the enzyme for the substrate. Additionally, differences in the metabolism of NAD+ and NMN were observed due to the type of cell culture conducted (primary or immortalized cells) and due to their species and tissue origin.

Extracellular NAD+ is a known signaling molecule that can act directly or indirectly through its metabolites at purinergic receptors and modulates many functions, including inflammatory processes and calcium signaling [19,20]. NAD+ appears in the extracellular space through lytic and non-lytic mechanisms, such as connexin 43 [21]. The major metabolites of NAD+ conversion are NMN and AMP (CD38, CD73, and CD203a activity), Nam and ADPR (CD38 activity), and NR, as a result of further metabolism of NMN (CD73 activity) [22]. In our research, we confirmed the presence of these pyridine derivatives. Furthermore, AMP was cleaved to Ado (CD73 activity) and further to Ino in PIEC CD73 cells. This is consistent with what we observed in endothelial cells when investigating the extracellular metabolism of adenine nucleotides [17]. In LECs, there were Nam and ADPR, and then Ado production. NMN, Nam, and NR metabolites could restore the intracellular NAD+ pool by the NAD+ salvage pathway [23]. Furthermore, studies about NAD+ metabolism and its precursors have great value [24]. There is ample evidence that the administration of NAD+ precursors has systemic benefits. For example, administered NR restores the tissue levels of NAD+ and increases autophagy, possibly serving as a protective response in acute kidney injury (AKI) [25]. NR supplementation and CD38 inhibition directly suppress neuroinflammation in the brain by boosting NAD+ [26]. NMN supplementation, likewise, prevents age-associated gene expression changes in essential metabolic organs and improves mitochondrial oxidative metabolism in skeletal muscle in mice [27]. Nowadays, the importance of NMN has increased and it has been assigned the function of a signaling molecule that interacts between three organs: the hypothalamus, adipose tissue, and skeletal muscle [28,29]. Moreover, interest in this compound is growing since the plasma transporter for NMN has been recently discovered [30], although not all approve of this finding [31]. Research into the beneficial effects of NAD+ enhancement therapies has also been conducted in humans [32]. In the case of administration to Nam, it is less important due to the controlled process of resynthesis to NAD+ and shorter retainment in the body than NMN [33]. Moreover, Nam may cause hepatotoxicity or flushing in high-dose medications [34].

This study showed that primary cells isolated from the lungs of WT and CD73 KO mice presented active hydrolysis of NAD+ to Nam and ADPR by CD38 on their surface. Moreover, LECs were characterized by the highest degree of hydrolysis of both NAD+ and NMN among the tested cell lines. Regardless of the lower activity of CD73, CD38 is the predominant enzyme on the surface of the lung endothelium. This is consistent with the function of CD38 in the respiratory system. NAD+ metabolites formed by CD38 (ADPR and adenosine) play a role in intracellular calcium regulation in various cell types, including airway smooth muscle (ASM) cells [35]. They contribute to airway inflammation and hyperresponsiveness [36,37]. There are also reports that CD38 knockout suppresses tumorigenesis in mice and the clonogenic growth of human lung cancer cells [38,39]. In our study, we also proved high CD38 activity on primary lung endothelial cells.

On the other hand, studies on PIEC cells showed low activity of CD38 in these cells and confirmed that CD73 is the main enzyme responsible for the hydrolysis of NAD+ and NMN. The transfection of PIEC cells by the human NT5E gene increased the degrading activity of NAD+ and NMN by about three times. The extracellular conversion of NMN to NR by CD73 localized in the luminal surface of endothelial cells represents important vasoprotective mechanisms maintaining intracellular NAD+ and the healthy phenotype of endothelial cells [35]. Much research on CD73 activity has been aimed at exploring adenosine-dependent mechanisms (AMP to adenosine hydrolysis). In our experiments, we have also tried to show the important role of the transformation of NAD^+ and NMN to NR by CD73.

An interesting result was the observation of the lack of NAD^+-cleavage activity during the two-hour incubation and the remark of NMN metabolism under the same conditions. This may indicate different affinities of enzymes for NAD^+ and NMN substrates. Differences in the Michaelis constant and Vmax were observed in the work by Mateuszuk et al. [40]. They compared the metabolism of extracellular NMN by CD73 and CD38 on Eahy.926 cells. Moreover, they showed a greater contribution of CD73 to NMN conversion in human endothelial cells compared to CD38, previously identified as the major NMN-degrading enzyme in mouse tissues in vivo [41].

Despite the importance of NAD^+ metabolism to human health and diseases, determining the levels of NAD^+ remains a challenge. Our research may also have some limitations. The reagents used for cell culturing were aimed at maximizing the growth and survival of cells in cultures and did not fully reflect the biological processes taking place in the body in vivo. FBS is an example of a compound where the presence of nucleotide pyrophosphatase and 5-nucleotidase activities has been discovered [42]. In our research, we also used this reagent. Each of the tested lines had the same conditions, so the impact on the experiment was similar. Another limitation in comparing the results with other studies may be the way of measuring NAD^+ and its metabolism. In our research, we relied on the measurement of the concentrations of the resulting products using reversed-phase HPLC [16]. Other authors are used to NMN spectroscopy [43]; however, the gold standard is LCMS [44,45].

To summarize, NAD^+ and NMN metabolism has been the subject of many studies, including brain cells [46], stem cells [47], muscles [48], fibroblasts [49,50], inflammatory cells [51], and WAT and BAT [52]. Our group has also examined NAD^+ and NMN hydrolysis on the surface of human aortic valves and vessels [53]. Endothelial cells are a common model in the study of inflammation, circulation, leukocyte transport, angiogenesis, and cancer research [54–56]. Recently, many associations with endothelial dysfunction and NAD^+ deficits, as well as CD38 activity, in the development of COVID-19 have also been demonstrated [57–60]. The differences observed in this study emphasize that the extracellular metabolism of NAD^+ and NMN are heterogeneous and results may vary depending on the tested material. This is further evidence that research on changes in the extracellular NAD^+ levels has its limitations and there is a lack of consistent standardization.

5. Conclusions

The extracellular metabolism of NAD^+ and NMN is active on endothelial cells, but the exact rates differ depending on the cell type and conditions. Primary cultures are characterized by the highest degree of NAD^+ and NMN hydrolysis compared to cell lines. Moreover, the presence of enzymes varies between species. We have demonstrated the presence of CD73, which was present in abundance on human HMEC-1 cells, while the murine cells (H5V) had less of this enzyme's activity. In the case of CD38, the amount of this enzyme on cell lines was similar, but its activity was highest in primary cells isolated from mice lungs. This may be related to the role CD38 plays in the respiratory system. On the other hand, in experiments using cells with CD73 overexpression, a significant increase in the metabolism of NAD^+ and NMN was observed, leading to the formation of Ado, which can be used, for example, in antiarrhythmic therapy. Due to differences in NAD^+ and NMN metabolism, as well as in the way this metabolism is measured, the results obtained with cell culture experiments should be compared with caution.

Supplementary Materials: The following supporting information can be downloaded at: https://www.mdpi.com/article/10.3390/biology11050675/s1, Figure S1: Determination of the dose inhibiting the metabolism of NAD^+ and NMN by deamino NAD. Figure S2: Examples from the chromatographic analysis.

Author Contributions: Conceptualization, P.J. and E.M.S.; Validation, E.M.S. and R.T.S.; Formal Analysis, P.J.; Investigation, P.J. and A.K.; Resources, P.J., P.M., B.K.-Z., M.T., P.K., M.F. and A.B.; Data Curation, P.J.; Writing—Original Draft Preparation, P.J.; Writing—Review & Editing, E.M.S. and R.T.S.; Visualization, P.J.; Supervision, E.M.S. and R.T.S.; Project Administration, E.M.S.; Funding Acquisition, P.J. All authors have read and agreed to the published version of the manuscript.

Funding: This work was supported by the Polish Ministry of Science and Higher Education (MB 664/256/63/73-3312) and the National Science Centre of Poland (UMO-2020/37/N/NZ3/03658). The funder had no role in study design, data collection, analysis, decision to publish, or preparation of the manuscript.

Institutional Review Board Statement: All experimental protocols used in this study were approved by the local commission for animal experiments in Bydgoszcz, Poland (27/2016).

Informed Consent Statement: Not applicable.

Data Availability Statement: The authors declare that the data supporting the findings of the study are available within the article.

Conflicts of Interest: The authors declare no conflict of interest.

References

1. Koch-Nolte, F.; Fischer, S.; Haag, F.; Ziegler, M. Compartmentation of NAD+-dependent signalling. *FEBS Lett.* **2011**, *585*, 1651–1656. [CrossRef] [PubMed]
2. Anwar, M.; Aslam, H.M.; Anwar, S. PARP inhibitors. *Hered. Cancer Clin. Pract.* **2015**, *13*, 4. [CrossRef] [PubMed]
3. Horenstein, A.; Chillemi, A.; Quarona, V.; Zito, A.; Roato, I.; Morandi, F.; Marimpietri, D.; Bolzoni, M.; Toscani, D.; Oldham, R.; et al. NAD+-Metabolizing Ectoenzymes in Remodeling Tumor–Host Interactions: The Human Myeloma Model. *Cells* **2015**, *4*, 520–537. [CrossRef] [PubMed]
4. Samiei, N.; Hosseini, S.; Maleki, M.; Moradi, L.; Joghataei, M.T.; Arabian, M. Modulatory Role of SIRT1 and Resistin as Therapeutic Targets in Patients with Aortic Valve Stenosis. *Arch. Med. Res.* **2019**, *50*, 333–341. [CrossRef] [PubMed]
5. Nikiforov, A.; Dölle, C.; Niere, M.; Ziegler, M. Pathways and Subcellular Compartmentation of NAD Biosynthesis in Human Cells. *J. Biol. Chem.* **2011**, *286*, 21767–21778. [CrossRef] [PubMed]
6. Zimmermann, H.; Zebisch, M.; Sträter, N. Cellular function and molecular structure of ecto-nucleotidases. *Purinergic Signal.* **2012**, *8*, 437–502. [CrossRef]
7. Boslett, J.; Hemann, C.; Christofi, F.L.; Zweier, J.L. Characterization of CD38 in the major cell types of the heart: Endothelial cells highly express CD38 with activation by hypoxia-reoxygenation triggering NAD(P)H depletion. *Am. J. Physiol. Cell Physiol.* **2018**, *314*, C297–C309. [CrossRef]
8. De Giorgi, M.; Pelikant-Malecka, I.; Sielicka, A.; Slominska, E.M.; Giovannoni, R.; Cinti, A.; Cerrito, M.G.; Lavitrano, M.; Smolenski, R.T. Functional analysis of expression of human ecto-nucleoside triphosphate diphosphohydrolase-1 and/or ecto-5′-nucleotidase in pig endothelial cells. *Nucleosides Nucleotides Nucleic Acids* **2014**, *33*, 313–318. [CrossRef]
9. Carmeliet, P. Blood vessels and nerves: Common signals, pathways and diseases. *Nat. Rev. Genet.* **2003**, *4*, 710–720. [CrossRef]
10. Eelen, G.; de Zeeuw, P.; Simons, M.; Carmeliet, P. Endothelial Cell Metabolism in Normal and Diseased Vasculature. *Circ. Res.* **2015**, *116*, 1231–1244. [CrossRef]
11. Franses, J.W.; Drosu, N.C.; Gibson, W.J.; Chitalia, V.C.; Edelman, E.R. Dysfunctional endothelial cells directly stimulate cancer inflammation and metastasis. *Int. J. Cancer* **2013**, *133*, 1334–1344. [CrossRef] [PubMed]
12. Kutryb-Zajac, B.; Mateuszuk, L.; Zukowska, P.; Jasztal, A.; Zabielska, M.A.; Toczek, M.; Jablonska, P.; Zakrzewska, A.; Sitek, B.; Rogowski, J.; et al. Increased activity of vascular adenosine deaminase in atherosclerosis and therapeutic potential of its inhibition. *Cardiovasc. Res.* **2016**, *112*, 590–605. [CrossRef] [PubMed]
13. Koszalka, P.; Ozüyaman, B.; Huo, Y.; Zernecke, A.; Flögel, U.; Braun, N.; Buchheiser, A.; Decking, U.K.M.; Smith, M.L.; Sévigny, J.; et al. Targeted disruption of cd73/ecto-5′-nucleotidase alters thromboregulation and augments vascular inflammatory response. *Circ. Res.* **2004**, *95*, 814–821. [CrossRef] [PubMed]
14. Mierzejewska, P.; Zabielska, M.A.; Kutryb-Zajac, B.; Tomczyk, M.; Koszalka, P.; Smolenski, R.T.; Slominska, E.M. Impaired l-arginine metabolism marks endothelial dysfunction in CD73-deficient mice. *Mol. Cell. Biochem.* **2019**, *458*, 133–142. [CrossRef]
15. Zhang, J.; Wang, C.; Wu, D.; Ying, W. Extracellular Degradation into Adenosine and the Activities of Adenosine Kinase and AMPK Mediate Extracellular NAD+-produced increases in the Adenylate Pool of BV2 Microglia under Basal Conditions. *bioRxiv* **2018**, *343*, 334268. [CrossRef]
16. Smolenski, R.T.; Lachno, D.R.; Ledingham, S.J.M.; Yacoub, M.H. Determination of sixteen nucleotides, nucleosides and bases using high-performance liquid chromatography and its application to the study of purine metabolism in hearts for transplantation. *J. Chromatogr. B Biomed. Sci. Appl.* **1990**, *527*, 414–420. [CrossRef]

17. Kutryb-Zajac, B.; Jablonska, P.; Serocki, M.; Bulinska, A.; Mierzejewska, P.; Friebe, D.; Alter, C.; Jasztal, A.; Lango, R.; Rogowski, J.; et al. Nucleotide ecto-enzyme metabolic pattern and spatial distribution in calcific aortic valve disease; its relation to pathological changes and clinical presentation. *Clin. Res. Cardiol.* **2020**, *109*, 1495. [CrossRef]
18. Grozio, A.; Sociali, G.; Sturla, L.; Caffa, I.; Soncini, D.; Salis, A.; Raffaelli, N.; De Flora, A.; Nencioni, A.; Bruzzone, S. CD73 Protein as a Source of Extracellular Precursors for Sustained NAD+ Biosynthesis in FK866-treated Tumor Cells. *J. Biol. Chem.* **2013**, *288*, 25938–25949. [CrossRef]
19. Moreschi, I.; Bruzzone, S.; Nicholas, R.A.; Fruscione, F.; Sturla, L.; Benvenuto, F.; Usai, C.; Meis, S.; Kassack, M.U.; Zocchi, E.; et al. Extracellular NAD + Is an Agonist of the Human P2Y 11 Purinergic Receptor in Human Granulocytes. *J. Biol. Chem.* **2006**, *281*, 31419–31429. [CrossRef]
20. Warszta, D.; Nebel, M.; Fliegert, R.; Guse, A.H. NAD derived second messengers: Role in spontaneous diastolic Ca2+ transients in murine cardiac myocytes. *DNA Repair (Amst.)* **2014**, *23*, 69–78. [CrossRef]
21. Bruzzone, S.; Guida, L.; Zocchi, E.; Franco, L.; De Flora, A. Connexin 43 hemi channels mediate Ca 2+ -regulated transmembrane NAD + fluxes in intact cells. *FASEB J.* **2001**, *15*, 10–12. [CrossRef] [PubMed]
22. Grant, R.; Berg, J.; Mestayer, R.; Braidy, N.; Bennett, J.; Broom, S.; Watson, J. A Pilot Study Investigating Changes in the Human Plasma and Urine NAD+ Metabolome During a 6 Hour Intravenous Infusion of NAD. *Front. Aging Neurosci.* **2019**, *11*, 257. [CrossRef] [PubMed]
23. Ratajczak, J.; Joffraud, M.; Trammell, S.A.J.; Ras, R.; Canela, N.; Boutant, M.; Kulkarni, S.S.; Rodrigues, M.; Redpath, P.; Migaud, M.E.; et al. NRK1 controls nicotinamide mononucleotide and nicotinamide riboside metabolism in mammalian cells. *Nat. Commun.* **2016**, *7*, 13103. [CrossRef] [PubMed]
24. Diguet, N.; Trammell, S.A.J.; Tannous, C.; Deloux, R.; Piquereau, J.; Mougenot, N.; Gouge, A.; Gressette, M.; Manoury, B.; Blanc, J.; et al. Nicotinamide riboside preserves cardiac function in a mouse model of dilated cardiomyopathy. *Circulation* **2018**, *137*, 2256–2273. [CrossRef] [PubMed]
25. Morevati, M.; Egstrand, S.; Nordholm, A.; Mace, M.L.; Andersen, C.B.; Salmani, R.; Lewin, E. Effect of NAD+ boosting on kidney ischemia-reperfusion injury. *PLoS ONE* **2021**, *16*, 2554. [CrossRef] [PubMed]
26. Roboon, J.; Hattori, T.; Ishii, H.; Takarada-Iemata, M.; Nguyen, D.T.; Heer, C.D.; Hori, O. Inhibition of CD38 and supplementation of nicotinamide riboside ameliorate lipopolysaccharide-induced microglial and astrocytic neuroinflammation by increasing NAD. *J. Neurochem.* **2021**, *158*, 311–327. [CrossRef]
27. Mills, K.F.; Yoshida, S.; Stein, L.R.; Grozio, A.; Kubota, S.; Sasaki, Y.; Redpath, P.; Migaud, M.E.; Apte, R.S.; Uchida, K.; et al. Long-Term Administration of Nicotinamide Mononucleotide Mitigates Age-Associated Physiological Decline in Mice. *Cell Metab.* **2016**, *24*, 795–806. [CrossRef]
28. Zapata-Pérez, R.; Tammaro, A.; Schomakers, B.V.; Scantlebery, A.M.L.; Denis, S.; Elfrink, H.L.; Giroud-Gerbetant, J.; Cantó, C.; López-Leonardo, C.; McIntyre, R.L.; et al. Reduced nicotinamide mononucleotide is a new and potent NAD + precursor in mammalian cells and mice. *FASEB J.* **2021**, *35*, 1–17. [CrossRef]
29. Di Stefano, M.; Nascimento-Ferreira, I.; Orsomando, G.; Mori, V.; Gilley, J.; Brown, R.; Janeckova, L.; Vargas, M.E.; Worrell, L.A.; Loreto, A.; et al. A rise in NAD precursor nicotinamide mononucleotide (NMN) after injury promotes axon degeneration. *Cell Death Differ.* **2015**, *22*, 731–742. [CrossRef]
30. Grozio, A.; Mills, K.F.; Yoshino, J.; Bruzzone, S.; Sociali, G.; Tokizane, K.; Lei, H.C.; Cunningham, R.; Sasaki, Y.; Migaud, M.E.; et al. Slc12a8 is a nicotinamide mononucleotide transporter. *Nat. Metab.* **2019**, *1*, 47–57. [CrossRef]
31. Schmidt, M.S.; Brenner, C. Absence of evidence that Slc12a8 encodes a nicotinamide mononucleotide transporter. *Nat. Metab.* **2019**, *1*, 660–661. [CrossRef] [PubMed]
32. Martens, C.R.; Denman, B.A.; Mazzo, M.R.; Armstrong, M.L.; Reisdorph, N.; McQueen, M.B.; Chonchol, M.; Seals, D.R. Chronic nicotinamide riboside supplementation is well-tolerated and elevates NAD + in healthy middle-aged and older adults. *Nat. Commun.* **2018**, *9*, 7. [CrossRef] [PubMed]
33. Kawamura, T.; Mori, N.; Shibata, K. β-nicotinamide mononucleotide, an anti-aging candidate compound, is retained in the body for longer than nicotinamide in rats. *J. Nutr. Sci. Vitaminol. (Tokyo)* **2016**, *62*, 272–276. [CrossRef] [PubMed]
34. Knip, M.; Douek, I.F.; Moore, W.P.T.; Gillmor, H.A.; McLean, A.E.M.; Bingley, P.J.; Gale, E.A.M. Safety of high-dose nicotinamide: A review. *Diabetologia* **2000**, *43*, 1337–1345. [CrossRef]
35. Graeff, R.; Guedes, A.; Quintana, R.; Wendt-Hornickle, E.; Baldo, C.; Walseth, T.; O'Grady, S.; Kannan, M. Novel Pathway of Adenosine Generation in the Lungs from NAD +: Relevance to Allergic Airway Disease. *Molecules* **2020**, *25*, 4966. [CrossRef]
36. Guedes, A.G.; Dileepan, M.; Jude, J.A.; Deshpande, D.A.; Walseth, T.F.; Kannan, M.S. Role of CD38/cADPR signaling in obstructive pulmonary diseases. *Curr. Opin. Pharmacol.* **2020**, *51*, 29–33. [CrossRef] [PubMed]
37. Deshpande, D.A.; Guedes, A.G.P.; Graeff, R.; Dogan, S.; Subramanian, S.; Walseth, T.F.; Kannan, M.S. CD38/cADPR Signaling Pathway in Airway Disease: Regulatory Mechanisms. *Mediators Inflamm.* **2018**, *2018*, 2042. [CrossRef] [PubMed]
38. Bu, X.; Kato, J.; Hong, J.A.; Merino, M.J.; Schrump, D.S.; Lund, F.E.; Moss, J. CD38 knockout suppresses tumorigenesis in mice and clonogenic growth of human lung cancer cells. *Carcinogenesis* **2018**, *39*, 242–251. [CrossRef]
39. Gally, F.; Hartney, J.M.; Janssen, W.J.; Perraud, A.L. CD38 plays a dual role in allergen-induced airway hyperresponsiveness. *Am. J. Respir. Cell Mol. Biol.* **2009**, *40*, 433–442. [CrossRef]

40. Mateuszuk, Ł.; Campagna, R.; Kutryb-Zając, B.; Kuś, K.; Słominska, E.M.; Smolenski, R.T.; Chlopicki, S. Reversal of endothelial dysfunction by nicotinamide mononucleotide via extracellular conversion to nicotinamide riboside. *Biochem. Pharmacol.* **2020**, *178*, 4019. [CrossRef]
41. Camacho-Pereira, J.; Tarragó, M.G.; Chini, C.C.S.; Nin, V.; Escande, C.; Warner, G.M.; Puranik, A.S.; Schoon, R.A.; Reid, J.M.; Galina, A.; et al. CD38 Dictates Age-Related NAD Decline and Mitochondrial Dysfunction through an SIRT3-Dependent Mechanism. *Cell Metab.* **2016**, *23*, 1127–1139. [CrossRef] [PubMed]
42. Kulikova, V.; Shabalin, K.; Nerinovski, K.; Yakimov, A.; Svetlova, M.; Solovjeva, L.; Kropotov, A.; Khodorkovskiy, M.; Migaud, M.E.; Ziegler, M.; et al. Degradation of Extracellular NAD+ Intermediates in Cultures of Human HEK293 Cells. *Metabolites* **2019**, *9*, 293. [CrossRef]
43. Shabalin, K.; Nerinovski, K.; Yakimov, A.; Kulikova, V.; Svetlova, M.; Solovjeva, L.; Khodorkovskiy, M.; Gambaryan, S.; Cunningham, R.; Migaud, M.E.; et al. NAD Metabolome Analysis in Human Cells Using ^{1}H NMR Spectroscopy. *Int. J. Mol. Sci.* **2018**, *19*, 3906. [CrossRef] [PubMed]
44. Yaku, K.; Okabe, K.; Nakagawa, T. Simultaneous measurement of NAD metabolome in aged mice tissue using liquid chromatography tandem-mass spectrometry. *Biomed. Chromatogr.* **2018**, *32*, 4205. [CrossRef] [PubMed]
45. Trammell, S.A.J.; Brenner, C. Targeted, LCMS-based Metabolomics for Quantitative Measurement of NAD(+) Metabolites. *Comput. Struct. Biotechnol. J.* **2013**, *4*, e201301012. [CrossRef] [PubMed]
46. Pang, H.; Jiang, Y.; Li, J.; Wang, Y.; Nie, M.; Xiao, N.; Hu, Z. Aberrant NAD + metabolism underlies Zika virus-induced microcephaly. *Nat. Metab.* **2021**, *3*, 1109–1124. [CrossRef]
47. Wang, J.; Liu, L.; Ding, Z.; Luo, Q.; Ju, Y.; Song, G. Exogenous NAD+ postpones the d-gal-induced senescence of bone marrow-derived mesenchymal stem cells via sirt1 signaling. *Antioxidants* **2021**, *10*, 254. [CrossRef]
48. Ryu, D.; Zhang, H.; Ropelle, E.R.; Sorrentino, V.; Mázala, D.A.; Mouchiroud, L.; Auwerx, J. NAD+ repletion improves muscle function in muscular dystrophy and counters global PARylation. *Sci. Transl. Med.* **2016**, *8*, 5504. [CrossRef]
49. Shi, B.; Wang, W.; Korman, B.; Kai, L.; Wang, Q.; Wei, J.; Varga, J. Targeting CD38-dependent NAD + metabolism to mitigate multiple organ fibrosis. *iScience* **2020**, *24*, 1902. [CrossRef]
50. Aleo, M.F.; Giudici, M.L.; Sestini, S.; Danesi, P.; Pompucci, G.; Preti, A. Metabolic fate of extracellular NAD in human skin fibroblasts. *J. Cell. Biochem.* **2000**, *80*, 360–366. [CrossRef]
51. García-Rodríguez, S.; Rosal-Vela, A.; Botta, D.; Cumba Garcia, L.M.; Zumaquero, E.; Prados-Maniviesa, V.; Cerezo-Wallis, D.; Lo Buono, N.; Robles-Guirado, J.Á.; Guerrero, S.; et al. CD38 promotes pristane-induced chronic inflammation and increases susceptibility to experimental lupus by an apoptosis-driven and TRPM2-dependent mechanism. *Sci. Rep.* **2018**, *8*, 6. [CrossRef] [PubMed]
52. Benzi, A.; Sturla, L.; Heine, M.; Fischer, A.W.; Spinelli, S.; Magnone, M.; Bruzzone, S. CD38 downregulation modulates NAD + and NADP(H) levels in thermogenic adipose tissues. *Biochim. Biophys. Acta Mol. Cell Biol. Lipids* **2021**, *1866*, 8819. [CrossRef] [PubMed]
53. Jablonska, P.; Kutryb-Zajac, B.; Mierzejewska, P.; Jasztal, A.; Bocian, B.; Lango, R.; Rogowski, J.; Chlopicki, S.; Smolenski, R.T.; Slominska, E.M. The new insight into extracellular NAD + degradation-the contribution of CD38 and CD73 in calcific aortic valve disease. *J. Cell. Mol. Med.* **2021**, *25*, 5884–5898. [CrossRef] [PubMed]
54. Cutler, B.R.; Chua, J.S.; Balagurunathan, K.; Anandh Babu, P.V. Methods to Analyze the Effect of Diet-Derived Metabolites on Endothelial Inflammation and Cell Surface Glycosaminoglycans. *Methods Mol. Biol.* **2022**, *2303*, 469–476. [CrossRef]
55. Canet, F.; Iannantuoni, F.; Marañon AM, D.; Díaz-Pozo, P.; López-Domènech, S.; Vezza, T.; Víctor, V.M. Does Empagliflozin Modulate Leukocyte-Endothelium Interactions, Oxidative Stress, and Inflammation in Type 2 Diabetes? *Antioxidants* **2021**, *10*, 1228. [CrossRef]
56. Armani, G.; Pozzi, E.; Pagani, A.; Porta, C.; Rizzo, M.; Cicognini, D.; Ferraris, E. The heterogeneity of cancer endothelium: The relevance of angiogenesis and endothelial progenitor cells in cancer microenvironment. *Microvasc. Res.* **2021**, *138*, 4189. [CrossRef]
57. Al-Farabi, M.J.; Nugraha, R.A.; Marsudi, B.A.; Azmi, Y. Biomarkers of endothelial dysfunction and outcomes in coronavirus disease 2019 (COVID-19) patients: A systematic review and meta-analysis. *Microvasc. Res.* **2021**, *138*, 4224. [CrossRef]
58. Evans, P.C.; Ed Rainger, G.; Mason, J.C.; Guzik, T.J.; Osto, E.; Stamataki, Z.; Neil, D.; Hoefer, I.E.; Fragiadaki, M.; Waltenberger, J.; et al. Endothelial dysfunction in COVID-19: A position paper of the ESC Working Group for Atherosclerosis and Vascular Biology, and the ESC Council of Basic Cardiovascular Science. *Cardiovasc. Res.* **2020**, *116*, 2177–2184. [CrossRef]
59. Altay, O.; Arif, M.; Li, X.; Yang, H.; Aydın, M.; Alkurt, G.; Mardinoglu, A. Combined Metabolic Activators Accelerates Recovery in Mild-to-Moderate COVID-19. *Adv. Sci.* **2021**, *8*, 1222. [CrossRef]
60. Horenstein, A.L.; Faini, A.C.; Malavasi, F. CD38 in the age of COVID-19: A medical perspective. *Physiol. Rev.* **2021**, *101*, 1457–1486. [CrossRef]

biology

MDPI

Article

Trophism and Homeostasis of Liver Sinusoidal Endothelial Graft Cells during Preservation, with and without Hypothermic Oxygenated Perfusion

Francesco Vasuri [1], Giuliana Germinario [2,3], Carmen Ciavarella [4], Michele Carroli [1], Ilenia Motta [4], Sabrina Valente [4], Matteo Cescon [2,3], Antonia D'Errico [1], Gianandrea Pasquinelli [4] and Matteo Ravaioli [2,3,*]

1 Pathology Unit, IRCCS Azienda Ospedaliero-Universitaria di Bologna, Via Albertoni 15, 40138 Bologna, Italy
2 Department of General Surgery and Transplantation, IRCCS Azienda Ospedaliero-Universitaria di Bologna, Via Albertoni 15, 40138 Bologna, Italy
3 Department of Medical and Surgical Sciences (DIMEC), University of Bologna, 40138 Bologna, Italy
4 Clinical Pathology, Experimental, Diagnostic and Specialty Medicine Department (DIMES), University of Bologna, IRCCS Azienda Ospedaliero-Universitaria di Bologna, Via Albertoni 15, 40138 Bologna, Italy
* Correspondence: matteo.ravaioli6@unibo.it; Tel.: +39-051-214-4810

Simple Summary: Little is known about the functions and intracellular mechanisms of the endothelial cells in liver grafts. In particular, we still know little about the effect of the most recent machine perfusion techniques currently applied to improve liver transplant outcomes. In this study we analyzed different endothelial markers of both immunohistochemical and gene expression in two different biopsies (for each donor). We observed a severe depression of endothelial trophism in liver grafts, which is restored after reperfusion. This is interesting for further studies on liver grafts, especially considering that the execution of HOPE seems to improve this functional recovery. We propose that our results may help improve the knowledge on graft tissues in order to customize the perfusion techniques prior to transplant and, therefore, improve liver transplant outcomes.

Abstract: The aim of the present study was to evaluate the homeostasis and trophism of liver sinusoidal endothelial cells (LSECs) in vivo in different stages of liver graft donation, in order to understand the effects of graft ischemia and perfusion on LSEC activity in liver grafts. Special attention was paid to grafts that underwent hypothermic oxygenated perfusion (HOPE). Forty-seven donors were prospectively enrolled, and two distinct biopsies were performed in each case: one allocation biopsy (at the stage of organ allocation) and one post-perfusion biopsy, performed after graft implant in the recipients. In all biopsies, immunohistochemistry and RT-PCR analyses were carried out for the endothelial markers CD34, ERG, Nestin, and VEGFR-2. We observed an increase in CD34 immunoreactivity in LSEC during the whole preservation/perfusion period ($p < 0.001$). Nestin and ERG expression was low in allocation biopsies, but increased in post-perfusion biopsies, in both immunohistochemistry and RT-PCR ($p < 0.001$). An inverse correlation was observed between ERG positivity and donor age. Our results indicate that LSEC trophism is severely depressed in liver grafts, but it is restored after reperfusion in standard conditions. The execution of HOPE seems to improve this recovery, confirming the effectiveness of this machine perfusion technique in restoring endothelial functions.

Keywords: CD34; endothelial cells; ERG; hypothermic oxygenated perfusion; liver transplantation; machine perfusion; nestin; VEGFR

Citation: Vasuri, F.; Germinario, G.; Ciavarella, C.; Carroli, M.; Motta, I.; Valente, S.; Cescon, M.; D'Errico, A.; Pasquinelli, G.; Ravaioli, M. Trophism and Homeostasis of Liver Sinusoidal Endothelial Graft Cells during Preservation, with and without Hypothermic Oxygenated Perfusion. *Biology* **2022**, *11*, 1329. https://doi.org/10.3390/biology11091329

Academic Editor: Gaetano Santulli

Received: 20 July 2022
Accepted: 5 September 2022
Published: 8 September 2022

1. Introduction

Vascular endothelial cells represent the interface between blood flow and tissue, exerting several functions beyond a mere barrier, including tissue homeostasis, metabolite and nutrient cross-transport, regulation of inflammation and neoangiogenesis, as well as control of the muscular tone of the vessel wall [1]. Liver sinusoidal endothelial cells (LSECs) show an even higher specialization level, being characterized by discontinuous surfaces, diffuse fenestrae for the passage of macromolecules, and the absence of a basal membrane [2]. LSECs actively cross-talk with hepatocytes, stellate cells, and Kupffer cells, contributing to all liver physiological processes: for instance, LSECs inhibit the profibrotic and vasoconstrictive activity of stellate cells [2]. In chronic liver diseases, in response to direct or indirect harmful stimuli, LSECs undergo so-called "endothelization" (or "capillarization") [3], with loss of fenestrae and acquisition of a phenotype commonly seen in usual capillaries, among which the expression of CD34 is the most evident, as well as the most useful in histopathological practice [4]. During chronic damage and endothelization, LSEC lose the inhibitory function on stellate cells, moving towards profibrotic and vasoconstrictive activity [2].

In the few last decades, the poor availability of grafts for liver transplantation has led to the search for new strategies to increase the donor pool: one of these strategies is the use of organs from extended criteria donors (ECD). The ECD inclusion criteria are age >60–65 years, and/or the presence of other criteria, such as donors in cardiovascular death (DCD), high transaminase levels or hypernatremia, long intensive care, or prolonged cold ischemia time [5]. Organ preservation is crucial when ECD transplant grafts are utilized, which appear more vulnerable when treated with standard techniques, such as static cold storage (SCS) [6], leading to a high risk of acute and chronic liver injury caused by ischemia–reperfusion after liver transplantation of ECD [7]. The machine perfusion (MP) is a recent organ preservation strategy to increase the survival of organs from ECD: at present, dynamic MP can be performed in hypothermic, subnormothermic, and normothermic conditions [8]. The hypothermic oxygenated perfusion (HOPE) allows the redirection from anaerobic metabolism to aerobic metabolism under hypothermic conditions, and protects grafts from species-related oxidative damage. The superiority of HOPE preservation to simple cold storage was reported in clinical liver preservation studies [9–12]. LSECs play a key role in the regulation of the venous pressure gradient (at 4 mmHg in normal conditions), by producing vasodilating (nitric oxide synthase, NOS) and vasoconstricting (endothelin-1) factors in response to intrahepatic shear stress [13]: this is of utmost importance, since one of the protective functions of machine perfusion techniques is the maintenance of adequate shear stress, in order to avoid endothelial damages and the shift in balance toward thrombosis and vasoconstriction [14]. The preservation of LSECs during machine perfusion is strongly time-dependent, since it has been demonstrated that LSECs can bear 8–16 h of cold storage, versus the 72 h reported for hepatocytes in vitro, as well as the fact that ischemic damage to LSECs drastically impacts on hepatocyte functional response in vivo [15,16].

The ETS-related gene (ERG) is a transcription factor with an emerging role in the regulation of endothelial functions, during embryogenesis, but also in several adult tissues [17]. ERG expression is constitutively regulated by several genes involved in neoangiogenesis and response to hypoxia, such as NOS, VE-cadherin, and von Willebrand factor, and it is inhibited by proinflammatory stimuli such as TNF-α [18–21]. Nestin represents another interesting marker of endothelial trophism and intracellular homeostasis regulation: it is a type IV intermediate filament, originally described in nervous stem cells, where it regulates the radial axon growth [22]. Our group studied Nestin expression and localization in the liver, finding a basal immunohistochemical expression in the portal arterioles: sinusoids usually do not show Nestin positivity, with the noteworthy exception of neoplastic sinusoids during hepatocellular carcinoma progression [23,24].

The aim of the present study is to evaluate the expression of different markers of LSECs trophism and homeostasis in vivo at different stages of liver graft donation, before and after HOPE, in order to understand the effects of graft ischemia and perfusion on LSECs.

2. Materials and Methods

2.1. Donors' Enrolment

The present study was approved in advance by the local Ethical Committee (code number 65/2018/SPER/AOUBo), and follows the ethical guidelines of the 1975 Declaration of Helsinki (6th revision, 2008); donors are kept anonymous. The study includes donors previously published in the context of a larger open-label, prospective, single-center, randomized clinical trial, where patients were stratified based on the contemporary presence of ECD liver criteria and randomized in a 1:1 ratio to receive a liver preserved with either HOPE after SCS during transportation or with SCS alone [9,25]. Livers undergoing SCS were stored in sterile organ bags with Belzer or Celsior solution, and cooled on ice as previously described and according to the Center policy [9,10,25]. HOPE was performed through the portal vein at a pressure of 5 mmHg by flushing the organ at low flow values (30 mL/min) with new oxygenated perfusion fluid (Belzer MPS) during back-table preparation with the aim of removing waste products and residual microthrombi, and to provide oxygen. Successively, the organs were treated with continuous HOPE until the grafts were transplanted.

An eligibility criterion specific for the present study was the availability of two distinct biopsies:

1. Allocation biopsy (A-Biopsy), performed at the stage of organ allocation for the assessment of graft suitability, as usually performed in our Institution [26];
2. Post-perfusion biopsy (PP-Biopsy), performed for the purposes of the study after graft implant in the recipients. This biopsy was specifically performed for the study.

Of the 110 liver donors prospectively enrolled in the context of the monocentric clinical trial, 49 satisfied the eligibility criteria; two more donors were excluded due to the lack of sufficient tissue in A-Biopsies to perform IHC: 47 donors were therefore finally enrolled for the current study, 22 males and 25 females, with a mean age of 74.8 $\pm$ 10.1 years (range 42–87 years).

Of these 47 donors, 34 (72.3%) donors were enrolled for HOPE perfusion, while 13 (27.7%) were preserved in SCS.

Collected donor variables included tobacco and/or alcohol consumption, chronic use of medications; heart, lung, and liver disease; hypertension; diabetes; dyslipidemia; nephropathy; vasculopathies; dyslipidemia; and body mass index (BMI). Machine perfusion variables included flow, pressure, and resistance; gas analysis variables of the effluent perfusate included PH, pCO2, paO2, glucose, and lactate at start of perfusion, and then every 30 min.

Analysis of recipients' characteristics was not included in the present study due to the study aims and the small sample size; however, a follow-up of the recipients was preliminarily included after a mean follow-up of 305.6 $\pm$ 193.5 days (range 45–723 days). Follow-ups included early allograft dysfunction (EAD) and primary non-function (PNF) cases, post-transplant acute kidney injury (AKI) with or without the need of continuous venovenous hemofiltration (CVVH), graft survival, and recipient survival.

2.2. Histopathological Analysis and Immunohistochemistry

Different endothelial markers were applied to evaluate the trophism of LSECs before and after perfusion. The term "trophism" was used to indicate the activation of those intracellular pathways leading to cell protection and/or survival in course of hypoxic stresses.

A-Biopsies were routinely frozen for quick histopathological evaluation for graft suitability, and subsequently fixed in formalin and embedded in paraffin (FFPE); PP-Biopsies were directly FFPE at the stage of sampling. In all cases, FFPE tissue was routinely processed, and 2-μm-thick sections were cut for hematoxylin–eosin and reticulin silver staining, as well as for immunohistochemistry (IHC). The histopathological variables collected are the same of the donor's checklist usually applied in our institution [26]: portal fibrosis stage according to Ishak [27], lobular fibrosis (absent, focal, diffuse), portal and lobular inflammation according to Ishak, lobular necrosis according to Ishak, myointimal thickening of hepatic arteries and arterioles (absent, mild, moderate, and severe), bile duct regression (absent, present) with or without bile duct reaction, percentage of macrovesicular and microvesicular steatosis, cholestasis (absent, mild, moderate, severe).

IHC was automatically performed by means of the automated immunostainer Benchmark® ultra (Ventana Medical Systems, Inc, Roche group, Tucson, AZ, USA), following the manufacturer's instructions. The antibodies used in the present study included: CD34 (clone QBEnd/10), Nestin (clone 10C2), and ERG (clone EPR3864). CD34 immunoreactivity was semiquantitively assessed in LSECs as mild/focal, moderate, and severe/diffuse, based on the extension of sinusoids "endothelization" from the periportal zone to the whole lobule. Nestin immunoreactivity was evaluated at two levels: (1) semiquantitively at LSEC level (when Nestin-positive sinusoids were present), applying the same method as CD34; (2) quantitatively counting the highest number of periportal Nestin-positive capillaries, which we observed to be common in our series. ERG immunoreactivity was quantitatively assessed by counting the number of ERG-positive endothelial nuclei in 10 high-power fields (HPF, 40× magnification): the mean number of ERG-positive nuclei/10 HPF was therefore counted. As controls, two liver graft specimens from healthy living donors (age 31 and 39 years) were used: in these cases, no CD34-positive endothelization of LSEC and no significant Nestin immunoreactivity were observed (as expected); the mean number of ERG-positive nuclei was 13.3 and 14.5/10 HPF respectively.

2.3. RNA Extraction and Reverse Transcriptase-Polymerase Chain Reaction (RT-PCR)

Total RNA was extracted from 16 FFPE tissue samples, eight HOPE and eight SCS cases, from both A-Biopsies and PP-Biopsies, using the FFPE Recover All (Thermo Fisher Scientific, Waltham, MA, USA) according to the manufacturer's instructions. RNA quality and concentration was evaluated by a ND-1000 spectrophotometer (NanoDrop, Thermo Fisher Scientific, Waltham, MA, USA). Reverse transcription was performed from 0.5 µg of total RNA in 20 µL reaction volume using the High-Capacity cDNA Reverse Transcription Kit (Thermo Fisher Scientific). Real-Time PCR for the analysis of Nestin, ETS Transcription Factor (ERG) and Vascular Endothelial Growth Factor Receptor-2 (VEGFR-2) was assessed in a CFX Connect Real-Time PCR Detection System (Bio-Rad Laboratories. Hercules, CA, USA) using the SYBR green master mix (Bio-Rad Laboratories). Primer sequences were designed through the NCBI primer tool and were the following: Nestin FWD GACCCTGAAGGGCAATCACA; Nestin REV GGCCACATCATCTTCCACCA; ERG FWD TCGCATTATGGCCAGCACTA, ERG REV CGTTCCGTAGGCACACTCAA; VEGFR-2 FWD CGGTCAACAAAGTCGGGAGA; VEGFR-2 REV CAGTGCACCACAAAGACACG (Sigma-Aldrich. St. Louis, MO, USA). Reactions were performed in triplicate and target gene expression was normalized to glyceraldehyde 3-phosphate dehydrogenase (GAPDH). Relative quantification was assessed by the comparative $2^{-\Delta\Delta Ct}$ method and data were expressed as fold changes of mRNA expression relative to A-Biopsies.

2.4. Statistical Analysis

Variables were reported as means ± standard deviation, ranges, and frequencies. Statistical analysis was carried out using SPSS® software for Windows, ver. 20. When applicable, t-tests, ANOVA, and non-parametric Mann–Whitney and Kruskal–Wallis tests were used to compare the variables. p values <0.05 was considered statistically significant.

3. Results

3.1. Donor and Graft Characteristics

Mean donors' BMI was 26.8 ± 4.5 (range 19.9–46.7), tobacco use was recorded in 17 (36.2%) cases, history of alcohol abuse in one (2.1%) case, cardiopathies were recorded in 18 (38.3%) cases, pneumopathies in 10 (21.3%), chronic hepatopathies in two (4.3%), chronic nephropathies in four (8.5%), hypertension in 30 (63.8%), diabetes in six (12.8%) donors (one of which insulin-dependent), dyslipidemia in 16 (34.0%) cases, and arterial and venous vasculopathies in eight (17.0%) and four (8.5%) donors, respectively; 37 (78.7%) donors chronically used medications (mostly antihypertensives).

Recipient follow-ups recorded seven (14.9%) cases of EAD, no cases (0%) of PNF, and 15 (31.9%) AKI cases, with seven (14.9%) cases needing CVVH. Only two recipients died in the follow-up, and only one needed retransplantation.

The main histopathological characteristics of the 47 liver grafts enrolled are listed in Table 1. Both A-Biopsies and PP-Biopsies were separately revised by two dedicated pathologists, blind to each other: no substantial differences were observed between A-Biopsies and the corresponding PP-Biopsies as far as histopathology is concerned. In case of discordance between the two biopsies concerning single variables, PP-Biopsy characteristics were chosen, due to the better tissue quality and lack of freezing artifacts [28]. In the 34 grafts perfused with HOPE, all the perfusion variables concerning the overall procedure were recorded as well, including both the flushing and the perfusion phases, according to the study protocol (Supplementary Table S1) [9].

Table 1. Histopathological features of the 47 liver grafts evaluated.

Histopathological Variable	No. (Percentage)
Portal Fibrosis (Ishak's stage)	
0	3 (6.4%)
1	19 (40.4%)
2	23 (48.9%)
3	2 (4.3%)
Lobular Fibrosis	
Absent	31 (66.0%)
Mild/focal	12 (25.5%)
Severe/diffuse	4 (8.5%)
Portal Inflammation	
Absent	7 (14.9%)
Mild	35 (75.5%)
Moderate-to-severe	5 (10.6%)
Myointimal Thickening	
Absent/mild	22 (46.8%)
Moderate	17 (36.2%)
Severe	8 (17.0%)
Biliocyte/bile duct regression	
Absent	5 (10.6%)
Mild	38 (80.9%)
Severe	4 (8.5%)
Lobular Inflammation	
Absent	30 (63.8%)
Mild	15 (31.9%)
Moderate-to-severe	2 (4.3%)
Lobular NecrosisAbsent	33 (70.2%)
Mild	11 (23.4%)
Moderate-to-severe	3 (6.4%)
Cholestasis	
Absent	36 (76.6%)
Mild	11 (23.4%)
Microvesicular steatosis (mean)	6.5 ± 6.7% (0–20%)
Macrovesicular steatosis (mean)	4.2 ± 6.1% (0–25%)

3.2. Modifications of Endothelial Trophism in Allocation and Post-Reperfusion Biopsies

In A-Biopsies, sinusoidal CD34 expression was mild/focal in 25 (53.2%) cases, moderate in 13 (27.7%), and diffuse in nine (19.1%). In PP-Biopsies, LSEC CD34 expression was mild/focal in five (10.6%) cases, moderate in 20 (42.6%), and diffuse in 22 (46.8%). We therefore observed an increase in CD34 immunoreactivity in graft sinusoids, mean 0.7 ± 0.9 points, as confirmed by the fact that the "diffuse" cases increased from 19.1% to 46.8%. This difference was statistically significant ($p < 0.001$, *t*-test; Figure 1a,b).

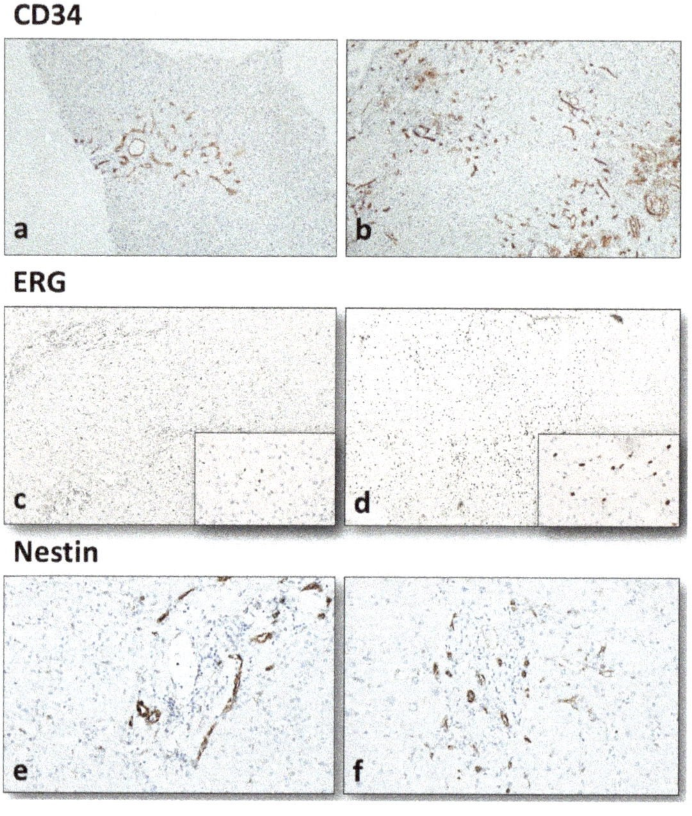

Figure 1. Immunohistochemistry results. CD34 immunoreactivity increased in liver sinusoids from allocation biopsy ((**a**), 2× magnification) to post-perfusion biopsy ((**b**), 2× magnification), as a progressively increasing periportal endothelization ($p < 0.001$). The increased expression of ERG was also statistically significant ($p < 0.001$) between allocation biopsy ((**c**), 2× magnification) and post-perfusion biopsy ((**d**), 2× magnification). The higher density of ERG-positive nuclei is highlighted by higher magnification (small squares, 20× magnification). Two portal tracts immunostained by Nestin ((**e**,**f**), 10× magnification), showing a variable number of periportal capillaries and a mild immunoreactivity in adjacent sinusoid.

The mean number of ERG-positive nuclei in A-Biopsy LSECs was 4.9 ± 7.5, with a maximum number ranging from 1 to 48 (mean maximum number 11.7 ± 12.5). The mean number of ERG-positive nuclei in PP-Biopsy sinusoidal endothelial cells was 22.0 ± 5.8, with a maximum number ranging from 13 to 47 (mean maximum number 28.3 ± 8.4). The increased expression of ERG was statistically significant ($p < 0.001$, Mann–Whitney test); however, an inverse correlation was observed between ERG positivity (on both biopsies) and donor age, as represented in Figure 2 ($p = 0.031$, ANOVA test; Figure 1c,d).

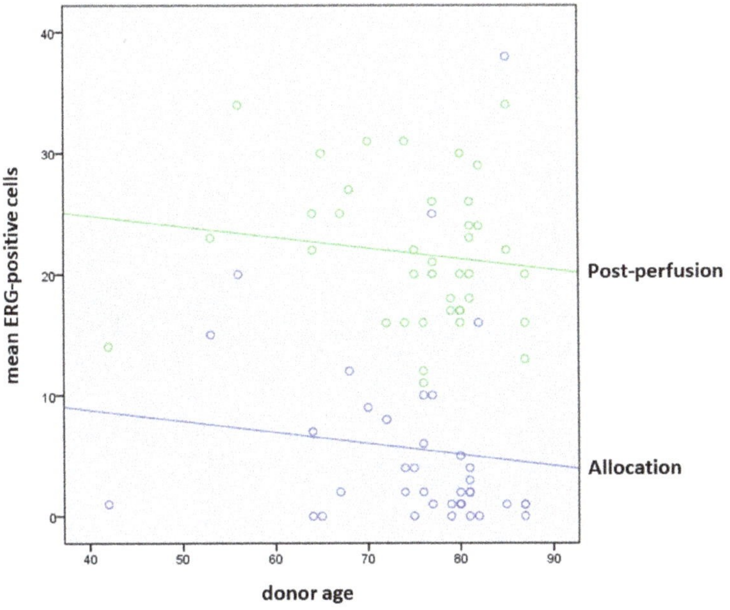

Figure 2. In post-perfusion biopsies (green dots and line), ERG significantly increased compared to allocation biopsies (blue dots and line), but in both groups, an inverse correlation was observed between ERG immunoreactivity and donor age.

Mild Nestin immunoreactivity was observed in LSECs in 6 (12.8%) A-Biopsies, with a low mean number of Nestin-positive periportal capillaries (mean 1.8 ± 2.6 positive periportal capillaries in hot spots). Nestin immunoreactivity in PP-Biopsy LSECs was absent in six (12.8%) cases, mild in 21 (44.6%), moderate in 14 (29.8%), and diffuse in six (12.8%). The mean number of Nestin-positive periportal capillaries was 7.4 ± 4.2 in hot spots (range 0–24). As for the first two markers, this increased Nestin expression from A-Biopsies to PP-Biopsies was statistically significant ($p < 0.001$, t-test; Figure 1e,f).

RT-PCR analysis confirmed the induction of the intracellular endothelial pathways suggested by IHC: the mean fold increases between A-Biopsies and PP-Biopsies for VEGFR-2, ERG and Nestin genes were 9.95 ± 17.77, 4.63 ± 7.83, and 2.04 ± 1.83, respectively (Figure 3). Notably, no differences in IHC expression were found between grafts perfused with HOPE and grafts preserved in SCS (see also Table 2), while in RT-PCR, the increase expression of VEGFR-2 was significantly higher after HOPE (mean 15.09) than after SCS (mean 1.36, $p = 0.025$, Kruskal–Wallis test), suggesting a beneficial role of HOPE in terms of endothelial cell homeostasis and survival.

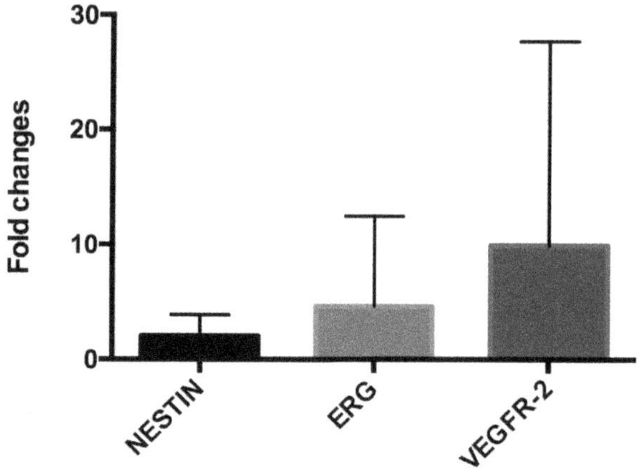

Figure 3. At RT-PCR analysis the mean fold increase from allocation biopsies to post-perfusion biopsies for was 2.04 for Nestin, 4.63 for ERG, and 9.95 for VEGFR-2.

Table 2. General, immunohistochemical and main follow-up of the patients enrolled in the HOPE and static cold storage (SCS) groups. * Chi-square test.

	HOPE (n = 34)	SCS (n = 13)	Sig.
Mean Age	73.7 years	78.8 years	n.s.
Sex (male)	17 (50%)	5 (38.5%)	n.s.
MELD score	14.4	23.0	n.s
Cold ischemia time (min)	411.2	361.8	n.s.
Post-perfusion CD34 (diffuse endothelization)	14 (41.2%)	7 (53.8%)	n.s.
Post-perfusion mean Nestin-positive capillaries	7.3	7.5	n.s.
Post-perfusion mean ERG-positive sinusoids	21.6	23.2	n.s.
Early allograft dysfunction	3 (8.8%)	4 (30.1%)	$p = 0.064$ *
Primary non-function	0	0	n.s.
Graft failure (retransplantation)	0	1	n.s.
Recipients' death	1	1	n.s

3.3. *Perfusion Characteristics Influencing Endothelial Trophism in Liver Grafts and Preliminary Follow-Up Results*

Among the 34 grafts who underwent to HOPE prior to liver transplantation, some perfusion characteristics correlated with the expression of the three endothelial markers. In particular, the increase in CD34 immunoreactivity in LSECs (endothelization) positively correlated with the time of the overall perfusion ($p = 0.047$, Spearman's test). The number of Nestin-positive periportal capillaries in PP-biopsies was positively correlated with the Lactate concentration ($p = 0.008$), and negatively correlated with the pH ($p = 0.005$) at the beginning of the re-cycle. Finally, the increase in the number of ERG-positive nuclei in PP-Biopsies positively correlated with the pO2 at the end of re-cycle ($p = 0.044$).

The only follow-up data that showed a correlation with graft endothelial trophism was the need for CVVH after transplant, which directly correlated with CD34 positivity in PP-Biopsies, suggesting a correlation between LSEC endothelization and kidney function in liver transplant recipients. In particular, all seven CVVH cases showed strong and diffuse positivity for CD34 at PP-Biopsy (p = 0.017, t-test).

4. Discussion

The present study aimed to evaluate the trophism in LSECs in vivo during liver graft procurement, preservation, and perfusion with HOPE. For this purpose, we carried out a morphological, IHC, and RT-PCR analyses on different endothelial markers on graft tissue before and after HOPE, as well as before and after SCS. Our results show an increase in LSEC CD34 immunoreactivity, which occurred invariably during the whole preservation/perfusion time: in the grafts perfused with HOPE, this endothelization is proportional to the overall perfusion time, but it does not seem to correlate to any of the other variables. CD34 immunoreactivity is a sign of so-called LSEC "endothelization", and even in the present context, it should be considered a sign of protection from a mechanic—rather than biologic—stimulus. Nonetheless, capillarized LSECs have been recently shown to produce extracellular matrix, thus demonstrating an indirect link between endothelization and so-called endothelial–mesenchymal transition (EndMT) [29]. EndMT is a model proposed to describe all those processes during which the endothelial cells lose their phenotype and normal functions in response to pathological noxae, to acquire a less differentiated mesenchymal phenotype, characterized by fusiform cell shape and extracellular matrix production, among others [30]. Hypoxia and inflammation have already been reported to stimulate EndMT; thus, this process is likely to play a role in regulating LSEC function during liver graft ischemia and reperfusion [30]. This setting fits well with the more "traditional" view of LSEC endothelization as a sort of metaplastic shift of endothelial cells towards a less specialized phenotype as a protection from high portal pressure, as it is demonstrated by the diffuse LSEC endothelization in course of cirrhosis [3,4]. The finding of a correlation between post-perfusion CD34 immunoreactivity and the need of CVVH in recipients reinforce further the importance of future studies on intrahepatic resistance and pressure of the grafts, both during HOPE and after transplantation.

In this study, we used Nestin as a more specific marker of EndMT: Nestin positivity is not normally observed in LSECs, as already reported by our group [23], as well as demonstrated by our controls. In A-Biopsies, the Nestin positivity generally remained very low, concerning both staining intensity and diffusion. Conversely, an increase in Nestin expression was observed in LSECs from PP-Biopsies with both IHC and gene expression: a periportal Nestin-positive microcirculation became visible in most cases, and RT-PCR demonstrated a twofold increase in Nestin mRNA from A-Biopsies to PP-biopsies. In addition, we observed a higher Nestin expression in HOPE grafts with higher lactate and lower pH at the beginning of the re-cycle. All these observations highlight a possible role of Nestin in EndMT as an adaptative mechanism of liver microcirculation during ischemia time and reperfusion.

ERG is a known mediator of endothelial trophism, essential for physiological and pathological neoangiogenesis [31], and it is generally expressed in normal endothelial cells [32], as also observed in our controls. ERG has been described to prevent the activation of the prothrombotic pathway during low shear stress (as in cold storage) [30], as well as to be involved in the regulation of the genes of the TGF-β family by enhancing Smad1 and reducing Smad2 and 3 activities [33]: through these mechanisms, ERG protect the endothelial cells from EndMT [34]. In addition, ERG inhibits EndMT by regulating several genes such as TGF-β, FLI1, Notch, and others [18–21,33]. ERG has been demonstrated to induce endothelial differentiation of non-vascular amniotic cell in vitro, actually reversing EndMT process [34]. An interesting finding in our study is the very low number of ERG-positive nuclei in A-Biopsies, which was further confirmed via gene expression with RT-PCR, followed by a total recovery after reimplantation (on PP-Biopsies). Unlike the former two markers, which increased as a response to mechanical and/or metabolic needs during ischemia and perfusion, the recovery of ERG expression is to be considered as restoring normal—and sometimes greater—endothelial trophism. This is also confirmed by our RT-PCR results, which showed a 4.6-fold increase in ERG mRNA in the PP-Biopsy, and a nearly 10-fold increase in VEGFR-2, which is a major vascular growth factor that works together with ERG in all neoangiogenetic processes, via the VEGF/MAPK/ERG pathway [17]. Notably, the VEGFR-2 increase was significantly higher after HOPE than after SCS: in the present study, the sample size and the low number of main events (two recipient deaths and one graft loss) prevents us from conducting a survival analysis (which was out of the study aims), but our results highlight once more the safety and the efficacy of HOPE at least in restoring LSEC functionality and cellular viability. The efficacy of HOPE in terms of short- and medium-term follow-up was reported in previous studies from our group [9,10].

Another interesting finding is that this "recovery" was inversely correlated to donor age, suggesting a physiological unbalance in ERG activities with ageing. This is in line with previous observations on the progressive lack of LSEC fenestrations, in which sinusoidal dysfunction and increased hepatic microvascular resistance correlated with age [35,36]. The comprehension of the effects of age in liver grafts is of utmost importance, since the use of increasingly older ECDs is becoming the rule in transplant centers.

5. Conclusions

In conclusion, our results showed that LSEC trophism is severely depressed in liver grafts, but it is restored after reperfusion in standard conditions. The execution of HOPE seems to improve this recovery, confirming the effectiveness of this MP technique in restoring endothelial functions. Whether this LSEC increases functionality impacts on graft survival and transplant outcomes will be the topic of further research.

Supplementary Materials: The following supporting information can be downloaded at: https://www.mdpi.com/article/10.3390/biology11091329/s1, Table S1: Perfusion variables.

Author Contributions: Conceptualization, F.V. and M.R.; methodology, G.G., C.C. and S.V.; validation, G.G. and C.C.; formal analysis, F.V. and G.G.; investigation, M.C. (Michele Carroli) and I.M.; resources, M.R.; data curation, M.C. (Michele Carroli) and F.V.; writing—original draft preparation, F.V.; writing—review and editing, A.D. and G.P.; supervision, A.D., G.P. and M.C. (Matteo Cescon); funding acquisition, M.R. All authors have read and agreed to the published version of the manuscript.

Funding: This research was funded by National Health System Research (RF-2016-02364732, Bando Ricerca Finalizzata 2016, Ministero della Salute–Rome, Italy).

Institutional Review Board Statement: This study was performed at the General Surgery and Transplant Unit of the IRCCS, Azienda Ospedaliero-Universitaria of Bologna in accordance with the Helsinki Declaration. All experimental procedures were approved by the Ethics in Research Committee in June 2016. The study underwent clinical trial registration (ClinicalTrials.gov ID: NCT03837197).

Informed Consent Statement: Informed consent was obtained from all subjects involved in the study.

Data Availability Statement: Not applicable.

Conflicts of Interest: The authors declare no conflict of interest.

References

1. Chiu, J.J.; Chien, S. Effects of disturbed blood flow on vascular endothelium: Pathophysiological basis and clinical perspectives. *Physiol. Rev.* **2011**, *91*, 327–387. [CrossRef] [PubMed]
2. Poisson, J.; Lemoinne, S.; Boulanger, C.; Durand, F.; Moreau, R.; Valla, D.; Rautou, P.E. Liver sinusoidal endothelial cells: Physiology and role in live disease. *J. Hepatol.* **2017**, *66*, 212–227. [CrossRef] [PubMed]
3. Schaffner, F.; Poper, H. Capillarization of hepatic sinusoids in man. *Gastroenterology* **1963**, *44*, 239–242. [CrossRef]

4. Lalor, P.F.; Lai, W.K.; Curbishley, S.M.; Shetty, S.; Adams, D.M. Human hepatic sinusoidal endothelial cells can be distiguished by expression of phenotypic markers related to their specialised functions in vivo. *World J. Gastroenterol.* **2006**, *12*, 5429–5439. [CrossRef]
5. Attia, M.; Silva, M.A.; Mirza, D.F. The marginal liver donor—An update. *Transpl. Int.* **2008**, *21*, 713–724. [CrossRef]
6. Schlegel, A.; Rougemont, O.; de Graf, R.; Clavien, P.A.; Dutkowski, P. Protective mechanisms of end-ischemic cold machine perfusion in DCD liver grafts. *J. Hepatol.* **2013**, *58*, 278–286. [CrossRef]
7. Feng, S.; Goodrich, N.P.; Bragg-Gresham, J.L.; Dykstra, D.M.; Punch, J.D.; DebRoy, M.A.; Greenstein, S.M.; Merion, R.M. Characteristics associated with liver graft failure: The concept of a donor risk index. *Am. J. Transpl.* **2006**, *6*, 783–790. [CrossRef]
8. Schlegel, A.; Muller, X.; Dutkowski, P. Machine perfusion strategies in liver transplantation. *HepatoBil. Surg. Nutr.* **2019**, *8*, 490–501. [CrossRef]
9. Ravaioli, M.; Germinario, G.; Dajti, G.; Sessa, M.; Vasuri, F.; Siniscalchi, A.; Morelli, M.C.; Serenari, M.; Del Gaudio, M.; Zanfi, C.; et al. Hypothermic Oxygenated Perfusion in Extended Criteria Donor Liver Transplantation—A Randomized Clinical Trial. *Am. J. Transpl. Publ. Online* **2022**. *Epub ahead of print*. [CrossRef]
10. Ravaioli, M.; de Pace, V.; Angeletti, A.; Comai, G.; Vasuri, F.; Baldassarre, M.; Maroni, L.; Odaldi, F.; Fallani, G.; Caraceni, P.; et al. Hypothermic Oxygenated New Machine Perfusion System in Liver and Kidney Transplantation of Extended Criteria Donors:First Italian Clinical Trial. *Sci. Rep.* **2020**, *10*, 6063. [CrossRef]
11. Schlegel, A.; Muller, X.; Dutkowski, P. Hypothermic Machine Preservation of the Liver: State of the Art. *Curr. Transplant. Rep.* **2018**, *5*, 93–102. [CrossRef] [PubMed]
12. Schlegel, A.; Porte, R.; Dutkowski, P. Protective mechanisms and current clinical evidence of hypothermic oxygenated machine perfusion (HOPE) in preventing post-transplant cholangiopathy. *J. Hepatol.* **2022**, *76*, 1330–1347. [CrossRef] [PubMed]
13. Parmar, K.M.; Larman, H.B.; Dai, G.; Zhang, Y.; Wang, E.T.; Moorthy, S.N.; Kratz, J.R.; Lin, Z.; Jain, M.K.; Gimbrone, M.A., Jr.; et al. Integration of flow-dependent endothelial phenotypes by Kruppel-like factor 2. *J. Clin. Investig.* **2006**, *116*, 49–58. [CrossRef]
14. Czigany, Z.; Lurje, I.; Tolba, R.H.; Neumann, U.P.; Tacke, F.; Lurje, G. Machine perfusion for liver transplantation in the era of marginal organs-New kids on the block. *Liver. Int.* **2019**, *39*, 228–249. [CrossRef]
15. Fernández, L.; Heredia, N.; Grande, L.; Gómez, G.; Rimola, A.; Marco, A.; Gelpí, E.; Roselló-Catafau, J.; Peralta, C. Preconditioning protects liver and lung damage in rat liver transplantation: Role of xanthine/xanthine oxidase. *Hepatology* **2002**, *36*, 562–572. [CrossRef]
16. Peralta, C.; Jiménez-Castro, M.B.; Gracia-Sancho, J. Hepatic ischemia and reperfusion injury: Effects on the liver sinusoidal milieu. *J. Hepatol.* **2013**, *59*, 1094–1106. [CrossRef]
17. Shah, A.V.; Birdsey, G.M.; Randi, A.M. Regulation of endothelial homeostasis, vascular development and angiogenesis by the transcription factor ERG. *Vasc. Pharmacol.* **2016**, *86*, 3–13. [CrossRef]
18. Birdsey, G.M.; Dryden, N.H.; Amsellem, V.; Gebhardt, F.; Sahnan, K.; Haskard, D.O.; Dejana, E.; Mason, J.C.; Randi, A.M. Transcription factor Erg regulates angiogenesis and endothelial apoptosis through VE-cadherin. *Blood* **2008**, *111*, 3498–3506. [CrossRef]
19. McLaughlin, F.; Ludbrook, V.J.; Cox, J.; von Carlowitz, I.; Brown, S.; Randi, A.M. Combined genomic and antisense analysis reveals that the transcription factor Erg is implicated in endothelial cell differentiation. *Blood* **2001**, *98*, 3332–3339. [CrossRef]
20. Laumonnier, Y.; Nadaud, S.; Agrapart, M.; Soubrier, F. Characterization of an upstream enhancer region in the promoter of the human endothelial nitric-oxide synthase gene. *J. Biol. Chem.* **2000**, *275*, 40732–40741. [CrossRef]
21. Sperone, A.; Dryden, N.H.; Birdsey, G.M.; Madden, L.; Johns, M.; Evans, P.C.; Mason, J.C.; Haskard, D.O.; Boyle, J.J.; Paleolog, E.M.; et al. The transcription factor Erg inhibits vascular inflammation by repressing NF-kappaB activation and proinflammatory gene expression in endothelial cells. *Arterioscler. Thromb. Vasc. Biol.* **2011**, *31*, 142–150. [CrossRef] [PubMed]
22. Michalczyk, K.; Ziman, M. Nestin structure and predicted function in cellular cytoskeletal organisation. *Histol. Histopathol.* **2005**, *20*, 665–671. [PubMed]
23. Vasuri, F.; Fittipaldi, S.; Giunchi, F.; Monica, M.; Ravaioli, M.; Degiovanni, A.; Bonora, S.; Golfieri, R.; Bolondi, L.; Grigioni, W.F.; et al. Facing the enigma of the vascular network in hepatocellular carcinomas in cirrhotic and non-cirrhotic livers. *J. Clin. Pathol.* **2016**, *69*, 102–108. [CrossRef] [PubMed]
24. Malvi, D.; de Biase, D.; Fittipaldi, S.; Grillini, M.; Visani, M.; Pession, A.; D'Errico, A.; Vasuri, F. Immunomorphology and molecular biology of mixed primary liver cancers: Is Nestin a marker of intermediate-cell carcinoma? *Histopathology* **2020**, *76*, 265–274. [CrossRef] [PubMed]
25. Ravaioli, M.; Maroni, L.; Angeletti, A.; Fallani, G.; de Pace, V.; Germinario, G.; Odaldi, F.; Corradetti, V.; Caraceni, P.; Baldassarre, M.; et al. Hypothermic Oxygenated Perfusion Versus Static Cold Storage for Expanded Criteria Donors in Liver and Kidney Transplantation: Protocol for a Single-Center Randomized Controlled Trial. *JMIR. Res. Protoc.* **2020**, *9*, e13922. [CrossRef]
26. D'Errico, A.; Riefolo, M.; Serenari, M.; de Pace, V.; Santandrea, G.; Monica, M.; de Cillia, C.; Ravaioli, M.; Cescon, M.; Vasuri, F. The histological assessment of liver fibrosis in grafts from extended criteria donors predicts the outcome after liver transplantation: A retrospective study. *Dig. Liver. Dis.* **2020**, *52*, 185–189. [CrossRef]
27. Ishak, K.; Baptista, A.; Bianchi, L.; Callea, F.; de Groote, J.; Gudat, F.; Denk, H.; Desmet, V.; Korb, G.; MacSween, R.N.; et al. Histological grading and staging of chronic hepatitis. *J. Hepatol.* **1995**, *22*, 696–699. [CrossRef]

28. Fiorentino, M.; Vasuri, F.; Ravaioli, M.; Ridolfi, L.; Grigioni, W.F.; Pinna, A.D.; D'Errico-Grigioni, A. Predictive value of frozen-section analysis in the histological assessment of steatosis before liver transplantation. *Liver Transplant.* **2009**, *15*, 1821–1825. [CrossRef]
29. Ruan, B.; Duan, J.L.; Xu, H.; Tao, K.S.; Han, H.; Dou, G.R.; Wang, L. Capillarized Liver Sinusoidal Endothelial Cells Undergo Partial Endothelial-Mesenchymal Transition to Actively Deposit Sinusoidal ECM in Liver Fibrosis. *Front. Cell. Dev. Biol.* **2021**, *9*, 671081. [CrossRef]
30. Hong, L.; Du, X.; Li, W.; Mao, Y.; Sun, L.; Li, X. EndMT: A promising and controversial field. *Eur. J. Cell. Biol.* **2018**, *97*, 493–500. [CrossRef]
31. Birdsey, G.M.; Shah, A.V.; Dufton, N.; Reynolds, L.E.; Almagro, L.O.; Yang, Y.; Aspalter, I.M.; Khan, S.T.; Mason, J.C.; Dejana, E.; et al. The endothelial transcription factor ERG promotes vascular stability and growth through Wnt/β-catenin signaling. *Dev. Cell.* **2015**, *32*, 82–96. [CrossRef] [PubMed]
32. Peghaire, C.; Dufton, N.P.; Lang, M.; Salles-Crawley, I.I.; Ahnström, J.; Kalna, V.; Raimondi, C.; Pericleous, C.; Inuabasi, L.; Kiseleva, R.; et al. The transcription factor ERG regulates a low shear stress-induced anti-thrombotic pathway in the microvasculature. *Nat. Commun.* **2019**, *10*, 5014. [CrossRef] [PubMed]
33. Dufton, N.P.; Peghaire, C.R.; Osuna-Almagro, L.; Raimondi, C.; Kalna, V.; Chauhan, A.; Webb, G.; Yang, Y.; Birdsey, G.M.; Lalor, P.; et al. Dynamic regulation of canonical TGFβ signalling by endothelial transcription factor ERG protects from liver fibrogenesis. *Nat. Commun.* **2017**, *8*, 895. [CrossRef]
34. Ginsberg, M.; James, D.; Ding, B.S.; Nolan, D.; Geng, F.; Butler, J.M.; Schachterle, W.; Pulijaal, V.R.; Mathew, S.; Chasen, S.T.; et al. Efficient direct reprogramming of mature amniotic cells into endothelial cells by ETS factors and TGFβ suppression. *Cell* **2012**, *151*, 559–575. [CrossRef]
35. Wilkinson, A.L.; Qurashi, M.; Shetty, S. The Role of Sinusoidal Endothelial Cells in the Axis of Inflammation and Cancer Within the Liver. *Front. Physiol.* **2020**, *11*, 990. [CrossRef]
36. Maeso-Díaz, R.; Ortega-Ribera, M.; Fernández-Iglesias, A.; Hide, D.; Muñoz, L.; Hessheimer, A.J.; Vila, S.; Francés, R.; Fondevila, C.; Albillos, A.; et al. Effects of aging on liver microcirculatory function and sinusoidal phenotype. *Aging Cell* **2018**, *17*, e12829. [CrossRef]

biology

Article

Anticancer Drugs Paclitaxel, Carboplatin, Doxorubicin, and Cyclophosphamide Alter the Biophysical Characteristics of Red Blood Cells, In Vitro

Elisaveta Skverchinskaya [1], Nadezhda Levdarovich [2], Alexander Ivanov [2], Igor Mindukshev [1] and Anton Bukatin [2,3,*]

1 Sechenov Institute of Evolutionary Physiology and Biochemistry of the Russian Academy of Sciences, 194223 Saint-Petersburg, Russia
2 Laboratory of Renewable Energy Sources, Alferov University, 194021 Saint-Petersburg, Russia
3 Institute for Analytical Instrumentation of the Russian Academy of Sciences, 198095 Saint-Petersburg, Russia
* Correspondence: antbuk.fiztek@gmail.com

Simple Summary: An important issue in cancer chemotherapy is minimizing its side effects. The extreme toxicity of chemotherapy drugs is due to their task of preventing the multiplication of cancer cells and causing cancer cell death. One of their most common undesirable side effects is anemia, which is caused by a decrease in the number of red blood cells (RBCs) circulating in blood, which in turn results in a lack of oxygen in tissues. The manifestation of anemia is associated not only with the inhibition of the hematopoietic function of bone marrow but also with direct damage to RBCs during the drugs' infusion and circulation. Here, we investigated how frequently used chemotherapy drugs directly affect RBCs. Our results show that chemotherapeutic drugs, whose main task is to damage the DNA of cancer cells and prevent their division, have a noticeable toxic effect on RBCs. However, this effect is lower than the effect caused by drugs, which disrupt the dynamics of the cytoskeleton during cell division. Direct simulation of RBCs' transport in microchannels of a microfluidic device was allowed to integrally assess the cells' functionality and the capability of passing through microcapillaries where gas transport mainly occurs. We demonstrate that after exposure to drugs, regardless of their type, the number of damaged cells did not exceed 10%, which indicates the balance of the drugs' therapeutic doses. Our data along with the developed research method could be used to work out an effective combination of chemotherapeutic drugs as well as to calculate the efficient therapeutic drug doses for cancer treatment to reduce anemia side effects.

Abstract: Red blood cells (RBCs) are the most numerous cells in the body and perform gas exchange between all tissues. During the infusion of cancer chemotherapeutic (CT) agents, blood cells are the first ones to encounter aggressive cytostatics. Erythrocyte dysfunction caused by direct cytotoxic damage might be a part of the problem of chemotherapy-induced anemia—one of the most frequent side effects. The aim of the current study is to evaluate the functional status of RBCs exposed to mono and combinations of widely used commercial pharmaceutical CT drugs with different action mechanisms: paclitaxel, carboplatin, cyclophosphamide, and doxorubicin, in vitro. Using laser diffraction, flow cytometry, and confocal microscopy, we show that paclitaxel, having a directed effect on cytoskeleton proteins, by itself and in combination with carboplatin, caused the most marked abnormalities—loss of control of volume regulation, resistance to osmotic load, and stomatocytosis. Direct simulations of RBCs' microcirculation in microfluidic channels showed both the appearance of a subpopulation of cells with impaired velocity (slow damaged cells) and an increased number of cases of occlusions. In contrast to paclitaxel, such drugs as carboplatin, cyclophosphamide, and doxorubicin, whose main target in cancer cells is DNA, showed significantly less cytotoxicity to erythrocytes in short-term exposure. However, the combination of drugs had an additive effect. While the obtained results should be confirmed in in vivo models, one can envisioned that such data could be used for minimizing anemia side effects during cancer chemotherapy.

Citation: Skverchinskaya, E.; Levdarovich, N.; Ivanov, A.; Mindukshev, I.; Bukatin, A. Anticancer Drugs Paclitaxel, Carboplatin, Doxorubicin, and Cyclophosphamide Alter the Biophysical Characteristics of Red Blood Cells, In Vitro. *Biology* **2023**, *12*, 230. https://doi.org/10.3390/biology12020230

Academic Editors: Olivier Barbier and Erxi Wu

Received: 18 November 2022
Revised: 29 January 2023
Accepted: 30 January 2023
Published: 31 January 2023

Keywords: red blood cells; chemotherapy; anemia; microcirculation; laser diffraction; osmotic fragility test; microfluidics

1. Introduction

According to estimates from the World Health Organization, cancer is the first/second leading cause of death [1]. Cancer treatment is a multi-component process in which, along with immunotherapy, surgery, and radiotherapy, chemotherapy still occupies a key role [2]. Chemotherapy (CT) is a powerful and aggressive treatment for cancer patients, the main goal of which is to destroy cancer cells. They are highly proliferative cells; therefore, the main mechanisms of CT drugs' action include DNA strand bonding and blocking DNA replication, blocking of metabolic pathways, and preventing replication. Additionally, for chemotherapeutic drugs' action, cell membranes represent either a possible target or an obstacle to the drugs' effect—suppression of malignization.

Currently, the main CT regimens for solid tumours include anthracyclines and taxanes accompanied by alkylating agents. The therapeutic effect of taxanes (paclitaxel (TAX), docetaxel) is due to blocking the depolymerization of microtubule tubulin, which leads to the inhibition of cell division [3]. The antiproliferative effect of platinum-based drugs (carboplatin (PLAT), cisplatin, oxaliplatin, etc.) is DNA–DNA and DNA–protein crosslinking, which leads to blockage of DNA replication and/or reparation [4]. These drugs work similarly to alkylating antineoplastic agents (anthracyclines), such as doxorubicin (RUBI). These agents' actions include interaction with DNA by intercalation, inhibition of protein kinases, and stabilization of the topoisomerase IIα complex after DNA cleavage, which in turn stops the replication process [5]. An exception to anthracycline drugs is cyclophosphamide (PHOS), whose action is mainly due to its metabolite phosphoramide mustard. Along with immunomodulation, this metabolite performs DNA alkylation, which causes cytotoxic apoptosis of tumour cells [6].

Although circulating erythrocytes (red blood cells, RBCs) have never been a target for cancer chemotherapy, they are the first to encounter aggressive cytostatics during infusion of CT drugs. RBCs lack a nucleus; therefore, the effect of CT drugs may be primarily due to the interaction with the lipid part of cell membrane and with the cytoskeleton. It has been shown that anthracyclines inhibit actin polymerization [6], which may play an important role in reducing the mechanical strength of erythrocytes and have a significant effect on reducing their rigidity. Moreover, interaction of anthracyclines only with the cell membrane without penetration into cells is sufficient for cancer cell death [7]. Thus, the target of action of anthracyclines can be both the lipid membrane and the cytoskeleton of erythrocytes. Another target in RBCs was reported to be the inhibition of Na-K-ATPase activity, which leads to volume regulation disruption [8].

Paclitaxel is another widely used chemotherapy drug from the class of taxanes. Its antiproliferative effect is linked to the ability to block the depolymerisation of tubulin and, thus, to stabilise microtubules. Mature erythrocytes lack microtubules but, as was shown by proteomic analysis, possess tubulin [9]. The treatment of erythrocytes with another taxane, Taxol®, resulted in an increase in the membrane tubulin pool and a decrease in erythrocyte deformability [10]. Additionally, it has been established that paclitaxel in erythrocyte membranes can also act on actin, a component of the cytoskeleton structure [11]. It has been shown that paclitaxel induces ionic pores in planar lipid bilayers. The drug molecules bind randomly on the cell surface in the form of single particles or clusters, and the favourable energetics in the adsorbed drug clusters can lead to the creation of ion channels [12,13].

Platinum drugs are actively used in combination therapy for solid tumours [14,15]. It is believed that with respect to erythrocytes, the main negative effect of platinum drugs is due to a decrease in antioxidant protection [16]. However, not all platinum preparations have equal prooxidative capabilities [17]. For example, treatment of erythrocytes with cisplatin leads to their transformation into stomatocytes [18] and to the disruption of lipid asymmetry with release of phosphatidylserine to the outer surface, which in turn increases procoagulant potential [19].

Another widely used CT drug is cyclophosphamide, which is a synthetic alkylating cytostatic used as an antitumour and immunosuppressive agent [20]. Cyclophosphamide must be metabolised to form mustard phosphoramide in order to exert its antitumour effect [21]. Little is known about the direct effect of cyclophosphamide on erythrocytes. In in vitro experiments, cyclophosphamide was shown to reduce erythrocyte antioxidant protection through a decrease in glutathione levels and the activity of the following enzymes: glutathione-S-transferase, catalase, glutathione peroxidase, and glutathione reductase [22].

In summary, different CT drugs might have different action on RBCs, but anemia during cancer treatment is the major multifactorial side effect, which occurs in 30–90% of patients with solid tumors (breast, lung, colon/rectum, stomach, and ovarian cancer) [23]. It can be induced not only by the drug itself but also by its toxic solvent used to improve solubility in case of taxanes [24,25]. In addition to reduced quality of life, anemia contributes to a suboptimal response to treatment, both through disruption of the CT dosing regimen and through limitation of the cytotoxic effect of CT in the setting of tumor hypoxia [26]. Two main pathophysiological causes of anemia in chemotherapy of oncology are: decreased RBCs formation due to the suppression of hematopoietic organs and increased RBC destruction in the circulation due to the direct cytotoxic effects of CT drugs and the release of substandard RBCs into the bloodstream [27].

Standard laboratory methods to determine the condition of RBCs are hematology analysis, which assesses their mean corpuscular volume, hemoglobin concentration, red cells distribution width, etc.; an osmotic fragility test, which indirectly characterizes the RBC's deformability; different flow cytometry tests, which show membrane structure and composition, enzyme activity, etc.; spectrophotometry for calculation of the percentage of hemolyzed cells and hemoglobin species; and confocal microscopy, which evaluates a cell's morphology. The limitation of all these methods is that they can only indirectly predict changes in RBCs' functionality. In contrast, modern microfluidic technologies can be used to simulate microcirculation conditions, study microrheology, and assess erythrocyte behavior in fluid flow on a single cell level. Thus, it is possible to uncover the relationship between changes in the RBC membrane and the dynamics of microvasculature flow caused by these alterations [28].

Analysis of erythrocyte behavior in microfluidic devices can directly detect microcirculatory disorders [29], the duration of blood bank storage [30], and oxidative stress level [31]. In the field of oncology, microfluidics is a key technique for liquid biopsy and has been successfully used for label-free sorting and isolation of circulating tumour cells [32] and extracellular vesicles [33].

However, little is known about the direct influence of CT drugs on the microcirculation of RBCs, which might be one of the significant reasons for anemia during cancer chemotherapy. We could not find any investigation of the simulation of RBC behavior under the microcirculation conditions in microfluidic devices after exposure to CT drugs. Moreover, current treatment protocols use a combination of two or more drugs to enhance their antiproliferative activity [34]. Furthermore, there is even less data on the combined effect of several CT drugs on red blood cells. From this perspective, knowledge of the off-target effects of CT drugs is crucial to understand possible further erythrocyte transformation and to make informed clinical decisions regarding drug selection and treatment regimen adjustments.

The aim of this study is to provide insight into the transformation of erythrocytes during cytotoxic injury caused by commercial pharmaceutical CT drugs in the first hours after exposure and to evaluate the contribution of cytological abnormalities to the biomechanical parameters of erythrocytes. To this end, we have investigated how the cytological and biophysical parameters of RBCs are changed when the cells are subjected to basic chemotherapy drugs with different action mechanisms. For this task, we have used flow cytometry analysis to determine cell viability and asymmetry of lipids in the cell membrane;

small-angle light diffraction to figure out the cell shape and osmotic fragility; and confocal microscopy to analyse morphology. These methods were combined with microfluidic simulations of microcirculation in microcapillaries to complexly define changes in RBCs' biophysical characteristics. We expect that a developed in vitro testing comparative approach of the drugs' direct influence on erythrocytes will provide an opportunity to assess the cells' resistance to a peak drug load during the infusion in the first hours. This approach can additionally facilitate the optimization of drug loading and therapeutic dose in order to reduce chemotherapy side effects.

2. Materials and Methods

2.1. *Preparation of RBC's Suspensions and Their Treatment*

Volunteers' blood was collected by venipuncture of the anterior cubital vein into S-monovette tubes (9NC, Sarstedt, Nümbrecht, Germany) with the addition of 2 mM EGTA. All volunteers were healthy at the time of blood sampling and did not take any medication for more than two weeks before blood donation. Each participant signed an informed consent for blood sampling and the anonymous presentation of the results.

The basic HEPES buffer was isotonic and was prepared according to the protocols for clinically approved LORCA analyzer (Laser-assisted Optical Rotational Cell Analyzer, https://lorrca.com/, accessed on 1 January 2020) [35]. It had the following composition: 10 mM HEPES, 140 mM NaCl, 5 mM KCl, 2 mM $MgCl_2$, 5 mM D-glucose, 2 mM EGTA, pH = 7.4 (pH-meter Metler Toledo, Columbus, OH, USA), and 300 mOsmol/kg H_2O (mOsmol) controlled by cryoscopic osmometer Osmomat 3000 (Gonotec, Germany).

RBCs' suspensions were obtained by whole blood centrifugation at 400 g for 3 min (Centrifuge ELMI-50CM, Elmi, Latvia) and two subsequent washes in HEPES buffer using the same centrifugation parameters. Washed RBCs were resuspended in HEPES buffer to a concentration of 5×10^8 cells/mL (corresponding to hematocrit 4–4.5%) to ensure that the specific concentration of drugs from different donors was similar. The hematological parameters of the blood and RBC suspension were controlled by the hematological counter Medonic-M20 (Boule Medical A.B., P.O. Box 42056 SE-126 13, Stockholm, Sweden).

The suspensions of RBCs were incubated with commercial pharmaceutical CT drugs at 37 °C for 3 h: paclitaxel—Paclitaxel-Ebewe (Sandoz, Ebewe Pharma Ges.m.b.h.Nfg.KG A-4866 Unterach, Austria), carboplatin—Carboplatin-Teva (Pharma B.V., Pharmachemie B.V., Swensweg 5, P.O. Box 552, 2003 RN Haarlem The Netherlands), doxorubicin—Farmorubicin® (Pfizer, Western Australia - 6102, Australia), cyclophosphamide—Endoxan® (Baxter, Halle/Westfalen, Germany), and their combinations TAX_PLAT and RUBI_PHOS, which are widely used in CT practice. All concentrations of used CT drugs corresponded to the maximum recommended therapeutic doses: TAX 175 mg/m^2, PLAT 400 mg/m^2, PHOS 600 mg/m^2, and RUBI 150 mg/m^2. Given that the drugs act on a finite number of cells, we calculated their concentrations for RBC concentration 5×10^8 cells/mL. The final TAX concentration was 17.5 µg/mL, PLAT—32.5 µg/mL, PHOS—65 µg/mL, and RUBI—16.5 µg/mL. The CT drugs were prepared prior to use, following the manufacturer's recommendations. The concentration of each drug was calculated according to the statistical average body surface area S = 1.81 m^2 and an average circulating blood volume V = 4.5 L.

2.2. *Osmotic Fragility Test*

The degree of osmotic fragility is used as a surrogate sign of deformability disorders [36] because it is a composite indicator of RBC shape, hydration, and, within certain limits, susceptibility to fracture in vivo [37]. The osmotic fragility test (OFT) is based on laser diffraction at small scattering angles (0–12°) performed by a laser particle analyser "LaSca-TM" (LLC "BioMedSystem", Saint-Petersburg, Russia). This method is much quicker than the standard OFT [37] and provides an osmotic resistance curve, which is the number of lysed cells in the buffer with different osmolality. It is sensitive to cytological changes in cells and can be applied to different cell suspensions including RBCs [37,38]. The analyzer was calibrated using 3, 6, and 10 μm latex beads (Invitrogen, Molecular Probes, Eugene, OR, USA).

The procedure of OFT was as follows: we placed the treated or control RBCs (1×10^6 cells/mL) in the cuvette in HEPES buffer (300 mOsmol). During the registration of the scattered laser intensity (SLI), we manually changed the buffer osmolality in the range of 300–100 mOsmol adding to the sample aliquots of dH_2O together with additional RBCs to maintain their concentration constant. The cuvette with the sample was temperature-controlled at 37 °C and equipped with a magnetic microstirrer (1200 rpm), which ensured rapid mixing of the suspension. The intensity of scattered light from the RBCs was continuously detected by forward scattering. The laser diffraction at the selected angles of detection made it possible to register cell swelling (increase in scattered light intensity at 1–6° angles) and hemolysis (decrease in scattered light intensity at 1–12°) [39].

The following parameters of RBCs were evaluated:

1. H50, mOsmol—the buffer osmolality, at which 50% of cells were lysed.
2. W, mOsmol—the distribution width of the osmotic resistance curve, at which 90% and 10% lysis occurs (W = H90–H10, mOsmol). It is a characteristic of the heterogeneity of the RBCs' pool.
3. MCV_{osm}, fL—the hydrodynamic cell volume versus the buffer osmolality. The MCV_{osm} curve was normalised to the MCV value at 300 mOsmol measured by the hematological analyser.
4. The asphericity index, AI, %, which is the normalised amplitude of the SLI oscillations of the RBCs in the buffer with physiological osmolality (300 mOsmol). This index is proportional to the shape asymmetry of the cells and can be used to distinguish the normal discoid and spherical shape of RBCs.

2.3. Flow Cytometry Analysis

Flow cytometry analysis was performed on CytoFLEX (BeckmanCoulter, Brea, CA, USA), with analysis of 20,000 events. For RBC detection, we used the forward scattering and side scattering coordinates (FSC/SSC), which gave information about cell size and structure.

Calcein-acetoxymethyl ester (Calcein-AM, C-AM; Molecular probes, Eugene, Oregon, USA) was used to estimate cells' esterase activity. After the incubation of RBCs with anticancer drugs, the suspension was diluted to 5×10^6 cells/mL and stained with Calcein-AM (5 μM, 40 min, 37 °C) in 300 μL of HEPES buffer. Mean Fluorescence Intensity (MFI) values of the control cells were taken as 100% efficiency of intracellular esterases in each individual experiment. To obtain intracellular esterase activity data of treated RBCs, the MFI values were normalised to the MFI of the control.

Transformation of the cytoskeleton with the formation of band3 protein clusters characterizes the final stages of RBC life [40]. This transformation of the cytoskeleton by CT drug action was assessed using the eosin-5-maleimide test (EMA; Molecular probes (Eugene, Oregon, USA). We incubated RBCs (5×10^6 cells/mL) with 0.07 mM EMA in HEPES buffer for 40 min, 25 °C.

Lipid asymmetry was assessed by the externalization of phosphatidylserine (PS) to the outer side of the membrane by the Annexin V test. RBCs (5×10^6 cells/mL) were incubated with Annexin V-FITC (Biolegend, Amsterdam, The Netherlands) for 15 min, 25 °C, in HEPES buffer. EGTA in HEPES buffer was replaced by 2 mM Ca^{2+}. The concentration of Annexin V-FITC was taken according to the manufacturer's recommendations.

In all the experiments, the fluorescent intensity was registered in FITC-channel (excitation 490 nm, emission 530 nm). For all CT drugs, the intensity of the fluorescent signal of the negative control ($n = 3$) was checked. To correctly record the signal of RBCs exposed to RUBI, which has its own laser-induced fluorescence, the negative control (RUBI-treated RBCs without dye staining) was subtracted. For RUBI, MFI did not exceed 0.5–1.1% of the total intensity of control cells stained with C-AM or EMA, which is negligible. Furthermore, it did not give any signal in the gate of Annexin V-positive cells (Annexin+, Ann+). For other CT drugs, the MFI level of the negative control was even lower.

2.4. Confocal Microscopy

An inverted Leica TCS SP5 MP confocal laser microscope (Leica Microsystems *GmbH, Wetzlar,* Germany) was used to visualize the RBCs' morphology after incubation with the anticancer drugs. We diluted 10 μL of treated RBCs in 200 μL of HEPES buffer with 3.7% of bovine serum albumin to prevent echinocytosis and placed this sample in Petri dishes (35 mm) with a centre hole replaced by a coverslip (SPL Lifesciences, Pocheon South, Korea). The microphotographs were processed using ImageJ software (Public Domain).

2.5. Microfluidic Analysis

For an integrative assessment of changes in RBCs under the influence of CT drugs, we evaluated RBCs' transit velocities through 2.5 × 8 × 200 μm microchannels in custom-made microfluidic chips [31,41]. This is an integral method for assessing functional disorders accumulated by erythrocytes, which takes into account changes in their shape, volume, adhesion, elasticity, and deformation characteristics.

The preparation of the microfluidic chips started with fabrication of polydimethylsiloxane (PDMS) replicas. We degassed the mixture of Sylgard 184 Silicone Elastomer Base and the Curing Agent 10:1 (Dow Corning, Midland, Michigan, USA). Then we filled the silicon mold with this mixture and cured it at 65 °C for 4 h. After the curing, we separated the replica from the mold and cut out the inlet and outlet holes with a 1 mm biopsy puncher. Then the PDMS replica was covalently bonded with a 75 × 25 × 1 mm glass slide after oxygen plasma treatment (PINK GmbH Thermosystem, Wertheim, Germany).

Before the experiment, we filled the microchips with HEPES buffer to prevent RBC adhesion to PDMS. All samples (5×10^7 cells/mL) were introduced into the chip under constant hydrostatic pressure. Recording of RBC transit in microchannels was performed via XIMEA MC023MG-SY video camera (XIMEA Corp., Lakewood, CA, USA) with a 400 fps frame rate through a Leica DM4000B LED microscope (Leica Microsystems GmbH, Wetzlar, Germany) with an N PLAN L 20×/0.40 objective (Leica Microsystems, Wetzlar, Germany). For each sample, we recorded at least 4–7 different channels of a 16-channel microchip for obtaining statistically correct data.

Analysis of obtained images was carried out by a custom MATLAB (The MathWorks) script [42], which calculated RBC transit velocities. Then the obtained values were normalised to the average RBC's velocity in wide channels, which corresponded to the average velocity of the fluid flow. After that, we constructed the probability density functions (Origin 2021, OriginLab Corporation) for each experiment and averaged them over all experiments.

2.6. Free Hemoglobin and Hemoglobin Species Calculation

To measure free hemoglobin (Hb) in RBC suspension, incubated samples were centrifuged at 400 g for 3 min. Then an aliquot of supernatant RBCs was diluted 20-fold with dH2O and analysed with a spectrophotometer (SPECS SSP-715-M, Spectroscopic systems LTD, Moscow, Russia). The calculation of the free hemoglobin (Hb) was based on the optical density of the plasma solution and lysate at 540 nm, where the optical density of the whole blood lysate was taken as 100%. For the correct calculations, the absorbance was also assessed at 700 nm to eliminate the influence of the solution's turbidity. The free Hb% was calculated according to Tarasev [43].

To analyse the formation of Hb forms under the action of CT drugs, the hemolysate of each sample (n = 3 donors) was scanned at 560, 577, 630, and 700 nm. The percentage of oxidised Hb was calculated by molar extinction coefficients of Hb species according to the method of Kanias [44].

2.7. Statistics

Microfluidic analysis data such as cell velocities and percentage of microchannel occlusions is presented as mean ± SE; the error bars on velocity histograms are shaded. Other data is presented as mean ± SD. To analyse the flow cytometry data, CytExpert (BeckmanCoulter, Inc. Brea, CA 92821, USA) and FCS Express Flow 7 (De Novo Software, Pasadena, USA) were used. To assess the laser diffraction data, the original software of the laser analyser LaSca-TM was used. Statistical significance was evaluated by Excel 16 (Microsoft, Redmond, WA, USA), GraphPad Prism 9 (GraphPad Software, San Diego, CA, USA), and Origin 2021 (OriginLab Corporation).

The differences between the groups were analysed by GraphPad Prism 9. The normal distribution was tested with the D'Agostino and Pearson normality test (in accordance with the recommendations of the GraphPad Prism guidelines). For multiple comparisons, one-way ANOVA followed by Tukey's multiple comparisons test Tukey HSD post-hoc (passed normality test), or Dunn's multiple comparisons test (no passed normality test) were used; values of $p < 0.05$ were considered statistically significant. All the obtained data and statistics are presented in Table A1.

3. Results

3.1. Osmotic Fragility Test Based on Laser Diffractometry

Unlike other blood cells, RBCs placed in a medium with an osmolality below the physiological values of blood plasma (285–305 mOsmol) become swollen. With a further decrease in osmolality, they lyse. Therefore, the OFT can be used as an integral indicator of RBC membrane disruption by CT drugs. OFT allows evaluation of a wide range of cell parameters. The most important is osmotic fragility/rigidity, which is a composite parameter of biophysical and morphological properties of RBCs' membranes: shape, cytoskeleton rigidity, and hydration. It analyses the tendency of cells to lyse in circulation [45].

3.1.1. OFT: CT Drugs Impair RBCs' Osmotic Resistance

The study of RBCs' fragility/rigidity after exposure to CT drugs was performed using two methods—the percentage of hemolysis at each buffer osmolality was calculated (Figure 1a), as was the osmotic fragility H50 (Figure 1b). Analysis of the hemolysis curve showed that RBCs under the action of TAX and TAX_PLAT started lysing earlier than cells in the other groups including control cells (Figure 1a). This means that these cells became fragile. Additionally, at low osmolality (100 mOsmol), TAX- and TAX_PLAT-treated cells showed a higher degree of rigidity to hypotonic load. It was also shown that the combined action of RUBI_PHOS increased the osmotic fragility of RBCs at a low osmolality of 120 mOsmol. Additionally, the hemolysis in this case was higher than that of RUBI and PHOS alone (Figure 1a, Table A1).

To assess the fragility/rigidity of RBCs, we used the H50 value, which indirectly represents the deformability of the cells [40,45,46]. The higher the H50 value, the more fragile the cells are. Incubation with TAX for 3–4 h resulted in a decrease of the H50 value by 6.4 ± 1.1 mOsmol, $n = 13$ donors. Incubation of RBCs with TAX_PLAT had the same effect (Figure 1b), but PLAT itself did not cause changes in the resistance of RBCs to osmotic loading.

The combined action of RUBI_PHOS caused a decrease in osmotic resistance (Figure 1a). In this case, lysis of half of the cell pool occurred at osmolality values higher than in their separate action. Thus, the action of TAX and TAX_PLAT contributes to an increase in osmotic rigidity (decrease in H50 level, $p \leq 4.5 \times 10^{-5}$), while the combined action of RUBI_PHOS increases osmotic fragility (H50: t-test $p \leq 0.054$; corrected Bartlett's statistic $p \leq 0.0151$).

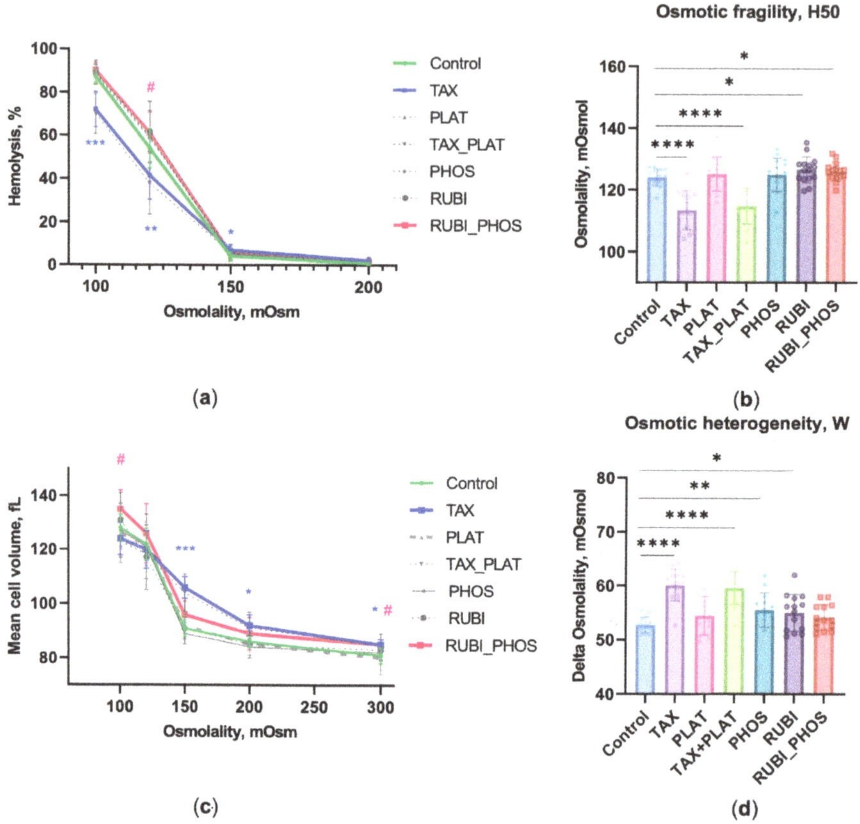

Figure 1. OFT demonstrates that CT drugs alter RBCs' osmotic resistance. (**a**) The osmotic lysis curves. TAX-treated cells started lysing earlier than the control ones, but were more rigid at 120 and 100 mOsmol; cells incubated with RUBI_PHOS were more sensitive to hemolysis at low osmolality 120 mOsmol; (**b**) H50 is the osmolality of the buffer at which 50% of the RBCs were lysed; CT drugs affect RBCs' membranes differently: TAX and TAX_PLAT caused increased osmotic stiffness of RBCs, and RUBI_PHOS caused osmotic fragility of RBCs; (**c**) Quantification of MCV during osmotic fragility test: MCV increased for TAX and TAX_PLAT earlier than for other drugs and control cells; (**d**) OFT revealed increased osmotic heterogeneity in the population of RBCs exposed to CT drugs. Data are presented as mean ± SD, n = 15 donors, one way ANOVA, Tukey HSD post-hoc; *, $p \leq 0.05$, **, $p \leq 0.01$, ***, $p \leq 0.001$, ****, and $p \leq 0.0001$ compared to control. Pink #, $p \leq 0.05$ refers to RUBI_PHOS compared to control, blue asterisks refer to TAX and TAX_PLAT compared to control.

3.1.2. OFT: CT Drugs Compromise the Ability of RBCs to Maintain Hydrodynamic Volume

During OFT, we recorded the entire range of scattering angles 0–12°, which made it possible to record MCV, known as the hydrodynamic volume of RBCs, under hypoosmotic loading. The use of laser detection in OFT allowed us to show for the first time that TAX action caused RBC swelling. This happened even with a slight decrease in the osmolality of the medium, meaning that high MCV levels are recorded throughout the 300–100 mOsmol range (Figure 1c). Thus, the MCV_{OSM} of TAX and TAX_PLAT at the osmolality of 200 mOsmol differed from the MCV at 300 mOsmol, 91.6 ± 4.6 vs. 84.6 ± 2.9 ($p \leq 0.008$). The observed phenomenon of swelling with slightly decreased medium osmolality is considered to be significant for microcirculation and the development of hemolytic anemia. At the same level of buffer osmolality (200 mOsmol), there was no increase in MCV_{OSM} for control, PLAT, and RUBI (Table A1).

3.1.3. OFT: CT Drugs Increase the Heterogeneity of the RBC Population

Due to the fact that the RBC pool consists of cells of different ages, the transformation of their membranes under the influence of CT drugs is not uniform. The index of osmotic heterogeneity W was used as another functional indicator of RBC anisocytosis (Figure 1d). All anticancer drugs were found to cause increased W levels, but TAX ($p \leq 9.2 \times 10^{-8}$) and the TAX_PLAT combination ($p \leq 1.55 \times 10^{-6}$) contributed the greatest degree of anisocytosis.

3.1.4. OFT: TAX and Its Combinations Disrupts the Discoid Shape of RBCs

Apart from recording cells stiffness/fragility, volume, and heterogeneity in mechanical characteristics, OFT based on laser diffractometry allows us to assess cell sphericity. Normal RBCs are diskocytes, so a signal of different amplitude (oscillations) is recorded during stochastic cell rotation in front of the detector during the measurement (Figure 2a). The amplitude of the signal from spherical particles decreases sharply, allowing this parameter to be used to estimate the shape of the RBCs (Figure 2a,b). To describe the RBCs' shape change, we introduced the asphericity index, which is the amplitude of the SLI oscillations normalised to the average SLI of the sample at the registration angle of 2.5°.

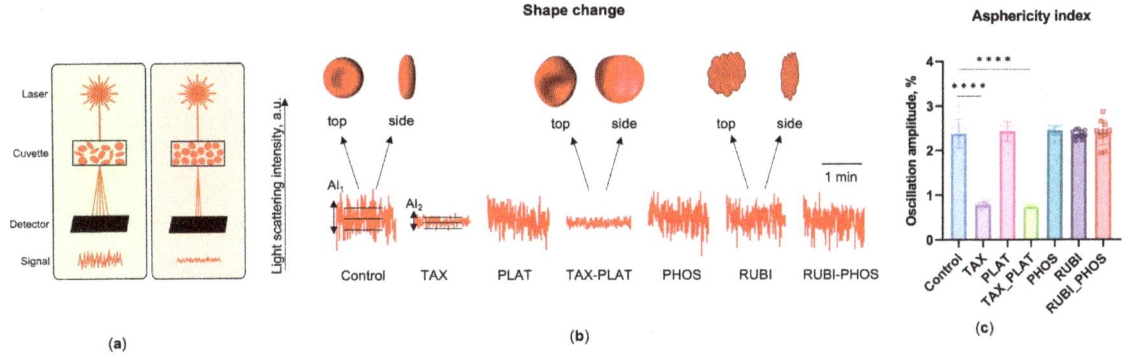

Figure 2. Disruption of native RBC shape under the action of CT drugs obtained by measuring the changes in amplitude of scattering light intensity (asphericity index) on laser diffractometer LaSca-TM. (**a**) At 300 mOsmol, discoid or flattened cells demonstrate a highly oscillated signal (left), while spherical cells demonstrate lower oscillation amplitude (right); (**b**) Representative amplitudes of light-scattering intensity oscillations demonstrate the shape changes of RBCs exposed to CT drugs. Three horizontal lines display the area of registration of the asphericity index (AI_1, AI_2). Above the graph, a schematic top and side view of RBCs are shown; (**c**) Asphericity index of RBCs, indicating TAX and its combination led to cell spherization. Data are presented as mean ± SD, n = 15 donors, one way ANOVA, Tukey HSD post-hoc; ****, $p \leq 0.0001$ compared to control.

The results show that TAX and TAX_PLAT reduced the asphericity index on average by three-fold (Figure 2c). Under the action of PLAT, PHOS, and RUBI, the flattened shape of RBCs did not change, and the amplitude of oscillations did not differ from controls ($p \leq 0.96$). We state "flattened" rather than "discoidal" because it could be other variants such as echinocytes. Changes in the asphericity index values were associated with impaired ability of RBCs to maintain their volume after incubation with TAX and TAX_PLAT, which was recorded by OFT (Figure 1c).

Summarizing, the results of OFT showed that the action of TAX and the combined action of TAX_PLAT led to a significant transformation of RBC membranes. Part of the cell population swelled and lysed rapidly, whereas the remaining portion of the cells showed increased resistance (rigidity) to the osmotic load. Incubation with TAX led to spherization of RBCs. The combined action of RUBI_PHOS, but not the individual drugs, contributed to the osmotic fragility of RBCs.

3.2. CT Drugs Change the Hematological and Morphometric Characteristics of RBCs

3.2.1. Hematological Analysis: CT Drugs Cause Increased Cell Volume and Population Heterogeneity

In addition to OFT, standard clinically approved hematological tests were performed. They showed that incubation of RBCs with TAX and TAX_PLAT resulted in a significant increase in mean corpuscular volume (MCV) levels (Figure 3a). For PHOS and RUBI, MCV levels did not differ from control. However, the MCV level was significantly higher when the cells were treated with their combination compared to the untreated control cells ($p \leq 0.0047$).

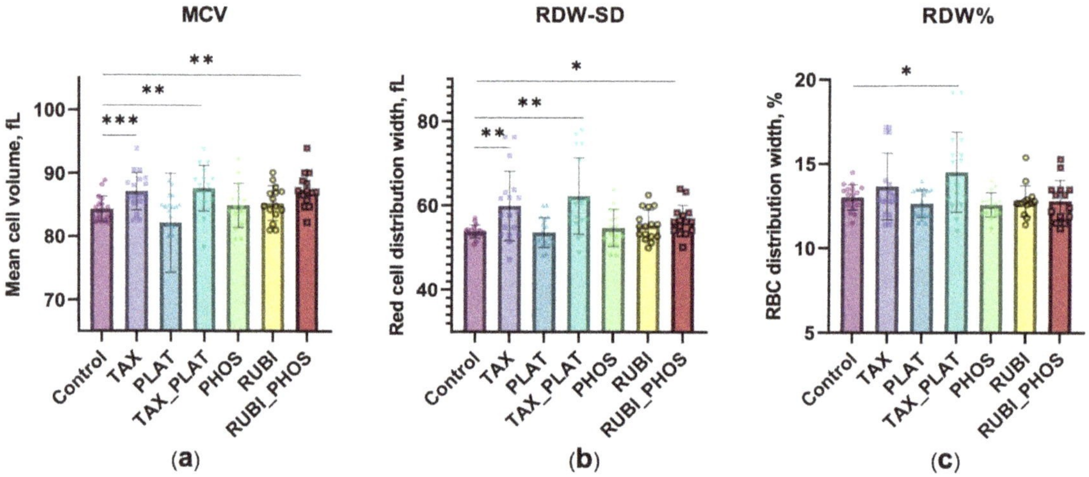

Figure 3. The action of CT drugs impaired the volume regulation of RBCs, resulting in increased MCV and volume heterogeneity. (**a**) MCV, (**b**) RDW-SD, and (**c**) RDW% calculated by Medonic-M hematology analyser. Data are presented as mean ± SD, $n = 15$ donors, one way ANOVA, Tukey HSD post-hoc; *, $p \leq 0.05$, **, $p \leq 0.01$, ***, $p \leq 0.001$ compared to control.

The cytological response may vary significantly within the same population during exposure to xenobiotics. Changes in RBC volume under the action of CT drugs did not occur in a uniform manner within the cell pool, which was recorded as an increase in red blood cell distribution width (RDW). In Figure 3b, we show RDW-SD, an index that is included in the routine blood count. In contrast to the more commonly used RDW% (Figure 3c), which calculates the coefficient of variation of the RBC's volume normalised by the average cell volume, RDW-SD shows a direct measurement of the width of the MCV histogram at 20% of its height. The results showed that RDW-SD was more sensitive to external factors than RDW% (Figure 3b,c): RDW-SD was significantly higher under the action of TAX, TAX_PLAT, and RUBI_PHOS, while RDW% revealed the differences only between control and TAX_PLAT. We believe that RDW-SD may be more useful in the individual assessment of anisocytosis in cancer patients as a more sensitive parameter for minor changes in RBC volume.

3.2.2. Confocal Microscopy: CT Drugs Change the Morphology of RBCs

To investigate the disorders detected during the OFT in more detail and to clarify the data of RBC shape changes, we performed 4 visualization experiments of nonfixed cells performed with confocal microscopy. Image analysis showed that all drugs influenced RBC morphology (Figure 4). Incubation of RBCs with TAX caused stomatocytosis, which is consistent with other studies [25,47]. The action of PLAT was gentler but led to the appearance of echinocytes I. Co-incubation with TAX_PLAT caused pronounced poikilocytosis—the appearance of stomatocytes III–IV. RBCs exposed to PHOS and RUBI mostly retained normal morphology. Nevertheless, it was noted that PHOS, RUBI, and the combination RUBI_PHOS caused the appearance of echinocytes I, eliptocytes, and a single appearance of shistocytes (fragmented RBCs). The classification of RBC forms is given according to [24,48].

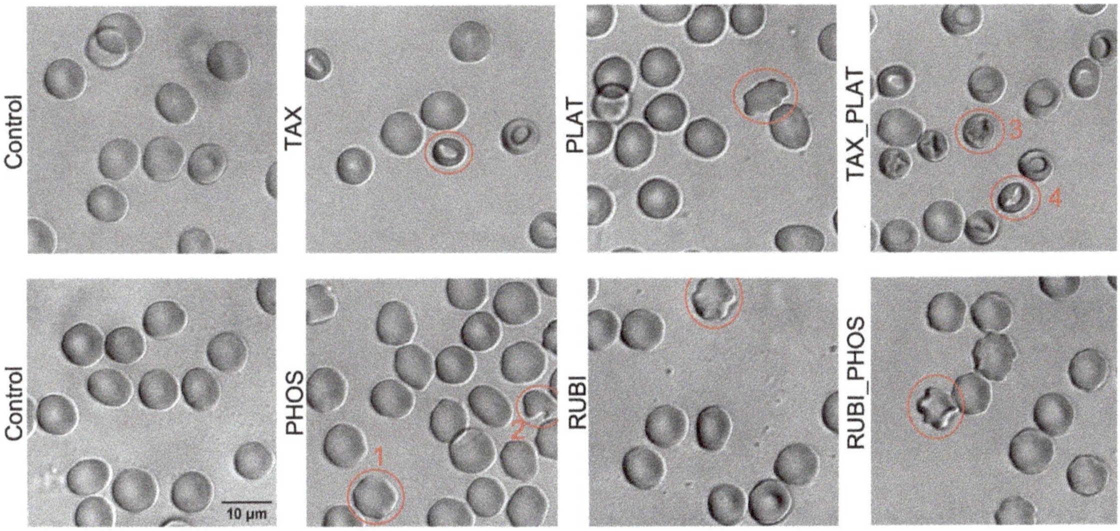

Figure 4. Disturbance of erythrocyte morphology under the action of CT drugs. Representative pseudocoloured confocal images (Leica TCS SP5 MP) of non-fixed RBCs show the appearance of anisocytosis and the disruption of RBCs' native morphology (red circled RBCs as an example): control cells had the normal shape of discocytes, TAX—discocytes and stomatocytes I–II, PLAT—discocytes and echinocytes I, PHOS—discocytes, echinocytes I (1) and schistocytes (2), RUBI—discocytes and echinocytes I; the combinations of drugs intensified the change in morphology of RBCs: TAX_PLAT demonstrated discocytes and stomatocytes at different stages (3, 4) while RUBI_PHOS—discocytes and echinocytes II.

Confocal imaging showed that contact of RBCs with TAX and TAX_PLAT caused stomatocytosis, while contact with PLAT, PHOS, RUBI, and RUBI_PHOS caused predominantly echinocytosis. The full statistics of RBC forms are presented in Figure A1. It should be emphasised that echinocytosis is a reversal process [49,50], whereas stomatocytosis is definitely a pathology.

3.3. *Flow Cytometry: Assessment of Membrane Transformation and Viability of RBCs*

Natural aging or induced accelerated RBC death is accompanied by physicochemical changes aimed to introduce recognition markers for macrophages. Such markers are the externalization of phosphatidylserine (PS), expression of neoantigens on the membrane surface, and impaired deformability [51]. We used two standard tests for RBC membrane transformation analysis to assess the degree of damage and readiness of RBCs for removal from microcirculation [52]. The first one is the PS binding to Annexin V, which characterises the relatively early stages of membrane transformation. The second one is a band3 clustering test with EMA labeling, characterising the final stages of RBCs' transformation before clearance. Additionally, we measured integral cell viability using the Calcein-AM test, which shows the activity of intracellular esterases.

3.3.1. Annexin-Positive Cells Test

According to the Annexin V test, the action of CT drugs on RBCs caused lipid asymmetry disruption, which was expressed in the phosphatidylserine exit to the outer side of the membrane (Figure 5a). The number of annexin-positive RBCs was significantly higher than in control for TAX ($p \leq 0.001$) and TAX_PLAT ($p \leq 0.04$). All the other investigated drugs also showed an increased number of annexin-positive cells.

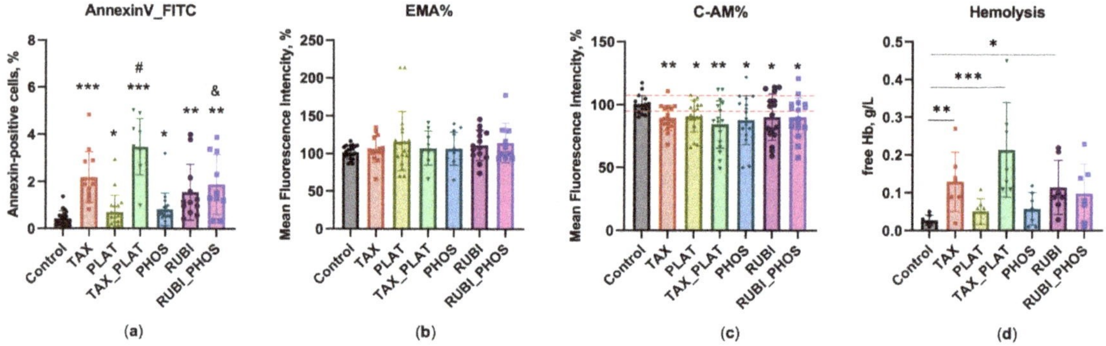

Figure 5. Effects of the anticancer drugs on RBCs' cytology parameters. (**a**) Annexin V test: all drugs induced the externalization of PS in RBCs (n = 10 donors); (**b**) EMA test revealed no membrane changes associated with Band3 translocation (n = 10 donors); (**c**) Calcein-AM test: CT drugs reduced RBCs' vitality. RUBI vs. control $p = 0.052$ (n = 13 donors); (**d**) Free Hb content in RBCs' incubation medium (n = 7 donors). Data are presented as mean ± SD, one way ANOVA, Tukey HSD post-hoc (C-AM, EMA, Hemolysis) and Dunn's post-hoc (AnnexinV); *, $p \leq 0.05$, **, $p \leq 0.01$, ***, $p \leq 0.001$ compared to control; #, $p = 0.0128$, compared to TAX, &, $p \leq 0.05$, compared to PHOS.

3.3.2. EMA Test: Action of Anticancer Drugs Does Not Cause Clustering of Band3 Membrane Complexes

Old or damaged RBCs are removed by phagocytosis in the spleen sinuses, as well as in the liver and bone marrow [53]. Macrophages recognize the erythrocytes to be removed by a number of aging markers, which include, among others, the exposure of new binding sites on the membrane surface. Since RBCs do not have the ability to synthesize new proteins, the creation of these binding sites occurs through the transformation of the existing membrane complexes with the band3 protein. These complexes form the core of the macrocomplexes of integrin and peripheral membrane proteins [54]. Profound changes in the structure of RBCs' membranes include the exposure of CD47 phagocytosis inhibitor sites located in the band3 complex, which switch them to the "eat me" signal of old/injured RBCs [55]. EMA dye binds predominantly with band3 protein. Increased intensity of EMA-dyed cells signifies the transformation of band3 complexes, which accompany the final stages of a RBC's life.

After the incubation of RBCs with CT drugs, no significant disturbances of band3 transmembrane complexes were detected (Figure 5b). Only a slight increase in MFI values for RUBI_PHOS ($p \leq 0.054$) was noted. The negative result obtained may be related to insufficient time for the deep transformation of the membranes. Previously, we found [38] that a significant increase in MFI while staining RBCs exposed to tert-butyl hydroperoxide requires at least 3 h of incubation with relatively high oxidant concentrations. Maximum MFI levels of band3 clusters were observed after 24 h exposure.

For a deeper analysis of the transformation of the RBC membrane and cytoskeleton under the influence of chemotherapy drugs, we investigated the hemichrome and methemoglobin (metHb) content in our samples. Hemichromes trigger prolonged band3 phosphorylation, which causes disruption of interactions between the cytoskeleton and membrane protein, which in turn contributes to shape disruption and enhances hemolysis [56]. MetHb leads to increased lateral diffusion of band3 complexes and cluster formation due to its high-affinity binding to the cytoplasmic band3 domain [52]. Spectroscopy of lysed RBCs pre-incubated with CT drugs showed that after 3–4 h of exposure there is an increase in the formation of reversible metHb (Hb(FeIII)) and nonreversible hemichrome (Hb(FeIV)) (Figure A2), which may also contribute to anemia during chemotherapy. Since RBC transformation, deformability, and band3 clustering strongly depends on oxidant concentration/time and the ratio of oxidant to cell count, we can conclude that the level of formation of oxidised forms of hemoglobin was not sufficient for the deep membrane transformation that is observed under severe oxidative stress.

3.3.3. Calcein-AM Vitality Test

Most of the intracellular esterases can hydrolyze a wide range of substrates [45], such as nonfluorescent calcein-AM, converting it into fluorescent calcein. Thus, the decreased MFI in the calcein-AM test reflects a complex disruption of the enzyme activity that supports the cell's vitality. This test is used as a quick option to assess the viability of nucleus-free cells [38]. Apart from RUBI ($p \leq 0.0502$), all the CT drugs and their combinations caused a decrease in the activity of intracellular esterases (Figure 5c), which is considered as a decrease in overall cell viability.

A significant decrease in MFI under the action of TAX ($p \leq 0.0079$) and TAX_PLAT ($p \leq 0.0013$) may be associated not only with inhibition of esterase activity but also with dye wash-out from cells when the integrity of membranes is disrupted. The analysis of free hemoglobin content (n = 7 donors) showed a significant increase in hemolysis of RBCs (Figure 5d). So, the decrease of calcein MFI for TAX- and TAX_PLAT-treated RBCs may be connected with both a decrease in intracellular esterase activity and a disruption of membrane integrity.

In addition, we should note that the analysis of cell distribution in FSC-SSC coordinates showed that TAX treatment stimulated microparticle formation in the samples, which was recorded by the accumulation of events in the gate calibrated on latex beads 3 μm in diameter (Figure A3).

In summary, the action of CT drugs leads to a decrease in the activity of intracellular esterases and the externalization of phosphatidylserine in RBCs. The most severe transformation of the cells occurred under the action of TAX: formation of microparticles and disruption of membrane integrity with the release of free hemoglobin into the incubation medium occurred.

3.4. Microfluidic Analysis: CT Drugs Degrade RBCs' Ability to Pass through the Microchannels

As an integrative test of the damage received by erythrocytes during incubation with CT drugs, we performed a microfluidic analysis of cell transport in $2.5 \times 8 \times 200$ μm microchannels [31,41]. The microfluidic analysis is a single-cell high-throughput method (100–1000 cells in one experiment) that can estimate cell functionality under microcirculation conditions.

All circulating RBCs have different ages; therefore, velocity distribution of untreated cells moving in the microchannels (Figure 6, blue histogram) can be approximated with a Gaussian function. Previously it was shown that when the cells are under oxidative stress, their population splits into two subpopulations of normal, undamaged cells and slow-damaged cells according to the capability of their built-in antioxidant defense system [31]. This leads to changing the velocity distribution from single modal to bimodal. For RBCs treated with CT drugs, similar to oxidative stress conditions, the velocity histogram transforms into bi- or trimodal (Figure 6, orange histograms).

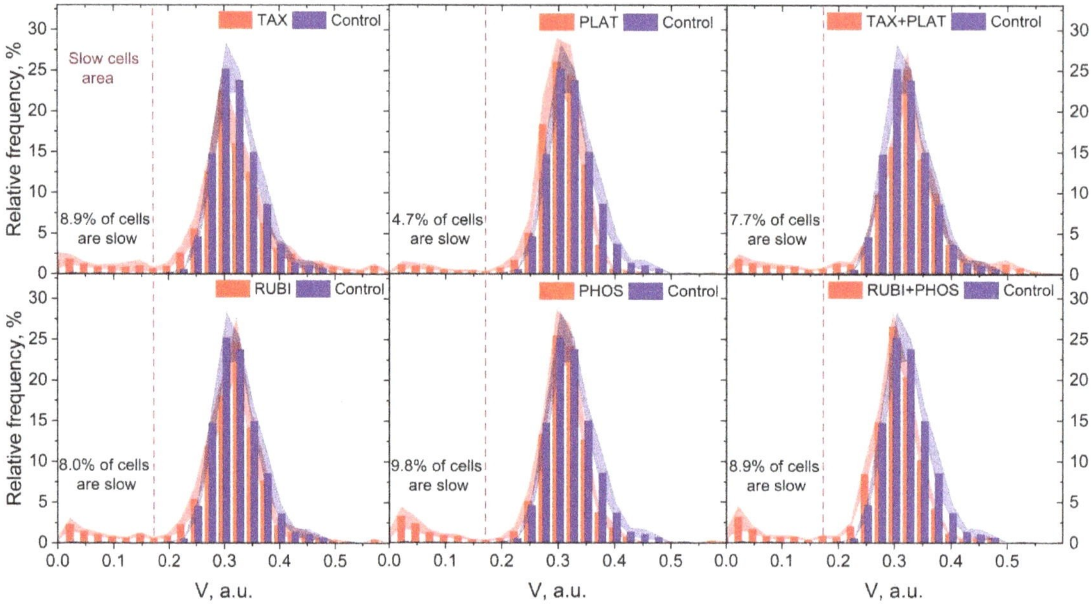

Figure 6. Distributions of RBCs' relative velocities in the microchannels of the microfluidic device. The action of CT drugs did not cause a significant shift in the main peak of cells' velocities but led to the appearance of a subpopulation of slow cells for all used chemotherapy drugs. The data are presented as mean ± SE, and the error bars are shaded. n = 10 donors.

The distribution changes can be separated into two variants. The first one is the appearance of the subpopulation of slow cells, which is typical for all investigated drugs (Figure 6, areas on the left of the graphs). These slow cells are unable to maintain the optimal speed in the microchannels inherent to the control cells due to impaired volumetric regulation and/or due to decreased deformation capacity. The second one is the appearance of fast cells (found only for TAX and TAX_PLAT) associated with the formation of microparticles (Figure A3). It is important to note that both transit abnormalities indicate pathology in the RBC population: decreased transit capacity carries the risk of microvascular blockage, while its increase leads to insufficient time for erythrocytes to be near the endothelial cells for efficient gas exchange.

The cutoff for counting slow cells was assumed to be 0.175 a.u. (Figure 6, dashed red line) resulting from the fact that the region below these values accounts for less than 0.45 ± 0.20% of control cells. Figure 6 displays the relative number of slow RBCs that had impaired microchannel transit characteristics. Although the appearance of the slow cell subpopulation was recorded for all CT drugs used (Control 0.45 ± 0.20%, TAX 8.9% ± 3.29%, PLAT 4.7% ± 1.88%, TAX_PLAT 7.7% ± 2.39%, RUBI 8.0% ± 2. 10%, PHOS 9.8% ± 3.76%, RUBI_PHOS 8.9% ± 3.30%), the data analysis showed no statistically significant differences between the number of cells in these subpopulations when different CT drugs were used. We attribute this to the fact that microfluidic analysis data are an integral indicator of the above-mentioned disorders in CT-drug-treated RBCs. At the same time, the main part of the population of RBCs under the action of CT drugs moved with a velocity inherent to the control cells: the position of the main velocity peak of the distribution histograms had no significant changes compared to the control cells.

Experimental results also demonstrated increased cell adhesion to the microchannels walls and to each other. Together with an increase in cell volume, shape changes, and changes in deformational properties under TAX, TAX_PLAT, PHOS, and RUBI_PHOS, this resulted in microchannels occlusions (Figures A4 and A5).

To summarize, all chemotherapy drugs had a damaging effect on the ability of several RBCs (4.7–9.8%) to pass microchannels. Moreover, it is important to note that the action of TAX and TAX_PLAT on RBCs resulted in a high number of microchannel occlusions.

4. Discussion

The aim of the study is to evaluate the disorders of RBCs exposed to CT drugs. We analysed the effects of the four main pharmacological compositions of CT drugs and their combinations included in current treatment regimens for solid tumours. The drugs differed in their action mechanism, and, as expected, TAX and its combination with PLAT had the most aggressive effect on RBCs. Such a drug combination decreased RBCs' total viability (Figure 5c), caused vesiculation (Figure A3), stomatocytosis (Figures 4 and A1 [24,25,47]), and hemolysis (Figure 5d) and led to cells' swelling (Figure 3a).

The main mechanism of TAX therapeutic action is related to disturbance of tubulin microtubules polymerization, which halts entry to anaphase-preventing cell division [3]. In the early stages of maturation, microtubules are present in pro-erythroblasts, but during maturation tubulin shifts to a disorganised structure. It continues to play a structural role in the RBC sedimentable fraction [57]. However, in RBCs there is no tubulin, but there is another class of cytoskeleton structures—actin microfilaments, on which TAX can also act. Its action on the membrane structures of RBCs leads to an increase in the transverse bonds between actin subunits and other proteins, such as tropomyosin, protein 4.1R, and dematin, which through adducin are connected with band3 membrane proteins [11,58]. Therefore, actin transformation, caused by TAX, affects membrane structural components and their links with the cell's cytoskeleton, which was indirectly shown with the Annexin V test (Figure 5a). This leads to violations of RBC shape and volume, which we observed in the osmotic fragility test (Figures 1c and 2c), hematological analysis (Figure 3a), and confocal microscopy (Figures 4 and A1). Hypothetically, TAX may also interfere with the molecular apparatus responsible for the perception and regulation of plasma membrane curvature, the nonmuscle myosin II and actin complex. Additionally, through actin polymerization TAX can distort signal transduction from transmembrane mechanosensors to cell volume regulation mechanisms [59].

In clinical treatment, TAX is usually used with platinum-based drugs. Our data show that PLAT itself has a low toxicity effect on RBCs because its main target in cancer cells is DNA. Therefore, in its combination with TAX, cell transformation mainly occurred due to TAX action. This follows from the close values of all measured parameters of cells treated with TAX and TAX_PLAT. Additionally, it should be mentioned that comparing our data with previously published results we can say that carboplatin has a much lower influence on RBCs than the older platinum-based drug cisplatin [18,19,60,61]. This might be due to the lower affinity of carboplatin for proteins [62]. Thus, carboplatin is preferred for therapy for preventing anemia.

Unlike taxanes, whose target is the cell cytoskeleton, the influence of anthracycline drugs (RUBI) did not lead to strong changes in RBCs' shape and volume. It is of note, that one of the targets of RUBI in cancer cells along with DNA is the lipid part of the membrane [7]. Therefore, we saw membrane transformations manifested in the higher Annexin V signal and hemolysis rate (Figure 5a,d). Nevertheless, we did not observe large changes in cell volume and shape (Figure 3a). Confocal microscopy revealed the appearance of the first order echinocytes, which is a reversible violation (Figures 4 and A1). The evidence is insufficient to link these disruptions of RBCs that we have identified to any mechanism of membrane transformation under RUBI action. The only conclusion possible is that RUBI interacts with RBC membrane structures, and its effects could be enhanced by combination with PHOS. By itself, PHOS has a low toxicity effect due to the fact that it should be transformed in cells into active forms by intracellular phosphatases. Upon such transformation, its main target is DNA, which is absent in RBCs.

Microfluidic functional analysis of CT-treated RBCs demonstrated the appearance of a subpopulation of slow-damaged cells (Figure 6) with a lower ability to move in microcapillaries than the healthy ones. Accordingly, these cells are removed from the blood and destroyed in the spleen. However, the number of slow cells did not exceed 10% for all investigated drugs and their combinations. Still, it should be noted that even this small number of cells with an altered biophysical phenotype could lead to profound consequences for blood rheology in cancer patients [63].

Despite the significant differences in the cytological and morphological status of RBCs treated with different drugs, the number of slow cells was approximately equal in all groups. We hypothesize that this uniform response is related to the application of all the drugs in therapeutic concentrations, which provides a balance between drugs' antiproliferative activities and cytotoxic effects. Additionally, the treatment of RBCs with TAX, TAX_PLAT, and RUBI_PHOS drugs resulted in occlusion of the microchannels due to cell adhesion to the walls and to each other. This effect might also lead to violations in microcirculation and anemia.

Interpretation of the obtained results should be performed keeping in mind several limitations. In vitro conditions during experimental studies were quite different from in vivo conditions in cancer patients. To evaluate the effects of direct exposure to pharmaceuticals drugs on human RBCs in vitro, washed cells were used. This approach revealed the negative effects of CT drugs, but it cannot be equivalent to the response of RBCs to these drugs in vivo. The response in patients may be more complex and depend on the condition of the red cell pool (impaired erythropoiesis and abnormal erythrocyte survival). Additionally, blood plasma composition is significantly different from the HEPES buffer used in the study. Many CT drugs in blood circulation can bind to plasma proteins, such as albumin, which may protect RBCs from direct cytotoxic injury. However, it is possible that the degree of negative effects of CT drugs will accumulate with several cycles of chemotherapy.

5. Conclusions

Our work shows that the response of RBCs to the direct action of commercial pharmaceutical CT drugs (impaired volume regulation, reduced cell vitality, shape change, and imbalance resistance to osmotic loading) depended upon the pharmacological group of the drugs. Drug combinations had additive effects. The greatest cytonegative effect was that of paclitaxel as an agent targeting cytoskeleton proteins. Drugs whose main target is DNA showed significantly lower toxicity. Microfluidic analysis showed that despite the differences in the level of cytological abnormalities, the microrheological behavior of erythrocytes was similar to and independent of the pharmacology of the drug. This can be attributed to the action of the drugs within therapeutic concentrations, which provides the balance between cytotoxicity in tumour and healthy cells.

Author Contributions: Conceptualization, E.S. and N.L.; methodology, A.B. and I.M.; software, A.B. and A.I.; validation, E.S., N.L. and A.B.; formal analysis, E.S. and N.L.; investigation, E.S. and N.L.; resources, A.B. and I.M.; data curation, E.S., N.L. and A.B.; writing—original draft preparation, E.S. and N.L.; writing—review and editing, A.B.; visualization, E.S. and N.L.; supervision, A.B. and I.M.; project administration, A.B.; funding acquisition, N.L. All authors have read and agreed to the published version of the manuscript.

Funding: This research was funded by the Russian Science Foundation, project number 22-24-00998.

Institutional Review Board Statement: The study was conducted in accordance with the guidelines of the Declaration of Helsinki and was approved by the Ethics Committee of the Sechenov Institute of Evolutionary Physiology and Biochemistry of the Russian Academy of Sciences (protocols no. 3–03 from 2 March 2021, and no. 1–04 from 7 April 2022).

Informed Consent Statement: Informed consent was obtained from all subjects involved in the study. A written agreement for the collection of blood samples and their deanonymised use was obtained from all donors before the experiments began.

Data Availability Statement: The data presented in this study are available on request from the corresponding author.

Acknowledgments: The authors would like to thank the IEFB Shared-use Centre for the equipment provided: the CytoFLEX (BeckmanCoulter, Brea, CA, USA) flow cytometer and the Leica TCS SP5 MP (Leica Mycrosystems, Germany) confocal station. The authors would like to thank Andrey Sorokin, professor of medicine (nephrology), microbiology and immunology from Medical College of Wisconsin, for revising the manuscript and language editing.

Conflicts of Interest: The authors declare no conflict of interest.

Appendix A

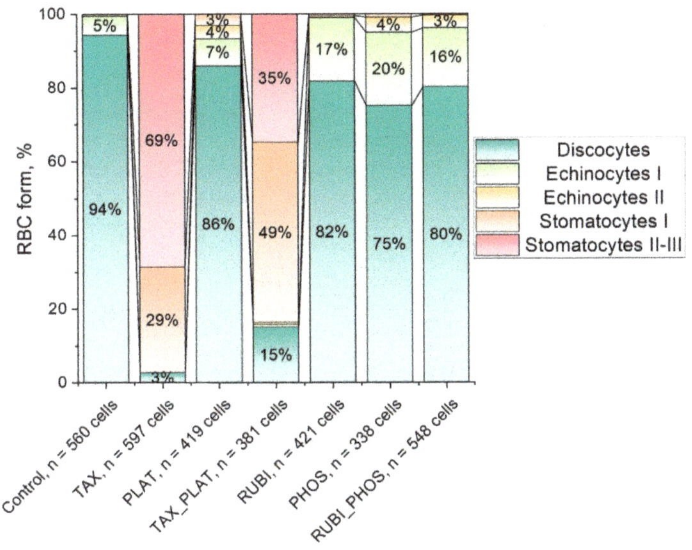

Figure A1. Statistics of RBCs forms according to the confocal microscopy images obtained after the incubation with CT drugs.

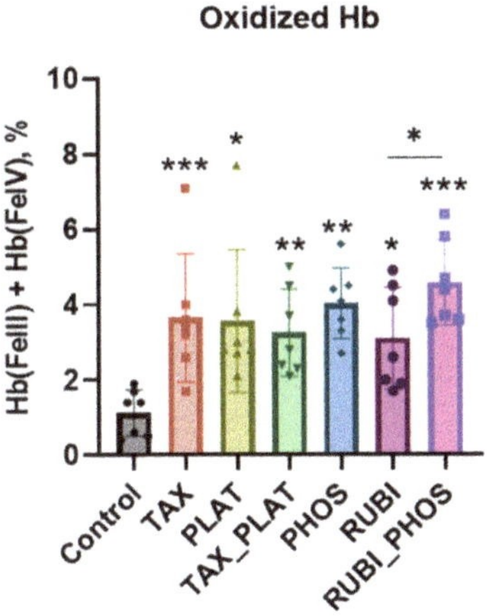

Figure A2. The action of CT drugs has contributed to the accumulation of oxidised forms of hemoglobin: the formation of reversible metHb and non reversible hemichrome. Spectral scans of Hb from hypoosmotically lysed RBCs after 3 h treatment with the CT drugs showed the elevated levels of oxidised Hb. Spectral range was from 450 to 700 nm. Data are presented as Mean ± SD, $n = 7$; one way ANOVA, Tukey HSD post hoc, *, $p \leq 0.05$, **, $p \leq 0.01$, ***, $p \leq 0.001$ compared to control.

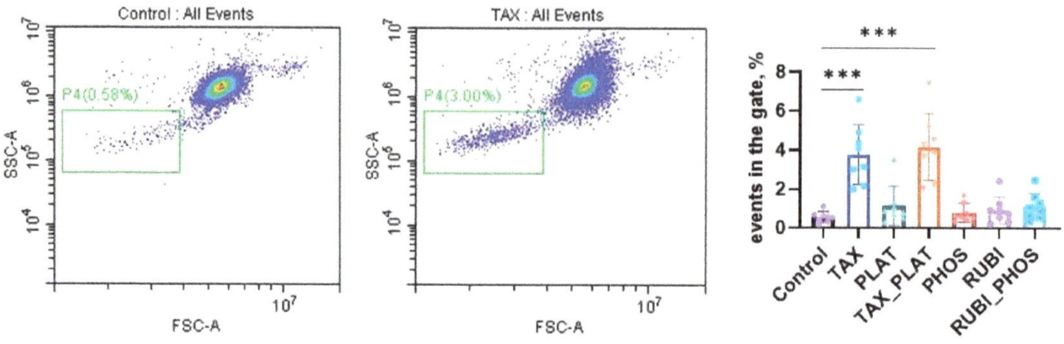

Figure A3. CT drugs induced RBCs transformation and microparticle formation. Representative SSC/FSC dot plots of control RBCs and RBCs under the action of TAX. The data shows the microparticle formation (left, green gate). The statistics of the formation of microparticles by CT drugs is shown on the right panel. Template and gating are corresponded to the left scan. Data are presented as Mean ± SD, n = 7 donors; one way ANOVA, Tukey HSD post hoc, ***, $p \leq 0.001$ compared to control.

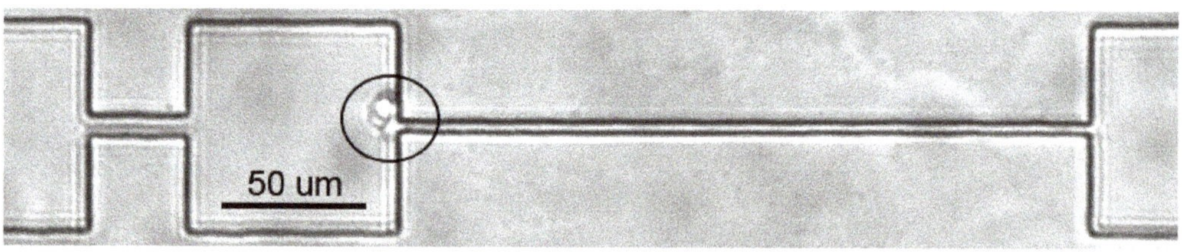

Figure A4. Representative image of a microchannel occlusion by RBCs treated with TAX drug.

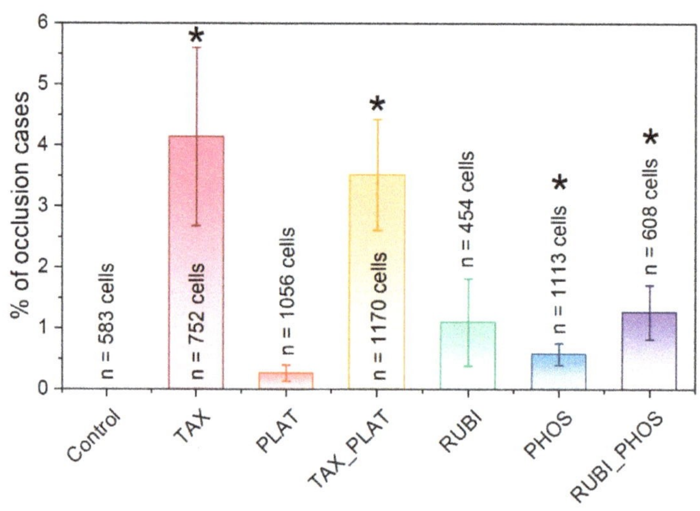

Figure A5. Percentage of microchannels occlusion cases. The number of cases of microchannel occlusions was counted manually; an occlusion was recorded if erythrocyte retention was at least 10 s. Data presented as mean ± SE, n = 3 donors, 2-tailed non-paired t-test; *, $p \leq 0.05$ compared to control.

Appendix B

Table A1. The hematology, OFT, spectrophotometry and microfluidics indicators of RBCs treated by the CT drugs; one way ANOVA, post hoc Dunnett's multiple comparisons test with adjusted p value. $p \leq$ compared to corresponding control.

	Control	TAX	PLAT	TAX_PLAT	PHOS	RUBI	RUBI_PHOS
				Hematology analysis			
MCV, fL $n = 15$	83.8 ± 3.1	87.1 ± 2.9 $p \leq 0.0321$	82.7 ± 7.6 $p \leq 0.416$	87.0 ± 4.7 $p \leq 0.0106$	84.9 ± 3.5 $p \leq 0.326$	85.2 ± 2.8 $p \leq 0.156$	86.7 ± 2.3 $p \leq 0.0345$
RDW-SD, fL $n = 15$	53.9 ± 1.5	59.9 ± 8.2 $p \leq 0.007$	53.7 ± 3.6 $p \leq 0.824$	62.3 ± 9.0 $p \leq 0.0001$	54.7 ± 4.4 $p \leq 0.465$	55.4 ± 3.6 $p \leq 0.140$	56.5 ± 3.6 $p \leq 0.047$
RDW-CV, % $n = 15$	13.0 ± 0.8	13.7 ± 2.0 $p \leq 0.208$	12.7 ± 2.4 $p \leq 0.178$	14.5 ± 0.7 $p \leq 0.023$	12.6 ± 0.9 $p \leq 0.090$	12.8 ± 0.9 $p \leq 0.487$	12.9 ± 1.2 $p \leq 0.592$
				Osmotic Fragility Test			
H50, mOsmol $n = 15$	123.9 ± 2.9	119.5 ± 3.9 $p \leq 0.001$	124.6 ± 4.4 $p \leq 0.521$	115.1 ± 5.7 $p \leq 0.000$	124.6 ± 5.1 $p \leq 0.610$	125.9 ± 3.5 $p \leq 0.037$	126.1 ± 3.0 $p \leq 0.054$
W, mOsmol $n = 15$	52.8 ± 1.5	60.2 ± 2.9 $p \leq 0.000$	54.3 ± 3.7 $p \leq 0.341$	59.7 ± 3.0 $p \leq 0.000$	55.7 ± 3.0 $p \leq 0.0345$	54.7 ± 3.7 $p \leq 0.0492$	55.2 ± 4.4 $p \leq 0.0324$
Asphericity Index, a.u. $n = 15$	2.36 ± 0.32	0.79 ± 0.06 $p \leq 0.000$	2.37 ± 0.33 $p \leq 0.967$	0.76 ± 0.04 $p \leq 0.000$	2.39 ± 0.23 $p \leq 0.817$	2.32 ± 0.17 $p \leq 0.628$	2.34 ± 0.32 $p \leq 0.868$
Hemolysis, % 100 mOsmol. $n = 15$	86.5 ± 3.1	72.1 ± 8.3 $p \leq 0.000$	89.2 ± 5.4 $p \leq 0.753$	70.0 ± 9.4 $p \leq 0.000$	88.2 ± 4.7 $p \leq 0.938$	88.6 ± 4.2 $p \leq 0.879$	90.6 ± 3.1 $p \leq 0.0352$
				Spectrophotometric analysis			
Free Hb, % $n = 7$	0.03 ± 0.01	0.13 ± 0.08 $p \leq 0.0208$	0.05 ± 0.03 $p \leq 0.09$	0.21 ± 0.13 $p \leq 0.001$	0.06 ± 0.04 $p \leq 0.10$	0.12 ± 0.07 $p \leq 0.02$	0.10 ± 0.08 $p \leq 0.0493$
				Microfluidic analysis			
Slow cells, %	0.45 ± 0.20	8.9 ± 3.29	4.7 ± 1.88	7.7 ± 2.39	9.8 ± 3.76	8.0 ± 2.10	8.9 ± 3.30
Number of analysed cells	5386	4379	6389	5188	5010	5420	4936
Occlusions, % $n = 3$	0 ± 0	4.14 ± 1.46 $p \leq 0.0472$	0.26539 ± 0.13 $p \leq 0.1193$	3.51702 ± 0.91 $p \leq 0.0179$	1.10036 ± 0.71 $p \leq 0.0305$	0.58389 ± 0.18 $p \leq 0.1747$	1.27221 ± 0.44 $p \leq 0.0452$

References

1. Sung, H.; Ferlay, J.; Siegel, R.L.; Laversanne, M.; Soerjomataram, I.; Jemal, A.; Bray, F. Global Cancer Statistics 2020: GLOBOCAN Estimates of Incidence and Mortality Worldwide for 36 Cancers in 185 Countries. *CA A Cancer J. Clin.* **2021**, *71*, 209–249. [CrossRef] [PubMed]
2. Abbas, Z.; Rehman, S. An Overview of Cancer Treatment Modalities. In *Neoplasm*; Shahzad, H.N., Ed.; InTech: Rang-Du-Fliers, France, 2018; ISBN 978-1-78923-777-1.
3. Jordan, M.A.; Wilson, L. Microtubules as a Target for Anticancer Drugs. *Nat. Rev. Cancer* **2004**, *4*, 253–265. [CrossRef] [PubMed]
4. Kelland, L. The Resurgence of Platinum-Based Cancer Chemotherapy. *Nat. Rev. Cancer* **2007**, *7*, 573–584. [CrossRef]
5. Minotti, G.; Menna, P.; Salvatorelli, E.; Cairo, G.; Gianni, L. Anthracyclines: Molecular Advances and Pharmacologic Developments in Antitumor Activity and Cardiotoxicity. *Pharmacol. Rev.* **2004**, *56*, 185–229. [CrossRef] [PubMed]
6. Colombo, R.; Necco, A.; Vailati, G.; Milzani, A. Dose-Dependence of Doxorubicin Effect on Actin Assembly in Vitro. *Exp. Mol. Pathol.* **1988**, *49*, 297–304. [CrossRef] [PubMed]
7. Alves, A.C.; Ribeiro, D.; Nunes, C.; Reis, S. Biophysics in Cancer: The Relevance of Drug-Membrane Interaction Studies. *Biochim. Biophys. Acta (BBA)—Biomembr.* **2016**, *1858*, 2231–2244. [CrossRef]
8. Shinohara, K.; Tanaka, K.R. The Effects of Adriamycin (Doxorubicin HCL) on Human Red Blood Cells. *Hemoglobin* **1980**, *4*, 735–745. [CrossRef]

9. Goodman, S.R.; Daescu, O.; Kakhniashvili, D.G.; Zivanic, M. The Proteomics and Interactomics of Human Erythrocytes. *Exp. Biol. Med.* **2013**, *238*, 509–518. [CrossRef]
10. Amaiden, M.R.; Monesterolo, N.E.; Santander, V.S.; Campetelli, A.N.; Arce, C.A.; Pie, J.; Hope, S.I.; Vatta, M.S.; Casale, C.H. Involvement of Membrane Tubulin in Erythrocyte Deformability and Blood Pressure. *J. Hypertens.* **2012**, *30*, 1414–1422. [CrossRef]
11. Chavez, J.D.; Keller, A.; Zhou, B.; Tian, R.; Bruce, J.E. Cellular Interactome Dynamics during Paclitaxel Treatment. *Cell Rep.* **2019**, *29*, 2371–2383.e5. [CrossRef]
12. Ashrafuzzaman, M.; Tseng, C.-Y.; Tuszynski, J.A. Regulation of Channel Function Due to Physical Energetic Coupling with a Lipid Bilayer. *Biochem. Biophys. Res. Commun.* **2014**, *445*, 463–468. [CrossRef] [PubMed]
13. Ashrafuzzaman, M.; Khan, Z.; Alqarni, A.; Alanazi, M.; Alam, M.S. Cell Surface Binding and Lipid Interactions behind Chemotherapy-Drug-Induced Ion Pore Formation in Membranes. *Membranes* **2021**, *11*, 501. [CrossRef] [PubMed]
14. Harper, B.W.; Krause-Heuer, A.M.; Grant, M.P.; Manohar, M.; Garbutcheon-Singh, K.B.; Aldrich-Wright, J.R. Advances in Platinum Chemotherapeutics. *Chemistry* **2010**, *16*, 7064–7077. [CrossRef] [PubMed]
15. Jeon, J.; Lee, S.; Kim, H.; Kang, H.; Youn, H.; Jo, S.; Youn, B.; Kim, H.Y. Revisiting Platinum-Based Anticancer Drugs to Overcome Gliomas. *IJMS* **2021**, *22*, 5111. [CrossRef] [PubMed]
16. Marković, S.D.; Djačić, D.S.; Cvetković, D.M.; Obradović, A.D.; Žižić, J.B.; Ognjanović, B.I.; Štajn, A.Š. Effects of Acute In Vivo Cisplatin and Selenium Treatment on Hematological and Oxidative Stress Parameters in Red Blood Cells of Rats. *Biol. Trace. Elem. Res.* **2011**, *142*, 660–670. [CrossRef] [PubMed]
17. Tolan, D.; Gandin, V.; Morrison, L.; El-Nahas, A.; Marzano, C.; Montagner, D.; Erxleben, A. Oxidative Stress Induced by Pt(IV) Pro-Drugs Based on the Cisplatin Scaffold and Indole Carboxylic Acids in Axial Position. *Sci. Rep.* **2016**, *6*, 29367. [CrossRef]
18. Suwalsky, M.; Hernández, P.; Villena, F.; Sotomayor, C.P. The Anticancer Drug Cisplatin Interacts with the Human Erythrocyte Membrane. *Z. Für Nat. C* **2000**, *55*, 461–466. [CrossRef]
19. Lü, C.; Yu, H.; Hou, J.; Zhou, J. Increased Procoagulant Activity of Red Blood Cells in the Presence of Cisplatin. *Chin. Med. J.* **2008**, *121*, 1775–1780. [CrossRef]
20. Cavallasca, J.A.; Costa, C.A.; del Maliandi, M.R.; Contini, L.E.; Fernandez de Carrera, E.; Musuruana, J.L. Severe Infections in Patients With Autoimmune Diseases Treated With Cyclophosphamide. *Reumatol. Clínica (Engl. Ed.)* **2015**, *11*, 221–223. [CrossRef]
21. Mills, K.A.; Chess-Williams, R.; McDermott, C. Novel Insights into the Mechanism of Cyclophosphamide-Induced Bladder Toxicity: Chloroacetaldehyde's Contribution to Urothelial Dysfunction in Vitro. *Arch. Toxicol.* **2019**, *93*, 3291–3303. [CrossRef]
22. Akamo, A.J.; Akinloye, D.I.; Ugbaja, R.N.; Adeleye, O.O.; Dosumu, O.A.; Eteng, O.E.; Antiya, M.C.; Amah, G.; Ajayi, O.A.; Faseun, S.O. Naringin Prevents Cyclophosphamide-Induced Erythrocytotoxicity in Rats by Abrogating Oxidative Stress. *Toxicol. Rep.* **2021**, *8*, 1803–1813. [CrossRef]
23. Xu, H.; Xu, L.; Page, J.; Cannavale, K.; Sattayapiwat, O.; Rodriguez, R.; Chao, C. Incidence of Anemia in Patients Diagnosed with Solid Tumors Receiving Chemotherapy, 2010–2013. *CLEP* **2016**, *8*, 61. [CrossRef] [PubMed]
24. Mark, M.; Walter, R.; Meredith, D.O.; Reinhart, W.H. Commercial Taxane Formulations Induce Stomatocytosis and Increase Blood Viscosity: Taxanes and Blood Rheology. *Br. J. Pharmacol.* **2001**, *134*, 1207–1214. [CrossRef] [PubMed]
25. Vader, P.; Fens, M.H.; Sachini, N.; van Oirschot, B.A.; Andringa, G.; Egberts, A.C.; Gaillard, C.A.; Rasmussen, J.T.; van Wijk, R.; van Solinge, W.W.; et al. Taxol ® -Induced Phosphatidylserine Exposure and Microvesicle Formation in Red Blood Cells Is Mediated by Its Vehicle Cremophor ® EL. *Nanomedicine* **2013**, *8*, 1127–1135. [CrossRef] [PubMed]
26. Bryer, E.; Henry, D. Chemotherapy-Induced Anemia: Etiology, Pathophysiology, and Implications for Contemporary Practice. *IJCTM* **2018**, *6*, 21–31. [CrossRef]
27. Rodgers, G.M.; Becker, P.S.; Blinder, M.; Cella, D.; Chanan-Khan, A.; Cleeland, C.; Coccia, P.F.; Djulbegovic, B.; Gilreath, J.A.; Kraut, E.H.; et al. Cancer- and Chemotherapy-Induced Anemia. *J. Natl. Compr. Canc. Netw.* **2012**, *10*, 628–653. [CrossRef]
28. Tomaiuolo, G. Biomechanical Properties of Red Blood Cells in Health and Disease towards Microfluidics. *Biomicrofluidics* **2014**, *8*, 51501. [CrossRef]
29. Aich, A.; Lamarre, Y.; Sacomani, D.P.; Kashima, S.; Covas, D.T.; de la Torre, L.G. Microfluidics in Sickle Cell Disease Research: State of the Art and a Perspective Beyond the Flow Problem. *Front. Mol. Biosci.* **2021**, *7*, 558982. [CrossRef]
30. Barshtein, G.; Pajic-Lijakovic, I.; Gural, A. Deformability of Stored Red Blood Cells. *Front. Physiol.* **2021**, *12*, 722896. [CrossRef]
31. Besedina, N.A.; Skverchinskaya, E.A.; Shmakov, S.V.; Ivanov, A.S.; Mindukshev, I.V.; Bukatin, A.S. Persistent Red Blood Cells Retain Their Ability to Move in Microcapillaries under High Levels of Oxidative Stress. *Commun. Biol.* **2022**, *5*, 659. [CrossRef]
32. Zhu, S.; Jiang, F.; Han, Y.; Xiang, N.; Ni, Z. Microfluidics for Label-Free Sorting of Rare Circulating Tumor Cells. *Analyst* **2020**, *145*, 7103–7124. [CrossRef] [PubMed]
33. Gholizadeh, S.; Shehata Draz, M.; Zarghooni, M.; Sanati-Nezhad, A.; Ghavami, S.; Shafiee, H.; Akbari, M. Microfluidic Approaches for Isolation, Detection, and Characterization of Extracellular Vesicles: Current Status and Future Directions. *Biosens. Bioelectron.* **2017**, *91*, 588–605. [CrossRef]
34. Mokhtari, R.B.; Homayouni, T.S.; Baluch, N.; Morgatskaya, E.; Kumar, S.; Das, B.; Yeger, H. Combination Therapy in Combating Cancer. *Oncotarget* **2017**, *8*, 38022–38043. [CrossRef] [PubMed]
35. Dobbe, J.G.G.; Hardeman, M.R. Red Blood Cell Aggregation as Measured with the LORCA. *Int. J. Artif. Organs* **2006**, *29*, 641–642. [CrossRef] [PubMed]
36. Clark, M.R.; Mohandas, N.; Shohet, S.B. Osmotic Gradient Ektacytometry: Comprehensive Characterization of Red Cell Volume and Surface Maintenance. *Blood* **1983**, *61*, 899–910. [CrossRef]

37. Walski, T.; Chludzińska, L.; Komorowska, M.; Witkiewicz, W. Individual Osmotic Fragility Distribution: A New Parameter for Determination of the Osmotic Properties of Human Red Blood Cells. *BioMed. Res. Int.* **2014**, *2014*, 1–6. [CrossRef]
38. Sudnitsyna, J.; Skverchinskaya, E.; Dobrylko, I.; Nikitina, E.; Gambaryan, S.; Mindukshev, I. Microvesicle Formation Induced by Oxidative Stress in Human Erythrocytes. *Antioxidants* **2020**, *9*, 929. [CrossRef]
39. Mindukshev, I.; Gambaryan, S.; Kehrer, L.; Schuetz, C.; Kobsar, A.; Rukoyatkina, N.; Nikolaev, V.O.; Krivchenko, A.; Watson, S.P.; Walter, U.; et al. Low Angle Light Scattering Analysis: A Novel Quantitative Method for Functional Characterization of Human and Murine Platelet Receptors. *Clin. Chem. Lab. Med.* **2012**, *50*, 1253–1262. [CrossRef]
40. Nobre, C.S.; Silva, J.A.; Jácomo, R.H.; Nery, L.F.A.; Barra, G.B. Flow Cytometric Analysis of Erythrocytes Osmotic Fragility in Hereditary Spherocytosis: A Case-Controlled Study Evaluating the Best Anticoagulant, Sample Pre-Treatment and NaCl Concentration for Reliable Screening of This Red Blood Cell Membrane Disorder. *Cytometry* **2018**, *94*, 910–917. [CrossRef]
41. Besedina, N.A.; Skverchinskaya, E.A.; Ivanov, A.S.; Kotlyar, K.P.; Morozov, I.A.; Filatov, N.A.; Mindukshev, I.V.; Bukatin, A.S. Microfluidic Characterization of Red Blood Cells Microcirculation under Oxidative Stress. *Cells* **2021**, *10*, 3552. [CrossRef]
42. Besedina, N.A.; Skverchinskaya, E.A.; Shmakov, S.V.; Ivanov, A.S.; Mindukshev, I.V.; Bukatin, A.S. Matlab Scripts for Calculation of Cell Velocity in Microfluidic Channel. *Zenodo* **2022**. [CrossRef]
43. Tarasev, M.; Chakraborty, S.; Alfano, K. RBC Mechanical Fragility as a Direct Blood Quality Metric to Supplement Storage Time. *Mil. Med.* **2015**, *180*, 150–157. [CrossRef] [PubMed]
44. Kanias, T.; Acker, J.P. Mechanism of Hemoglobin-Induced Cellular Injury in Desiccated Red Blood Cells. *Free. Radic. Biol. Med.* **2010**, *49*, 539–547. [CrossRef] [PubMed]
45. Orbach, A.; Zelig, O.; Yedgar, S.; Barshtein, G. Biophysical and Biochemical Markers of Red Blood Cell Fragility. *Transfus. Med. Hemother.* **2017**, *44*, 183–187. [CrossRef]
46. Nemeth, N.; Kiss, F.; Miszti-Blasius, K. Interpretation of Osmotic Gradient Ektacytometry (Osmoscan) Data: A Comparative Study for Methodological Standards. *Scand. J. Clin. Lab. Investig.* **2015**, *75*, 213–222. [CrossRef]
47. Huisjes, R.; Bogdanova, A.; van Solinge, W.W.; Schiffelers, R.M.; Kaestner, L.; van Wijk, R. Squeezing for Life—Properties of Red Blood Cell Deformability. *Front. Physiol.* **2018**, *9*, 656. [CrossRef]
48. Bessis, M. Red cell shapes. An illustrated classification and its rationale. In *Red Cell Shape*; Bessis, M., Weed, R.I., Leblond, P.F., Eds.; Springer: Berlin/Heidelberg, Germany, 1973; pp. 1–25. ISBN 978-3-642-88064-3.
49. Mrowietz, C.; Hiebl, B.; Franke, R.P.; Park, J.-W.; Jung, F. Reversibility of Echinocyte Formation after Contact of Erythrocytes with Various Radiographic Contrast Media. *Clin. Hemorheol. Microcirc.* **2008**, *39*, 281–286. [CrossRef]
50. Wong, P. An Explanation of the Reversal of Erythrocyte Echinocytosis by Incubation and Storage by Serum Albumin. *Clin. Hemorheol. Microcirc.* **2018**, *68*, 383–389. [CrossRef]
51. Thiagarajan, P.; Parker, C.J.; Prchal, J.T. How Do Red Blood Cells Die? *Front. Physiol.* **2021**, *12*, 655393. [CrossRef]
52. Arashiki, N.; Kimata, N.; Manno, S.; Mohandas, N.; Takakuwa, Y. Membrane Peroxidation and Methemoglobin Formation Are Both Necessary for Band 3 Clustering: Mechanistic Insights into Human Erythrocyte Senescence. *Biochemistry* **2013**, *52*, 5760–5769. [CrossRef]
53. Gottlieb, Y.; Topaz, O.; Cohen, L.A.; Yakov, L.D.; Haber, T.; Morgenstern, A.; Weiss, A.; Chait Berman, K.; Fibach, E.; Meyron-Holtz, E.G. Physiologically Aged Red Blood Cells Undergo Erythrophagocytosis in Vivo but Not in Vitro. *Haematologica* **2012**, *97*, 994–1002. [CrossRef] [PubMed]
54. Bruce, L.J. A Band 3-Based Macrocomplex of Integral and Peripheral Proteins in the RBC Membrane. *Blood* **2003**, *101*, 4180–4188. [CrossRef] [PubMed]
55. Burger, P.; de Korte, D.; van den Berg, T.K.; van Bruggen, R. CD47 in Erythrocyte Ageing and Clearance—the Dutch Point of View. *Transfus. Med. Hemother.* **2012**, *39*, 348–352. [CrossRef] [PubMed]
56. Pantaleo, A.; Ferru, E.; Pau, M.C.; Khadjavi, A.; Mandili, G.; Mattè, A.; Spano, A.; De Franceschi, L.; Pippia, P.; Turrini, F. Band 3 Erythrocyte Membrane Protein Acts as Redox Stress Sensor Leading to Its Phosphorylation by p ^{72} Syk. *Oxidative Med. Cell. Longev.* **2016**, *2016*, 1–11. [CrossRef] [PubMed]
57. Nigra, A.D.; Santander, V.S.; Dircio-Maldonado, R.; Amaiden, M.R.; Monesterolo, N.E.; Flores-Guzmán, P.; Muhlberger, T.; Rivelli, J.F.; Campetelli, A.N.; Mayani, H.; et al. Tubulin Is Retained throughout the Human Hematopoietic/Erythroid Cell Differentiation Process and Plays a Structural Role in Sedimentable Fraction of Mature Erythrocytes. *Int. J. Biochem. Cell Biol.* **2017**, *91*, 29–36. [CrossRef]
58. Gokhin, D.S.; Fowler, V.M. Feisty Filaments: Actin Dynamics in the Red Blood Cell Membrane Skeleton. *Curr. Opin. Hematol.* **2016**, *23*, 206–214. [CrossRef]
59. Barvitenko, N.; Aslam, M.; Lawen, A.; Saldanha, C.; Skverchinskaya, E.; Uras, G.; Manca, A.; Pantaleo, A. Two Motors and One Spring: Hypothetic Roles of Non-Muscle Myosin II and Submembrane Actin-Based Cytoskeleton in Cell Volume Sensing. *IJMS* **2021**, *22*, 7967. [CrossRef]
60. Dasari, S.; Bernard Tchounwou, P. Cisplatin in Cancer Therapy: Molecular Mechanisms of Action. *Eur. J. Pharmacol.* **2014**, *740*, 364–378. [CrossRef]
61. Tikhomirova, I.A.; Muravyov, A.V.; Petrochenko, E.P.; Kislov, N.V.; Cheporov, S.V.; Peganova, E.V. Alteration of Red Blood Cell Microrheology by Anti-Tumor Chemotherapy Drugs. *Biochem. Mosc. Suppl. Ser. A* **2016**, *10*, 135–141. [CrossRef]

62. Wagstaff, A.J.; Ward, A.; Benfield, P.; Heel, R.C. Carboplatin: A Preliminary Review of Its Pharmacodynamic and Pharmacokinetic Properties and Therapeutic Efficacy in the Treatment of Cancer. *Drugs* **1989**, *37*, 162–190. [CrossRef]
63. Kuck, L.; McNamee, A.P.; Simmonds, M.J. Impact of Small Fractions of Abnormal Erythrocytes on Blood Rheology. *Microvasc. Res.* **2022**, *139*, 104261. [CrossRef] [PubMed]

biology

MDPI

Article

IDH1 Promotes Foam Cell Formation by Aggravating Macrophage Ferroptosis

Ben Li [1,†], Chufan Wang [1,†], Peng Lu [1,†], Yumeng Ji [1], Xufeng Wang [1], Chaoyang Liu [1], Xiaohu Lu [1], Xiaohan Xu [1,2,*] and Xiaowei Wang [1,*]

1 Department of Cardiovascular Surgery, The First Affiliated Hospital, Nanjing Medical University, Nanjing 210000, China
2 The Friendship Hospital of Ili Kazakh Autonomous Prefecture Ili, Jiangsu Joint Institute of Health, Yining 835000, China
* Correspondence: zyc2001@stu.njmu.edu.cn (X.X.); wangxiaowei@njmu.edu.cn (X.W.)
† These authors contributed equally to this work.

Simple Summary: In our study, the involvement of IDH1 in atherosclerotic foam cells was explored. Inhibiting macrophage ferroptosis and foam cell formation by knocking down IDH1 is a promising study direction for better understanding the occurrence and progression of atherosclerosis, as well as the treatment targets for atherosclerosis.

Abstract: A distinctive feature of ferroptosis is intracellular iron accumulation and the impairment of antioxidant capacity, resulting in a lethal accumulation of lipid peroxides leading to cell death. This study was conducted to determine whether inhibiting isocitrate dehydrogenase 1 (IDH1) may help to prevent foam cell formation by reducing oxidized low-density lipoprotein (ox-LDL)-induced ferroptosis in macrophages and activating nuclear factor erythroid 2-related factor 2 (NRF2). Gene expression profiling (GSE70126 and GSE70619) revealed 21 significantly different genes, and subsequent bioinformatics research revealed that ferroptosis and IDH1 play essential roles in foam cell production. We also confirmed that ox-LDL elevates macrophage ferroptosis and IDH1 protein levels considerably as compared with controls. Ferrostatin-1 (Fer-1), a ferroptosis inhibitor, reduced ox-LDL-induced elevated Fe^{2+} levels, lipid peroxidation (LPO) buildup, lactate dehydrogenase (LDH) buildup, glutathione (GSH) depletion, glutathione peroxidase 4 (GPX4), ferritin heavy polypeptide 1 (FTH1), and solute carrier family 7 member 11 (SLC7A11) protein downregulation. More crucially, inhibiting IDH1 reduced Fe^{2+} overload, lipid peroxidation, LDH, and glutathione depletion, and elevated GPX4, FTH1, and SLC7A11 protein expression, resulting in a reduction in ox-LDL-induced macrophage ferroptosis. IDH1 inhibition suppressed ox-LDL-induced macrophage damage and apoptosis while raising NRF2 protein levels. We have demonstrated that inhibiting IDH1 reduces ox-LDL-induced ferroptosis and foam cell formation in macrophages, implying that IDH1 may be an important molecule regulating foam cell formation and may be a promising molecular target for the treatment of atherosclerosis.

Keywords: IDH1; foam cell; macrophage; ferroptosis; NRF2

Citation: Li, B.; Wang, C.; Lu, P.; Ji, Y.; Wang, X.; Liu, C.; Lu, X.; Xu, X.; Wang, X. IDH1 Promotes Foam Cell Formation by Aggravating Macrophage Ferroptosis. *Biology* **2022**, *11*, 1392. https://doi.org/10.3390/biology11101392

Academic Editor: Gaetano Santulli

Received: 25 July 2022
Accepted: 20 September 2022
Published: 23 September 2022

1. Introduction

Atherosclerosis is a chronic inflammatory condition of the artery wall in which macrophages play a significant role [1]. The production of foam cells is a key step in the pathogenesis of atherosclerosis, and inhibiting foam cell formation could be a viable way to prevent atherosclerotic lesions from developing [2]. Lipid peroxy radicals and hydroperoxides produced by free radical oxidation of polyunsaturated fatty acids (PUFAs), known as LPO, play a significant role in atherosclerosis [3,4].

A distinctive feature of ferroptosis is intracellular iron accumulation and the impairment of antioxidant capacity, resulting in a lethal accumulation of lipid peroxides leading to cell death [5]. Many pathogenic processes have been linked to ferroptosis, including cancer, ischemic organ damage, neurological illnesses, autoimmune diseases, and atherosclerosis [6,7]. Advanced human atherosclerotic plaques have lipid peroxidation, intraplaque bleeding, and iron deposition [8]. The mechanism of ferroptosis in atherosclerosis, however, is unknown.

IDH1 is a protease that may oxidize isocitrate to oxalosuccinate, which is then converted to α-ketoglutarate. It is found in the cytoplasm and peroxisomes. In studies, IDH1 mutations have been associated with malignant tumors such as secondary glioblastoma, cholangiocarcinoma, and periosteal cartilage tumors [9]. IDH1 mutations in tumors accelerate lipid reactive oxygen species (ROS) accumulation, leading to glutathione depletion through the downregulation of GPX4 [10]. However, it is unknown whether IDH1 enhances ferroptosis and thus worsens atherosclerosis progression.

In our study, the involvement of IDH1 in atherosclerotic foam cells was explored. Inhibiting macrophage ferroptosis and foam cell formation by knocking down IDH1 is a promising study direction for better understanding the occurrence and progression of atherosclerosis, as well as the treatment targets for atherosclerosis.

2. Materials and Methods

2.1. Data Mining from the Gene Expression Omnibus (GEO) Database

Gene expression profiles can be searched and downloaded using the GEO database. Appropriate microarray datasets were screened from the GEO database to analyze differential genes between foam cells and macrophages. The GEO database provided the gene expression profiles of GSE70126 and GSE70619. The platforms GPL6885 and GPL1261 provided these RNA profiles. GSE70126 is a genomic profile of foam cell macrophages isolated from fat-fed ApoE null mice and non-foamy macrophages isolated from control mice fed a normal diet. GSE70619 is the expression data from foam cells of apolipoprotein E-deficient mice. GEO2R is an online tool provided by the GEO database, which is mainly used for the differential analysis of expression profile chips. GEO2R was used to research the two groups of samples and find genes that were significantly differentially expressed at $p < 0.05$ and $|\log FC| > 1$ experimental conditions.

2.2. Analysis of Functional Enrichment

We used DAVID (https://david.ncifcrf.gov) (accessed on 12 August 2021) and Bioinformatics (http://www.bioinformatics.com.cn) (accessed on 12 August 2021) to perform Gene Ontology (GO) and Kyoto Encyclopedia of Genes and Genomes (KEGG) enrichment analysis on differentially expressed genes (DEGs). Bioinformatics is a free online platform for data analysis and visualization [11]. p values less than 0.05 are considered statistically significant.

2.3. FerrDb Database

FerrDb is the first manually maintained ferroptosis database, as well as a database of ferroptosis-related disorders. Thus far, the database has extracted and sorted 259 regulated genes, 111 markers, and 95 ferroptosis-related disorders from publications about ferroptosis in the PubMed database [12].

2.4. Cells and Reagents

The cell bank of the Chinese Academy of Sciences provided RAW264.7 mouse macrophages. Ox-LDL was provided by Guangzhou Yiyuan Biotechnology Co., Ltd., YB-002, Guangzhou, China. Proteintech (Wuhan, China) provided the primary antibodies against IDH1, NRF2, GPX4, SLC7A11, and GAPDH. Abcam (Cambridge, UK) provided the primary antibodies against FTH1. Fer-1 is a potent inhibitor of ferroptosis. Fer-1 was obtained from Sigma-Aldrich (St. Louis, MO, USA).

2.5. Treatment and Transfection of Cells

RAW264.7 macrophages were cultivated at 37 °C in 5% CO_2 using DMEM containing 10% fetal bovine serum (FBS). We divided cells into the following groups at random: control, ox-LDL, ox-LDL + si-IDH1, ox-LDL + Fer-1, ox-LDL + si-IDH1 + Fer-1. RAW264.7 cells were pretreated with 5 μmol/L Fer-1 (24 h), followed by stimulation of ox-LDL for 24 h. Genechem (Shanghai, China) designed and synthesized LV-Idh1-RNAi to target IDH1. CON313 was used as a control, transfected with shRNA to reduce IDH1 levels (targeting sequence: CCGGGCCAAATTAGCTCAGGCCAAACTCGAGTTTGGCCTGAGCTAATTTG-GCTTTTTG). To carry out the transfection, we followed the manufacturer's instructions.

2.6. Analysis of Cell Viability

RAW264.7 macrophages were seeded in 96-well plates (3×10^3 cells/well) and treated with several experimental treatments 24 h after cell adhesion, followed by the addition of cell counting kit 8 (CCK-8) reagents (10 μL/well). A microplate reader (Multiskan FC, Waltham, MA, USA) was used to measure the absorbance at 450 nm. Representative images of cell death in each group were collected using an optical microscope (Nikon TS2, Tokyo, Japan).

2.7. Western Blotting Analysis

Proteins were separated using 10% SDS-PAGE gels and transferred onto PVDF membranes. After blocking with 5% bovine serum albumin (BSA) for 1 h at room temperature, the membranes were incubated at 4 °C overnight with specific primary antibodies: anti-IDH1 (1:1000, Proteintech, 12332-1-AP), anti-GAPDH (1:1000, Proteintech, 60004-1-Ig), anti-NRF2 (1:1000, Proteintech, 16396-1-AP), anti-GPX4 (1:1000, Proteintech, 67763-1-Ig), anti-SLC7A11 (1:1000, Proteintech, 26864-1-AP), or anti-FTH1 (1:1000, Abcam, ab65080), followed by the addition of the corresponding secondary antibodies. The signal was detected using a chemiluminescence imaging system (Tanon 5200 Multi 4600SF, Shanghai, China).

2.8. Reverse Transcription-Quantitative PCR (RT-qPCR)

The purified DNA finally produced by CHIP was used as template DNA for the PCR reaction. qPCR analysis was performed using SYBR-Green I (Takara Bio, Inc., Kusatsu, Japan). GAPDH was used as an endogenous control. The primers used in this study were provided by Realgene (Realgene, Nanjing, China) (Supplementary Table S2): IDH1 forward, 5′-AGGTGGGCTGAGGAGGC-3′ and reverse, 5′-ACAGGGTAGGTCCGAGCTTT-3′; GAPDH forward, 5′-GAAGGTGAAGGTCGGAGTC-3′ and reverse, 5′-GAAGATGGTGATGGGATTTC-3′. Finally, the period threshold (Ct) value was analyzed by the $2^{-\Delta\Delta Ct}$ method. Each individual experiment was performed a total of three times.

2.9. Measurement of Intracellular GSH Level

RAW264.7 macrophages were seeded in 96-well plates (3×10^3 cells/well) and treated with several experimental treatments 24 h after cell adhesion. After stimulation, a total glutathione detection kit (Beyotime, S0052, Shanghai, China) was used to detect the intracellular glutathione levels, and process the data, according to the manufacturer's instructions. A microplate reader (Multiskan FC, Waltham, MA, USA) was used to measure the absorbance at 412 nm.

2.10. Iron Assay

RAW264.7 macrophages were seeded in 96-well plates (3×10^3 cells/well) and treated with several experimental treatments 24 h after cell adhesion. After stimulation, an iron ion colorimetric detection kit (E1042, Applygen, Beijing, China) was used to analyze the iron concentration in each group. A microplate reader (Multiskan FC, Waltham, MA, USA) was used to measure the absorbance at 550 nm.

2.11. Lipid Peroxidation Assay

RAW264.7 macrophages were seeded in 96-well plates (3×10^3 cells/well) and treated with several experimental treatments 24 h after cell adhesion. After stimulation, a lipid peroxide test kit (A106-1, Jiancheng, Nanjing, China) was used to evaluate the amount of lipid peroxide in each group. A microplate reader (Multiskan FC, Waltham, MA, USA) was used to measure the absorbance at 586 nm.

RAW264.7 macrophages were seeded in 6-well plates (1×10^6 cells/well) and treated with several experimental treatments 24 h after cell adhesion. Lipid peroxidation levels were detected using a C11BODIPY probe (D3861, Thermo Fisher Scientific, Waltham, MA, USA). Cells were incubated with 2 μM C11BODIPY for 30 min at 37 °C, then washed three times with PBS. Finally, cells were imaged using a fluorescence microscope (THUNDER DMi8, LEICA, Munich, Germany). The mean fluorescence intensity of C11BODIPY Green, which measures the amount of lipid ROS in the cell, was calculated using ImageJ software (NIH, Bethesda, MD, USA).

2.12. Annexin V-APC/7-AAD Staining and Flow Cytometry

The Annexin V-APC/7-AAD apoptosis detection kit (KGA1025, KeyGen Biotech, Nanjing, China) was used to detect cell death. Cells were collected, dyes were applied, and the cells were treated for 20 min. Flow cytometry (Beckman Coulter Cytoflex S, Brea, CA, USA) was then used to examine the samples.

2.13. Oil Red O Staining

RAW264.7 macrophages were seeded in 6-well plates (1×10^6 cells/well) and treated with several experimental treatments 24 h after cell adhesion, then washed three times with PBS and fixed in 4% paraformaldehyde for 30 min. We used oil red O to counterstain macrophages for an hour before being imaged under an optical microscope (Nikon TS2, Tokyo, Japan). Image J analysis software was further utilized to examine the experimental data.

2.14. LDH Release Assay

RAW264.7 macrophages were seeded in 6-well plates (1×10^6 cells/well) and treated with several experimental treatments 24 h after cell adhesion. A commercially available lactate dehydrogenase assay kit (C0016, Beyotime, Shanghai, China) was used to detect LDH release. A microplate reader (Multiskan FC, Waltham, MA, USA) was used to measure the absorbance at 490 nm.

2.15. α-KG Assay

RAW264.7 macrophages were seeded in 6-well plates (1×10^6 cells/well) and treated with several experimental treatments 24 h after cell adhesion. α-Ketoglutarate was collected using an α-ketoglutarate assay kit (MAK054, Sigma Aldrich, St. Louis, MO, USA), and a coupled enzyme assay was used to determine the α-ketoglutarate level. A microplate reader (Multiskan FC, Waltham, MA, USA) was used to measure the absorbance at 570 nm.

2.16. NADPH Assay

The NADPH/NADP ratio was determined using an NADP+/NADPH detection kit (WST-8 method) (S0179, Beyotime, Shanghai, China). Briefly, RAW264.7 macrophages (1×10^6 cells) were split into two aliquots. One aliquot was left on ice (containing both NADP and NADPH), and the other was incubated at 60 °C for 30 min to deplete NADP (only NADPH remained). The content of NADPH was determined by the absorbance at 450 nm (Multiskan FC, Waltham, MA, USA). The concentrations of NADPH and NADP were calculated separately from the standard curve. Finally, we calculated the NADPH/NADP ratio.

2.17. JASPAR Analysis

The binding sites of IDH1 and NRF2 were predicted by the JASPAR database (accessed on 27 August 2021) (https://jaspar.genereg.net). The sequences 2000 bp upstream and 100 bp downstream of the origin of the IDH1 gene were obtained from the NCBI database as potential promoter regions (Chr2: 208254972-208257071). The sequences were then compared with the transcription factors in the JASPAR database to predict the possible binding sites of IDH1 and target transcription factors. Transcription factor binding sites (TFBS) in JASPAR are represented by visualization techniques. The ordinate represents the information content, and the greater the information content, the greater the probability of occurrence. The abscissa aligns the positions of TFBS, each position is formed by the stacking of possible bases [13].

2.18. Chromatin Immunoprecipitation (ChIP) Assay

Binding of IDH1 and NRF2 was detected using a chromatin immunoprecipitation (ChIP) assay kit (P2078, Beyotime, Shanghai, China). Briefly, different groups of RAW264.7 macrophages were treated with 37% paraformaldehyde. Lysates were sonicated and centrifuged. The supernatant was immunoprecipitated with the antibody anti-NRF2 overnight at 4 °C (1:100, 16396-1-AP, Proteintech). The same amount of mouse IgG antibody (1:200, cat. no. 554002, BD Biosciences, Franklin Lakes, NJ, USA) was used as a negative control. The DNA–protein complexes were eluted using elution buffer. Purified DNA was used as template DNA for PCR reactions.

2.19. Statistical Analysis

GraphPad Prism 8.0 was used for statistical analysis, and all experiments were conducted independently in triplicate. The data are presented in the form of mean ± standard deviation (SD). An analysis of variance was used to examine the mean differences between groups, with p values < 0.05 considered to be statistically significant.

3. Results

3.1. Identification and Enrichment Analysis of DEGs in Foam Cells

Following GEO2R analysis of the GSE70126 and GSE70619 datasets, genes differentially expressed between foam cells and macrophages were discovered. Figures 1A and 2A show the volcano map of DEGs in each dataset. To further understand the biological functions of the DEGs, we utilized DAVID to perform GO functional and KEGG pathway enrichment analyses on both datasets. The DEGs were substantially enriched in response to lipopolysaccharides and transcription, DNA-templated, according to GO analysis findings for biological processes (BP). The DEGs were found in the extracellular matrix and nucleus for cellular components (CC). The DEGs were considerably enriched in extracellular matrix binding and protein binding for molecular functions (MF). Furthermore, KEGG pathway analysis indicated that these DEGs were mostly engaged in ECM-receptor interactions, as well as ubiquitin-mediated proteolysis (Figures 1B and 2B).

3.2. Identification and Enrichment Analysis of Target DEGs

The Venn diagram revealed 21 DEGs that overlapped between the two datasets (Figure 3A), including 12 upregulated genes and 9 downregulated genes, for which we used Bioinformatics to undertake GO function and KEGG pathway enrichment analyses. A list of related genes is provided in the Supplementary Materials (Supplementary Table S1). For BP, 21 overlapping DEGs were considerably enriched in the regulation of T-cell-mediated immunity. A further 21 overlapping DEGs were considerably enriched in fibrillar centers in the case of CC; while for MF, 21 overlapping DEGs were enriched in structural constituents of cytoskeleton. In addition, these DEGs were predominantly involved in glutathione metabolism (Figure 3B–E). It is well known that ferroptosis is mainly manifested by decreased GSH and expression of GPX4; thus, we looked for differential genes associated with ferroptosis in overlapping DEGs as a target gene. After intersecting with the 259 ferroptosis-related regulatory genes sorted by FerrDb, we obtained a target gene IDH1 (Figure 3F).

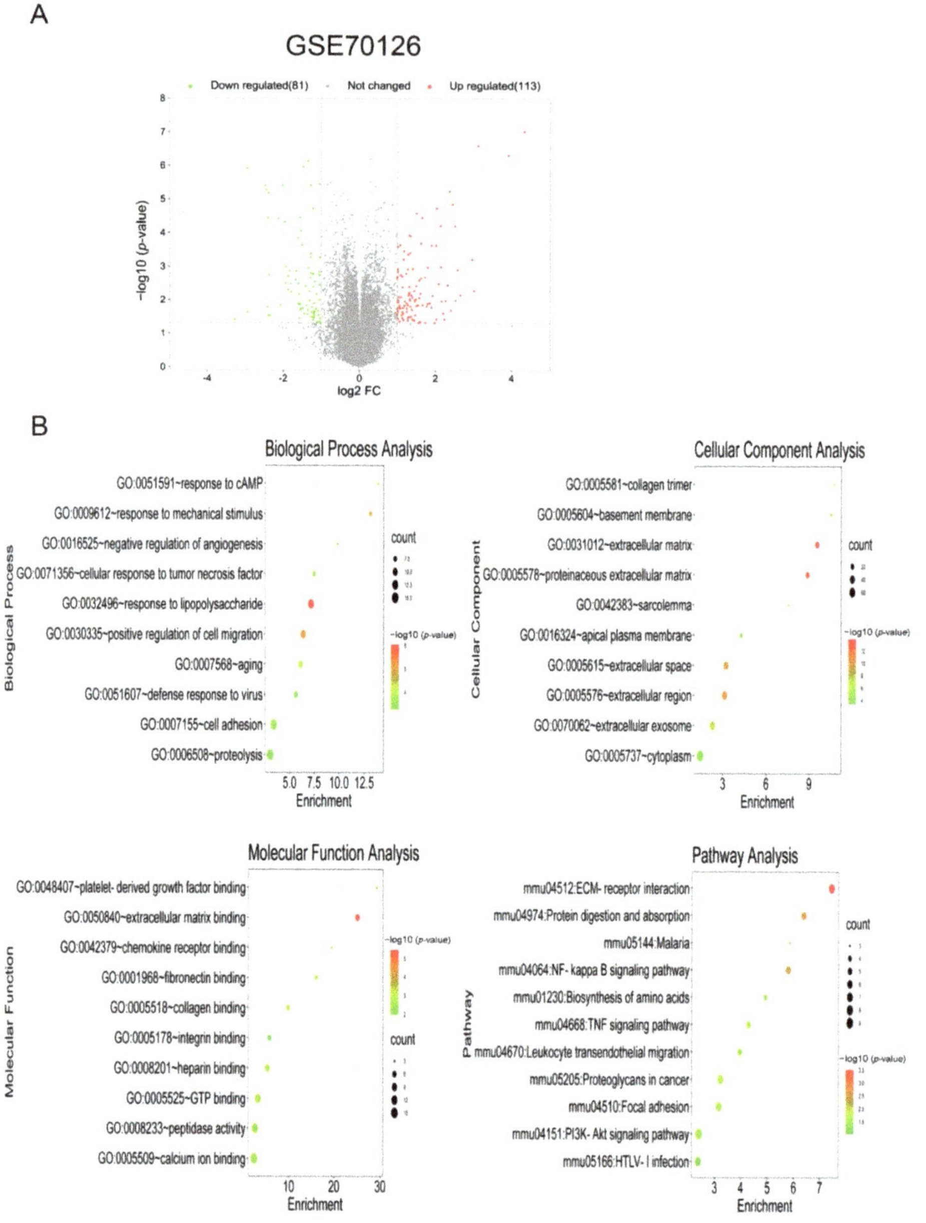

Figure 1. Identification and enrichment analysis of DEGs in foam cells. (**A**) Volcano plot of gene expression profile data between foam cells and macrophages in GSE70126. (**B**) Functional and signaling pathway analysis of DEGs between foam cells and macrophages.

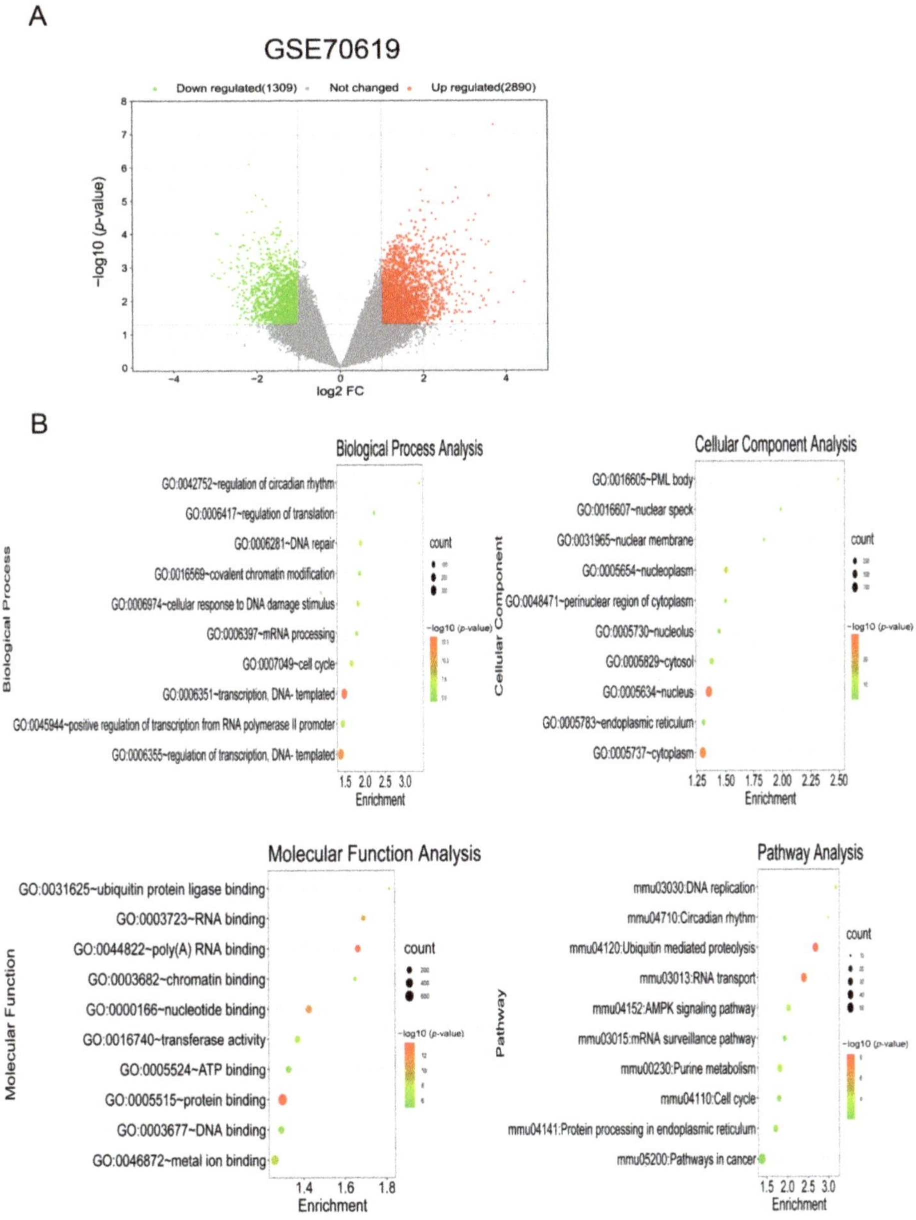

Figure 2. Identification and enrichment analysis of DEGs in foam cells. (**A**) Volcano plot of gene expression profile data between foam cells and macrophages in GSE70619. (**B**) Functional and signaling pathway analysis of DEGs between foam cells and macrophages.

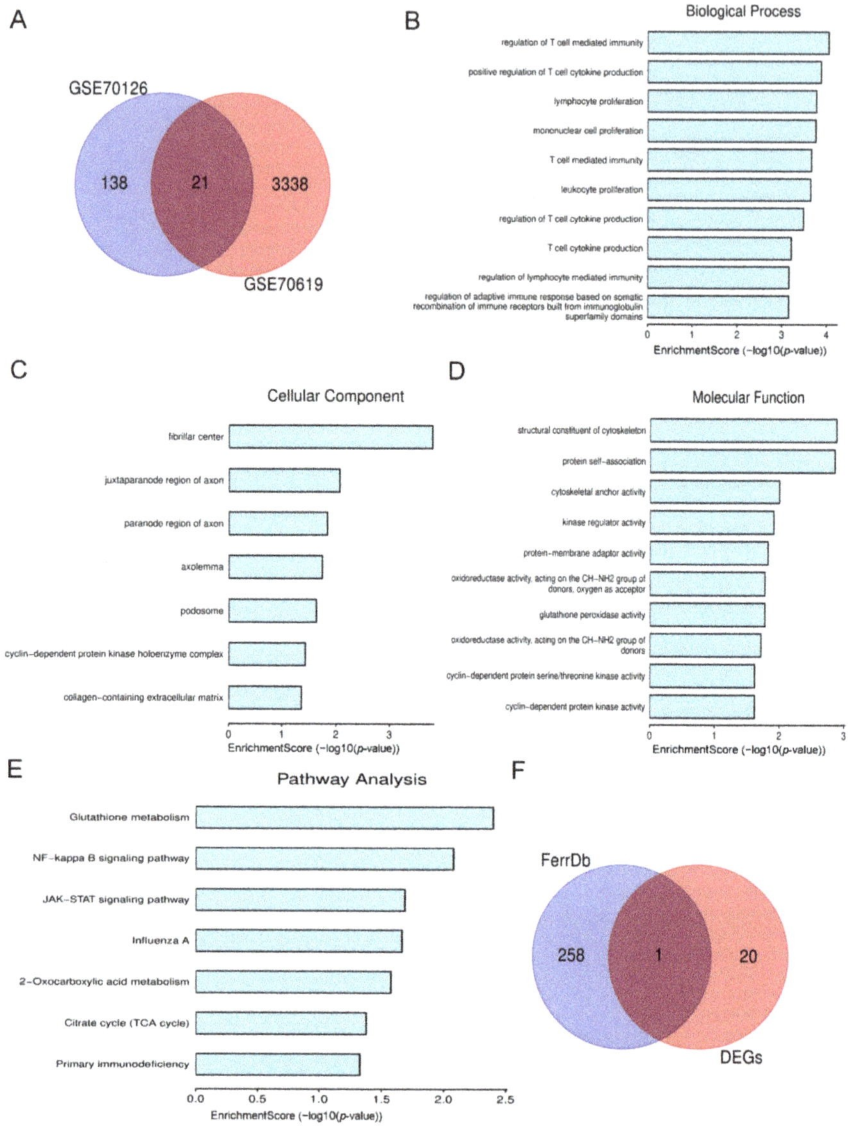

Figure 3. Identification and enrichment analysis of target DEGs. (**A**) Venn diagram of overlapping DEGs from GSE70126 and GSE70619 datasets. (**B**) Biological processes of 21 overlapping DEGs. (**C**)'Cellular components of 21 overlapping DEGs. (**D**) Molecular functions of 21 overlapping DEGs. (**E**) KEGG pathway of 21 overlapping DEGs. (**F**) Venn diagram of 21 overlapping DEGs and 259 ferroptosis-related regulatory genes.

3.3. Expression of IDH1 in ox-LDL-Induced Foam Cells

To develop a macrophage-derived foam cell model, different dosages of ox-LDL were used to stimulate RAW264.7 cells. In the CCK-8 experiment, 50 μg/mL ox-LDL significantly reduced cell viability (Figure 4A). As a result, in subsequent investigations, we used 50 μg/mL ox-LDL to enhance foam cell development, which was likewise the dosage used in previous findings [14,15].

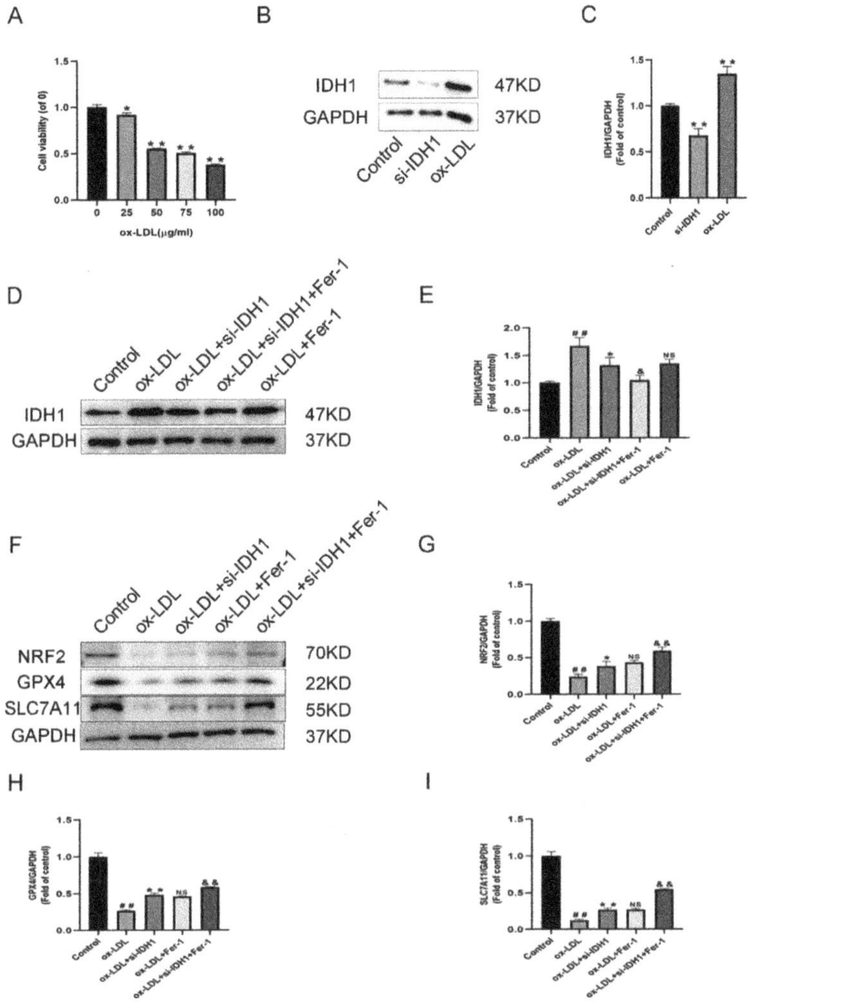

Figure 4. Expression of IDH1 in ox-LDL-induced foam cells. (**A**) Viability of RAW264.7 cells treated with ox-LDL (25, 50, 75, 100 μg/mL) for 24 h. * $p < 0.05$, ** $p < 0.01$ vs. control group. (**B**,**C**) Western blot was used to detect the levels of IDH1 protein expression. ## $p < 0.01$ si-IDH1 treatment group vs. control group; * $p < 0.05$, ** $p < 0.01$ ox-LDL-treated group vs. control group. (**D**,**E**) Western blot was used to detect the levels of IDH1 protein expression. ## $p < 0.01$ vs. control group; * $p < 0.05$, ** $p < 0.01$ vs. ox-LDL-treated group; & $p < 0.05$, && $p < 0.01$ vs. ox-LDL and si-IDH1 treatment group. NS vs. ox-LDL and si-IDH1 treatment group. (**F**–**I**) NRF2, GPX4, and SLC7A11 expression. ## $p < 0.01$ vs. control group; * $p < 0.05$, ** $p < 0.01$ vs. ox-LDL-treated group; NS vs. ox-LDL and si-IDH1 treatment group. & $p < 0.05$, && $p < 0.01$ vs. ox-LDL and Fer-1 treatment group. NS, not significant. Data are presented as mean ± SD.

In ox-LDL-induced foam cells, IDH1 was higher than in RAW264.7 cells, according to the Western blotting (WB) data (Figure 4B–E). The low IDH1 expression in the si-IDH1 group showed successful LV-Idh1-RNAi transfection (Figure 4B,C). Furthermore, for the ox-LDL + si-IDH1 group, low IDH1 expression could be successfully maintained. On the basis of IDH1 knockdown, Fer-1 was added, and the IDH1 protein level was further reduced. Interestingly, for IDH1 protein expression, the ox-LDL + si-IDH1 and ox-LDL

+ Fer-1 groups displayed no statistical difference. This suggests that the ferroptosis inhibitor Fer-1 and IDH1 knockdown have similar effects in reducing IDH1 expression.

3.4. *Inhibition of IDH1 Can Reduce the Production of ox-LDL-Induced Foam Cells*

The production of foam cells is a key step in the pathogenesis of atherosclerosis, and inhibiting foam cell formation could be a viable way to prevent atherosclerotic lesions from developing [2]. The oil red O staining findings showed that ox-LDL stimulation could enhance lipid accumulation, while the inhibition of IDH1 or addition of Fer-1 reduced the formation of foam cells. For ox-LDL + si-IDH1 + Fer-1 compared with ox-LDL + Fer-1, the experimental results showed that the formation of foam cells was further reduced, suggesting that the inhibition of IDH1 further ameliorated the formation of foam cells (Supplementary Figure S1A,B).

3.5. *Inhibition of IDH1 Ameliorates ox-LDL-Induced Ferroptosis*

Ferroptosis is dependent on the accumulation of intracellular iron causing an elevation of toxic LPO. In addition, it also revealed the decrease in GSH and the characteristic genes GPX4 and SLC7A11.

RAW264.7 cells induced by ox-LDL showed higher Fe^{2+} levels, accumulation of LPO, depletion of GSH, and downregulation of GPX4, FTH1, and SLC7A11 protein levels, indicating that ox-LDL promotes macrophage ferroptosis (Figure 4F,H,I, Figure 5A–C and Figure S1C,D). The knockdown of IDH1 showed reduced Fe^{2+} and LPO accumulation and increased levels of GSH, GPX4, FTH1, and SLC7A11, implying that the inhibition of IDH1 could reduce the ferroptosis induced by ox-LDL (Figure 4F,H,I, Figure 5A–C andm). The experimental results of the ox-LDL + si-IDH1 and ox-LDL + Fer-1 groups were similar, demonstrating that IDH1 inhibition can play a similar role in inhibiting ferroptosis as Fer-1 (Figure 4F,H,I, Figure 5A–C and Figure S1C,D). For ox-LDL + si-IDH1 + Fer-1 compared with ox-LDL + Fer-1, the IDH1 protein level was further reduced. The experimental results showed that the content of Fe^{2+} and LPO was further reduced, while the levels of GSH, GPX4, FTH1, and SLC7A11 were further increased, suggesting that the inhibition of IDH1 further ameliorated ox-LDL-induced ferroptosis (Figure 4F,H,I, Figure 5A–C and Figure S1C,D).

Lipid peroxidation was a critical biomarker associated with ferroptosis. We detected lipid peroxidation levels using C11BODIPY staining. The results showed that lipid peroxidation levels rose in ox-LDL-induced foam cells compared to the control group, but knocking down IDH1 reversed the increased lipid peroxidation levels. This suggested that ferroptosis was involved in ox-LDL-induced foam cell formation, and knocking down IDH1 relieved this by attenuating ferroptosis (Supplementary Figure S2A,B).

3.6. *Inhibition of IDH1 Suppresses α-KG and NADPH Levels*

It has been reported that the suppression of IDH1 inhibits α-KG and NADPH levels [16]. Therefore, we tested the α-KG and NADPH levels after generating the IDH1 KD macrophage cell line. The inhibition of IDH1 reduced the levels of α-KG and NADPH (Supplementary Figure S1E,F).

3.7. *Inhibition of IDH1 Ameliorates Macrophage Damage and Apoptosis Induced by ox-LDL*

We used the CCK8 test to assess cell viability, and flow cytometry to quantify the amount of apoptosis to see how IDH1 affected ox-LDL-induced macrophage damage and apoptosis. Macrophage viability was lowered and apoptosis was increased when ox-LDL was added. However, the inhibition of IDH1 expression before ox-LDL induction improved macrophage viability and inhibited apoptosis. Between the ox-LDL + si-IDH1 and ox-LDL + Fer-1 groups, there was no significant change in the cell survival or apoptosis rate, implying that IDH1 downregulation served a comparable function to the ferroptosis inhibitor Fer-1. In addition, the inhibition of IDH1 expression on the basis of ox-LDL + Fer-1 further ameliorated ox-LDL-induced macrophage damage and apoptosis (Figure 5D,E and Figure 6A,B).

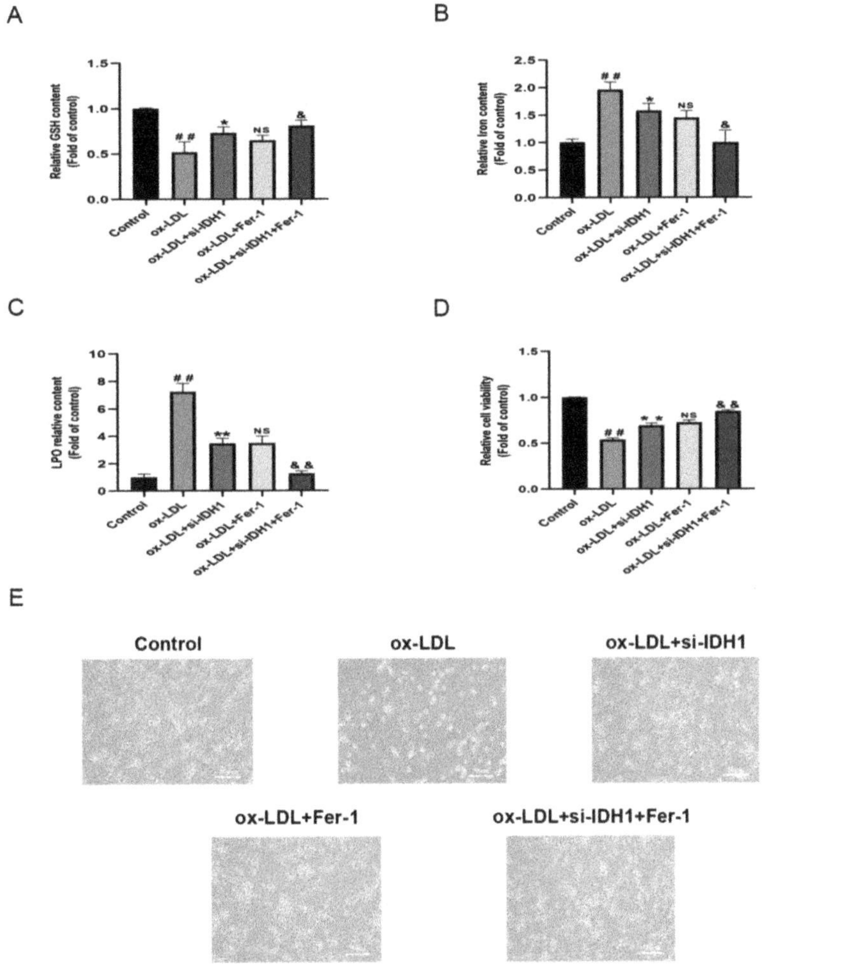

Figure 5. Inhibition of IDH1 ameliorates ox-LDL-induced ferroptosis. (**A**) The levels of GSH. (**B**) The level of iron content. (**C**) The total content of LPO. (**D**) Cell survival determined by CCK-8. (**E**) The brightfield images showing dead cells (Scale bars = 50 μm). $^{##}$ $p < 0.01$ vs. control group; * $p < 0.05$, ** $p < 0.01$ vs. ox-LDL-treated group; NS vs. ox-LDL and si-IDH1 treatment group. $^{\&}$ $p < 0.05$, $^{\&\&}$ $p < 0.01$ vs. ox-LDL and Fer-1 treatment group. NS, not significant. Data are presented as mean ± SD.

In addition, we complemented LDH experiments to investigate whether the knockdown of IDHI could protect macrophages from cell death at higher concentrations of ox-LDL (100 μg/mL). The experimental results showed that when ox-LDL was added, increased cell death resulted in increased LDH activity. However, the inhibition of IDH1 expression prior to ox-LDL induction suppressed LDH activity. The cellular levels of LDH did not change significantly between the ox-LDL + si-IDH1 and ox-LDL + Fer-1 groups, implying that IDH1 downregulation was functionally equivalent to the ferroptosis inhibitor Fer-1. Furthermore, the inhibition of IDH1 expression on the basis of ox-LDL + Fer-1 further ameliorated ox-LDL-induced macrophage damage. Therefore, we conclude that the knockdown of IDHI reduces macrophage death (Supplementary Figure S1G).

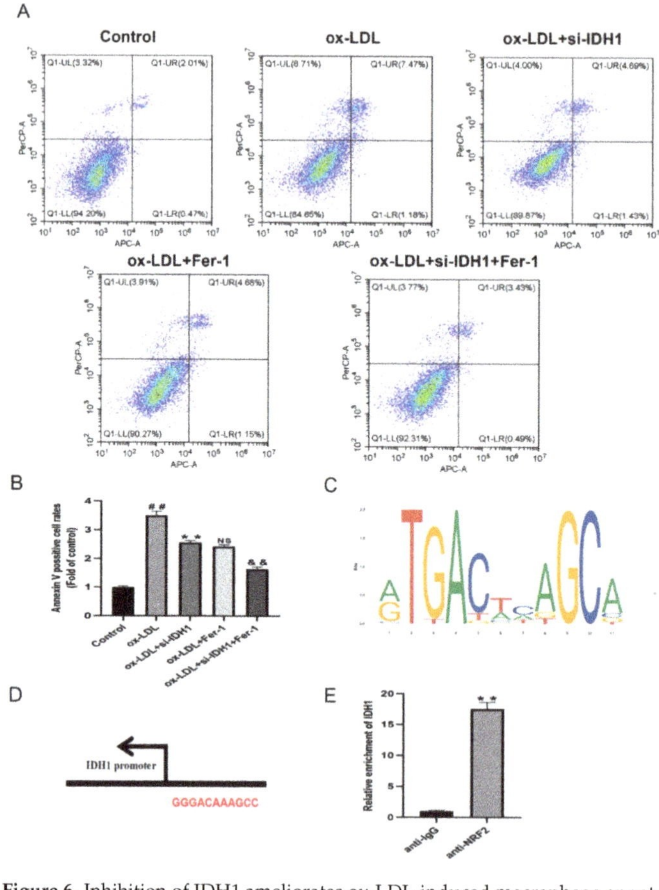

Figure 6. Inhibition of IDH1 ameliorates ox-LDL-induced macrophage apoptosis. (**A**,**B**) Apoptosis of RAW264.7 cells in different groups was examined via flow cytometry. (**C**) JASPAR-predicted NRF2 transcription factor binding sites. (**D**) Nrf2 and IDH1 potential binding sites. (**E**) Relative enrichment of IDH1, detected by ChIP. ## $p < 0.01$ vs. control group; ** $p < 0.01$ vs. ox-LDL-treated group; NS vs. ox-LDL and si-IDH1 treatment group. && $p < 0.01$ vs. ox-LDL and Fer-1 treatment group. NS, not significant. Data are presented as mean ± SD.

3.8. IDH1 May Regulate ox-LDL-Induced Macrophage Ferroptosis via NRF2

Ferroptosis is cell death caused by oxidative stress. By modulating gene expression in the GSH antioxidant system and iron metabolism, NRF2 can influence ferroptosis. Ferroptosis has been linked to cancer and neurological illnesses in a growing number of studies, and NRF2, which can regulate ferroptosis, is a promising pharmacological target. In our study, the online database JASPAR provided the TFBS of NRF2 (Figure 6C); the IDH1 and NRF2 potential binding sites are GGGACAAAGCC (Figure 6D). The presence of these binding sites on the IDH1 promoter suggests that IDH1 may interact with NRF2. The addition of ox-LDL reduced NRF2 levels when compared to controls, whereas the inhibition of IDH1 prior to ox-LDL induction upregulated the expression of NRF2, which was similar to the ox-LDL + Fer-1 group. In addition, further reducing the expression of IDH1 on the basis of ox-LDL + Fer-1 would further upregulate the protein level of NRF2

(Figure 4F,G). Further ChIP assays demonstrated that the IDH1 promoter sequence was significantly enriched by the anti-NRF2 antibody (Figure 6E). In summary, this means that IDH1 may regulate ox-LDL-induced macrophage ferroptosis and foam cell formation through NRF2.

4. Discussion

Foam cells are involved in all stages of atherosclerosis development [17]. The formation of macrophage-derived foam cells is caused by the dysregulation of lipid metabolism. When foam cells accumulate beneath the lipid-rich arterial endothelium, they begin to perform a variety of functions that promote the development of atherosclerosis, increasing the likelihood of plaque rupture [18]. Furthermore, foam cells do not exist in isolation and can contribute to the development of atherosclerosis through pro-inflammation of apoptotic cells. The formation of foam cells and apoptosis induced by ox-LDL are important in the pathogenesis of atherosclerosis [19]. After apoptosis and necrosis, foam cells die, and intracellular lipids accumulate on arterial walls, exacerbating atherosclerotic lesions [20]. Apoptotic cell removal has been shown to help with plaque stabilization [21]. Excess lipid accumulation in macrophages inhibits macrophage egress from plaques, resulting in macrophage apoptosis and secondary necrosis, which exacerbates foam cell formation and atherosclerotic progression [22]. Apoptosis of macrophages causes necrotic areas to enlarge in late-stage atherosclerotic plaques [23]. Numerous studies have shown that ox-LDL-induced macrophage apoptosis and foam cell formation play an important role in the onset and progression of atherosclerosis, providing information for clinical practice and drug development. The inhibition of ox-LDL-induced macrophage apoptosis and foam cell formation could be a potential treatment target for atherosclerosis [24,25].

Foam cells are essential to the formation of atherosclerosis lesions, and increasing data suggest that blocking foam cell formation is a possible treatment option [2,25]. In this study, we screened 21 potential target genes in foam cells through the GEO database. KEGG suggested that DEGs were mainly enriched in glutathione metabolism. The main manifestation of ferroptosis is the reduction in GSH and GPX4; thus, we evaluated the ferroptosis-related target gene IDH1 in 21 DEGs. In foam cells induced by ox-LDL, IDH1 was higher than in RAW264.7 cells. The inhibition of IDH1 ameliorated the foam cell formation induced by ox-LDL.

Ferroptosis is distinguished by the buildup of intracellular iron and the weakening of antioxidant ability, which results in programmed cell death [26]. Many pathogenic processes, including neurological diseases, myocardial I/R damage, and COVID-19 cardiac injury, have been linked to ferroptosis [27–29]. Existing studies have certified that ferroptosis is related to the progression of atherosclerosis [30–32]. IDH1 has the ability to convert isocitrate to oxalosuccinate, which is then converted to α-ketoglutarate. Tumor-derived IDH1 mutations promote lipid ROS accumulation and subsequent ferroptosis. The inhibition of IDH1 reduced the levels of α-KG and NADPH, as in previous reports [16].

IDH1 mutations decrease GPX4, a key enzyme in the ferroptosis and removal of lipid ROS, and accelerate glutathione depletion [10]. However, it is unclear whether IDH1 reduces ox-LDL-induced ferroptosis and, as a result, influences the progression of atherosclerosis. IDH1 inhibition increased the GPX4, FTH1, and SLC7A11 protein levels and repressed glutathione depletion, whereas downregulation of IDH1 decreased Fe^{2+}, LDH levels, and lipid ROS generation. This means that the inhibition of IDH1 ameliorated ox-LDL-induced ferroptosis.

Ox-LDL-induced macrophage damage and apoptosis are critical in the formation and development of atherosclerosis [33,34], and inhibiting IDH1 expression in macrophages prevented viability and apoptosis death in our work. By increasing the generation of GPX4, GSH, and NADPH, and preserving cellular iron homeostasis, the NRF2-Keap1 pathway decreases atherosclerosis-related ferroptosis [30]. IDH1 has binding sites with NRF2, according to database predictions, and the inhibition of IDH1 could upregulate the expression of NRF2. It has been reported that the promoter region of IDH1 binds significantly to NRF2 [35]. Similarly, in our study, NRF2 can bind to the promoter region of IDH1. By activating NRF2, the inhibition of IDH1 may relieve foam cell formation by ameliorating ox-LDL-induced macrophage ferroptosis.

By influencing the gene expression of the GSH antioxidant system and iron metabolism, NRF2 can impact AS-related ferroptosis [30,36]. NRF2 is decreased in AS patients. In addition, when ox-LDL stimulates cells, NRF2 is significantly reduced [7]. In our study, ChIP showed that IDH1 was able to bind to NRF2. In addition, IDH1 may also affect the expression of NRF2 by regulating oxidative metabolism-related products. This indicates that IDH1 may regulate ox-LDL-induced macrophage ferroptosis through NRF2, and we will explore the relevant mechanism in subsequent studies.

We observed that ox-LDL induces macrophage ferroptosis, while ox-LDL promotes the upregulation of IDH1 expression. Given that IDH1 mutation can lead to iron death in cells as reported in previous studies [10,37], we speculate that there may be a regulatory relationship between IDH1 and ferroptosis. Moreover, we found that the knockdown of IDH1 expression with specific IDH1 siRNA ameliorated the reduction in GSH, iron accumulation, decreased cell viability, and increased lipotoxicity. After adding ferroptosis inhibitors, these phenomena were further improved, indicating that the inhibition of IDH1 could partially ameliorate ox-LDL-induced ferroptosis. However, ox-LDL-induced ferroptosis and IDH1 upregulation cannot indicate whether ferroptosis promotes increased IDH1 expression or IDH1 promotes ferroptosis, and these two phenomena may be parallel rather than causal. Although our present results show that IDH1 expression was further downregulated in the si-IDH1 + Fer-1 group compared to the si-IDH1 only group, there was no difference when Fer-1 was used alone. This suggested that the inhibition of ferroptosis will further downregulate IDH1 in the presence of low levels of IDH1. Ferroptosis was improved after the inhibition of IDH1, which may suggest that there may be a relationship between ferroptosis and IDH1-dependent positive feedback regulatory mechanisms. Therefore, in-depth studies are needed to clarify the regulatory relationship between ferroptosis and IDH1, such as the detection of IDH1 expression changes after the induction of ferroptosis with ferroptosis agonists. In addition, careful design is necessary to determine whether the overexpression of IDH1 promotes ferroptosis, while adding ferroptosis inhibitors at the same time reverses the ferroptosis caused by the overexpression of IDH1. More importantly, the regulatory mechanism was found through deeper exploration. This research will aid in a better understanding of IDH1's numerous activities in ferroptosis and open the door for the discovery of new ferroptosis therapeutic targets.

For the ox-LDL + si-IDH1 + Fer-1 group, the phenotypes such as gene expression and cell death can only be partially rescued. Our study showed that neither the suppression of IDH1 levels nor the use of Fer-1 could completely alleviate ox-LDL stimulation-induced macrophage ferroptosis and foam cell formation. Similarly, adding Fer-1 on the basis of inhibiting IDH1 only partially alleviated the ferroptosis of macrophages and ox-LDL-induced foam cell formation, but could not completely reverse the cell damage caused by ox-LDL. In follow-up studies, we will try to enhance the efficiency of IDH1 intervention or increase the concentration of Fer-1 and explore the ratio between the two to explore whether ox-LDL stimulation-induced macrophage ferroptosis and foam cell formation can be completely rescued. The possible mechanism will be verified by deeper exploration.

Recent studies have discovered that Jak2V617F, ALOX5, and NCF2 are involved in the formation of atherosclerosis by regulating macrophage ferroptosis [38,39]. Furthermore, the important role of macrophage-mediated ferroptosis in cancer therapy has also been widely researched [40,41]. In colorectal cancer, there is a crosstalk between ferroptosis and three other programmed cell deaths, including necrosis, autophagy, and apoptosis [42]. An increasing number of ferroptosis inhibitors have been discovered in recent years [43], and ferroptosis has become a new therapeutic target for more disorders [44].

In our findings, knocking down IDH1 inhibited macrophage ferroptosis and foam cell formation, which is a potential research area for better understanding the occurrence and development of atherosclerosis. One of the study's shortcomings is the lack of human or in vivo data, as well as investigations on non-macrophage-derived foam cells. GSE70126 and GSE70619 were from separate platforms, we examined their respective differential genes and then took the intersection to find similar differential genes. One of the study's limitations might be the lack of batch effect correction. In further investigations, we will investigate the relevant mechanisms, potentially resulting in the discovery of novel treatment targets for ferroptosis.

5. Conclusions

Overall, in ox-LDL-induced macrophages, IDH1 was shown to be substantially expressed. The inhibition of IDH1 modulates the progression of atherosclerosis by ameliorating macrophage viability and apoptosis. Furthermore, inhibiting IDH1 may alleviate atherosclerosis by ameliorating ox-LDL-induced macrophage ferroptosis by activating NRF2.

Supplementary Materials: The following supporting information can be downloaded at: https://www.mdpi.com/article/10.3390/biology11101392/s1, Figure S1: Oil red O staining and the levels of FTH1, α-KG, NADPH, and LDH; Figure S2: Measurement of lipid peroxidation; Table S1: Differentially expressed genes; Table S2: Primer sequences.

Author Contributions: B.L., C.W. and P.L. designed the experiments. B.L., Y.J. and X.W. (Xufeng Wang) wrote the manuscript. C.L. and X.L. evaluated the data. X.X. and X.W. (Xiaowei Wang) oversaw the study. All authors have read and agreed to the published version of the manuscript.

Funding: The Jiangsu Provincial Department of Health provided funding for this project (no. ZDA2020004).

Institutional Review Board Statement: Not applicable.

Informed Consent Statement: Not applicable.

Data Availability Statement: The original contributions presented in the study are included in the article/Supplementary Materials. Further inquiries can be directed to the corresponding authors.

Conflicts of Interest: The authors declare no conflict of interest.

References

1. Orecchioni, M.; Kobiyama, K.; Winkels, H.; Ghosheh, Y.; McArdle, S.; Mikulski, Z.; Kiosses, W.B.; Fan, Z.; Wen, L.; Jung, Y.; et al. Olfactory receptor 2 in vascular macrophages drives atherosclerosis by NLRP3-dependent IL-1 production. *Science* **2022**, *375*, 214–221. [CrossRef]
2. Liu, X.; Wu, J.; Tian, R.; Su, S.; Deng, S.; Meng, X. Targeting foam cell formation and macrophage polarization in atherosclerosis: The Therapeutic potential of rhubarb. *Biomed. Pharmacother.* **2020**, *129*, 110433. [CrossRef] [PubMed]
3. Chen, X.; Li, X.; Xu, X.; Li, L.; Liang, N.; Zhang, L.; Lv, J.; Wu, Y.C.; Yin, H. Ferroptosis and cardiovascular disease: Role of free radical-induced lipid peroxidation. *Free Radic. Res.* **2021**, *55*, 405–415. [CrossRef]
4. Bai, T.; Li, M.; Liu, Y.; Qiao, Z.; Wang, Z. Inhibition of ferroptosis alleviates atherosclerosis through attenuating lipid peroxidation and endothelial dysfunction in mouse aortic endothelial cell. *Free Radic. Biol. Med.* **2020**, *160*, 92–102. [CrossRef]
5. Qiu, Y.; Cao, Y.; Cao, W.; Jia, Y.; Lu, N. The Application of Ferroptosis in Diseases. *Pharmacol. Res.* **2020**, *159*, 104919. [CrossRef]
6. Jiang, X.; Stockwell, B.R.; Conrad, M. Ferroptosis: Mechanisms, biology and role in disease. *Nat. Rev. Mol. Cell. Biol.* **2021**, *22*, 266–282. [CrossRef]
7. Yang, K.; Song, H.; Yin, D. PDSS2 Inhibits the Ferroptosis of Vascular Endothelial Cells in Atherosclerosis by Activating Nrf2. *J. Cardiovasc. Pharmacol.* **2021**, *77*, 767–776. [CrossRef] [PubMed]
8. Martinet, W.; Coornaert, I.; Puylaert, P.; De Meyer, G.R.Y. Macrophage Death as a Pharmacological Target in Atherosclerosis. *Front. Pharmacol.* **2019**, *10*, 306. [CrossRef]
9. Bruce-Brand, C.; Govender, D. Gene of the month: IDH1. *J. Clin. Pathol.* **2020**, *73*, 611–615. [CrossRef] [PubMed]
10. Wang, T.X.; Liang, J.Y.; Zhang, C.; Xiong, Y.; Guan, K.L.; Yuan, H.X. The oncometabolite 2-hydroxyglutarate produced by mutant IDH1 sensitizes cells to ferroptosis. *Cell Death Dis.* **2019**, *10*, 755. [CrossRef] [PubMed]
11. Huang, D.W.; Sherman, B.T.; Lempicki, R.A. Bioinformatics enrichment tools: Paths toward the comprehensive functional analysis of large gene lists. *Nucleic Acids Res.* **2009**, *37*, 1–13. [CrossRef]
12. Zhou, N.; Bao, J. FerrDb: A manually curated resource for regulators and markers of ferroptosis and ferroptosis-disease associations. *Database* **2020**, *2020*, baaa021. [CrossRef]
13. Wasserman, W.W.; Sandelin, A. Applied bioinformatics for the identification of regulatory elements. *Nat. Rev. Genet.* **2004**, *5*, 276–287. [CrossRef]

14. Xie, Z.; Wang, X.; Liu, X.; Du, H.; Sun, C.; Shao, X.; Tian, J.; Gu, X.; Wang, H.; Tian, J.; et al. Adipose-Derived Exosomes Exert Proatherogenic Effects by Regulating Macrophage Foam Cell Formation and Polarization. *J. Am. Heart Assoc.* **2018**, *7*, e007442. [CrossRef] [PubMed]
15. Geng, J.; Yang, C.; Wang, B.; Zhang, X.; Hu, T.; Gu, Y.; Li, J. Trimethylamine N-oxide promotes atherosclerosis via CD36-dependent MAPK/JNK pathway. *Biomed. Pharmacother.* **2018**, *97*, 941–947. [CrossRef]
16. Calvert, A.E.; Chalastanis, A.; Wu, Y.; Hurley, L.A.; Kouri, F.M.; Bi, Y.; Kachman, M.; May, J.L.; Bartom, E.; Hua, Y.; et al. Cancer-Associated IDH1 Promotes Growth and Resistance to Targeted Therapies in the Absence of Mutation. *Cell Rep.* **2017**, *19*, 1858–1873. [CrossRef]
17. Chistiakov, D.A.; Melnichenko, A.A.; Myasoedova, V.A.; Grechko, A.V.; Orekhov, A.N. Mechanisms of foam cell formation in atherosclerosis. *J. Mol. Med.* **2017**, *95*, 1153–1165. [CrossRef] [PubMed]
18. Maguire, E.M.; Pearce, S.; Xiao, Q. Foam cell formation: A new target for fighting atherosclerosis and cardiovascular disease. *Vasc. Pharmacol.* **2019**, *112*, 54–71. [CrossRef]
19. Zhang, H.; Liu, Q.; Lin, J.L.; Wang, Y.; Zhang, R.X.; Hou, J.B.; Yu, B. Recombinant Human Thioredoxin-1 Protects Macrophages from Oxidized Low-Density Lipoprotein-Induced Foam Cell Formation and Cell Apoptosis. *Biomol. Ther.* **2018**, *26*, 121–129. [CrossRef]
20. Zhang, J.; Zu, Y.; Dhanasekara, C.S.; Li, J.; Wu, D.; Fan, Z.; Wang, S. Detection and treatment of atherosclerosis using nanoparticles. *Wiley Interdiscip. Rev. Nanomed. Nanobiotechnol.* **2017**, *9*, e1412. [CrossRef]
21. Tabas, I. Macrophage apoptosis in atherosclerosis: Consequences on plaque progression and the role of endoplasmic reticulum stress. *Antioxid. Redox Signal.* **2009**, *11*, 2333–2339. [CrossRef] [PubMed]
22. Luo, Y.; Duan, H.; Qian, Y.; Feng, L.; Wu, Z.; Wang, F.; Feng, J.; Yang, D.; Qin, Z.; Yan, X. Macrophagic CD146 promotes foam cell formation and retention during atherosclerosis. *Cell Res.* **2017**, *27*, 352–372. [CrossRef] [PubMed]
23. Ren, J.L.; Chen, Y.; Zhang, L.S.; Zhang, Y.R.; Liu, S.M.; Yu, Y.R.; Jia, M.Z.; Tang, C.S.; Qi, Y.F.; Lu, W.W. Intermedin1-53 attenuates atherosclerotic plaque vulnerability by inhibiting CHOP-mediated apoptosis and inflammasome in macrophages. *Cell Death Dis.* **2021**, *12*, 436. [CrossRef] [PubMed]
24. Wang, D.; Yang, Y.; Lei, Y.; Tzvetkov, N.T.; Liu, X.; Yeung, A.; Xu, S.; Atanasov, A.G. Targeting Foam Cell Formation in Atherosclerosis: Therapeutic Potential of Natural Products. *Pharmacol. Rev.* **2019**, *71*, 596–670. [CrossRef]
25. Poznyak, A.V.; Wu, W.K.; Melnichenko, A.A.; Wetzker, R.; Sukhorukov, V.; Markin, A.M.; Khotina, V.A.; Orekhov, A.N. Signaling Pathways and Key Genes Involved in Regulation of foam Cell Formation in Atherosclerosis. *Cells* **2020**, *9*, 584. [CrossRef]
26. Yang, X.D.; Yang, Y.Y. Ferroptosis as a Novel Therapeutic Target for Diabetes and Its Complications. *Front. Endocrinol.* **2022**, *13*, 853822. [CrossRef]
27. Liu, Y.; He, L.; Liu, B.; Ying, Y.; Xu, J.; Yu, M.; Dang, J.; Liu, K. Pharmacological inhibition of sphingolipid synthesis reduces ferroptosis by stimulating the HIF-1 pathway. *iScience* **2022**, *25*, 104533. [CrossRef]
28. Chen, H.Y.; Xiao, Z.Z.; Ling, X.; Xu, R.N.; Zhu, P.; Zheng, S.Y. ELAVL1 is transcriptionally activated by FOXC1 and promotes ferroptosis in myocardial ischemia/reperfusion injury by regulating autophagy. *Mol. Med.* **2021**, *27*, 14. [CrossRef]
29. Jacobs, W.; Lammens, M.; Kerckhofs, A.; Voets, E.; Van San, E.; Van Coillie, S.; Peleman, C.; Mergeay, M.; Sirimsi, S.; Matheeussen, V.; et al. Fatal lymphocytic cardiac damage in coronavirus disease 2019 (COVID-19): Autopsy reveals a ferroptosis signature. *ESC Heart Fail.* **2020**, *7*, 3772–3781. [CrossRef]
30. Wang, Y.; Zhao, Y.; Ye, T.; Yang, L.; Shen, Y.; Li, H. Ferroptosis Signaling and Regulators in Atherosclerosis. Front. *Cell Dev. Biol.* **2021**, *9*, 809457.
31. Yu, Y.; Yan, Y.; Niu, F.; Wang, Y.; Chen, X.; Su, G.; Liu, Y.; Zhao, X.; Qian, L.; Liu, P.; et al. Ferroptosis: A cell death connecting oxidative stress, inflammation and cardiovascular diseases. *Cell Death Discov.* **2021**, *7*, 193. [CrossRef] [PubMed]
32. Guo, Y.; Lu, C.; Hu, K.; Cai, C.; Wang, W. Ferroptosis in Cardiovascular Diseases: Current Status, Challenges, and Future Perspectives. *Biomolecules* **2022**, *12*, 390. [CrossRef] [PubMed]
33. Negre-Salvayre, A.; Auge, N.; Camare, C.; Bacchetti, T.; Ferretti, G.; Salvayre, R. Dual signaling evoked by oxidized LDLs in vascular cells. *Free Radic. Biol. Med.* **2017**, *106*, 118–133. [CrossRef] [PubMed]
34. Farahi, L.; Sinha, S.K.; Lusis, A.J. Roles of Macrophages in Atherogenesis. *Front. Pharmacol.* **2021**, *12*, 785220. [CrossRef]
35. Bai, M.; Yang, L.; Liao, H.; Liang, X.; Xie, B.; Xiong, J.; Tao, X.; Chen, X.; Cheng, Y.; Chen, X.; et al. Metformin sensitizes endometrial cancer cells to chemotherapy through IDH1-induced Nrf2 expression via an epigenetic mechanism. *Oncogene* **2018**, *37*, 5666–5681. [CrossRef]
36. Dodson, M.; Castro-Portuguez, R.; Zhang, D.D. NRF2 plays a critical role in mitigating lipid peroxidation and ferroptosis. *Redox Biol.* **2019**, *23*, 101107. [CrossRef]
37. Su, L.; Huang, Y.; Zheng, L.; Zhu, Z.; Wu, Y.; Li, P. Isocitrate dehydrogenase 1 mutation in cholangiocarcinoma impairs tumor progression by sensitizing cells to ferroptosis. *Open Med.* **2022**, *17*, 863–870. [CrossRef]
38. Liu, W.; Östberg, N.K.; Yalcinkaya, M.; Dou, H.; Endo-Umeda, K.; Tang, Y.; Hou, X.; Xiao, T.; Fidler, T.P.; Abramowicz, S.; et al. Erythroid lineage Jak2V617F expression promotes atherosclerosis through erythrophagocytosis and macrophage ferroptosis. *J. Clin. Investig.* **2022**, *132*, e155724. [CrossRef]
39. Li, M.; Xin, S.; Gu, R.; Zheng, L.; Hu, J.; Zhang, R.; Dong, H. Novel Diagnostic Biomarkers Related to Oxidative Stress and Macrophage Ferroptosis in Atherosclerosis. *Oxidative Med. Cell. Longev.* **2022**, *2022*, 8917947. [CrossRef]

40. Zhang, Y.; Huang, Z.; Cheng, J.; Pan, H.; Lin, T.; Shen, X.; Chen, W.; Chen, Q.; Gu, C.; Mao, Q.; et al. Platelet-Vesicles-Encapsulated RSL-3 Enable Anti-Angiogenesis and Induce Ferroptosis to Inhibit Pancreatic Cancer Progress. *Front. Endocrinol.* **2022**, *13*, 865655. [CrossRef]
41. Wang, Y.; Fan, J.; Chen, T.; Xu, L.; Liu, P.; Xiao, L.; Wu, T.; Zhou, Q.; Zheng, Q.; Liu, C.; et al. A novel ferroptosis-related gene prognostic index for prognosis and response to immunotherapy in patients with prostate cancer. *Front. Endocrinol.* **2022**, *13*, 975623. [CrossRef] [PubMed]
42. Zhong, Y.; Zhang, W.; Yu, H.; Lin, L.; Gao, X.; He, J.; Li, D.; Chen, Y.; Zeng, Z.; Xu, Y.; et al. Multi-platform-based characterization of ferroptosis in human colorectal cancer. *iScience* **2022**, *25*, 104750. [CrossRef] [PubMed]
43. Hinder, L.; Pfaff, A.L.; Emmerich, R.E.; Michels, S.; Schlitzer, M.; Culmsee, C. Characterization of Novel Diphenylamine Compounds as Ferroptosis Inhibitors. *J. Pharmacol. Exp. Ther.* **2021**, *378*, 184–196. [CrossRef] [PubMed]
44. Cotticelli, M.G.; Xia, S.; Lin, D.; Lee, T.; Terrab, L.; Wipf, P.; Huryn, D.M.; Wilson, R.B. Ferroptosis as a Novel Therapeutic Target for Friedreich's Ataxia. *J. Pharmacol. Exp. Ther.* **2019**, *369*, 47–54. [CrossRef] [PubMed]

biology MDPI

Article

Recovery of Ischemic Limb and Femoral Artery Endothelial Function Are Preserved in Mice with Dextran Sodium Sulfate-Induced Chronic Colitis

Hao Wu [1,2], Qiang Zhu [1], Xuanyou Liu [1], Hong Hao [1], Zhe Sun [3], Meifang Wang [1], Michael A. Hill [3], Canxia Xu [2] and Zhenguo Liu [1,*]

1 Center for Precision Medicine and Division of Cardiovascular Medicine, University of Missouri School of Medicine, Columbia, MO 65212, USA
2 Department of Gastroenterology, Third Xiangya Hospital, Central South University, Changsha 410013, China
3 Dalton Cardiovascular Research Center, University of Missouri, Columbia, MO 65212, USA
* Correspondence: liuzheng@health.missouri.edu

Simple Summary: The present study examines the effect of experimental inflammatory bowel disease on femoral artery endothelial function and limb ischemia recovery in female mice using a chronic colitis model induced by dextran sodium sulfate exposure. As expected, plasma levels of proinflammatory cytokines, including interleukin-6, interleukin-17, tumor necrosis factor alpha, and chemokine ligand 1, were significantly increased in the chronic colitis model. However, ROS levels in the ischemic muscle tissues were not significantly increased in mice with colitis as compared to controls. There were no significant changes in endothelium-dependent or -independent vasodilation of femoral artery between the colitis model and the control. Recovery of function and blood flow of the ischemic limb and capillary density in the ischemic muscle were preserved in the colitis model as compared with the control.

Abstract: Inflammatory bowel disease (IBD) produces significant systemic inflammation and increases the risk of endothelial dysfunction and peripheral artery disease. Our recent study demonstrated that abdominal aortic endothelial cell function was impaired selectively in female mice with chronic colitis. This study aimed to test the hypothesis that experimental colitis leads to femoral artery endothelial cell dysfunction and impairs limb ischemia recovery in female mice. An experimental chronic colitis model was created in female C57BL/6 mice with dextran sodium sulfate (DSS) treatment. Unilateral hind limb ischemia was produced by femoral artery ligation. Limb blood perfusion, vascular density, tissue ROS levels, and plasma levels of proinflammatory cytokines were assessed. Femoral artery endothelium-dependent and -independent vasodilation of the contralateral limb were evaluated ex vivo using acetylcholine and nitroglycerin, respectively. As expected, the plasma levels of proinflammatory cytokines, including tumor necrosis factor alpha (TNF-α), interleukin (IL)-6, and IL-17, were significantly increased in the DSS-induced colitis model. However, ROS levels in the ischemic muscle tissues were not significantly increased in colitis model as compared to the controls. There were no significant changes in endothelium-dependent or -independent vasodilation of the femoral artery between colitis model and the control. Recovery of function and blood flow in the ischemic limb and capillary density in the ischemic gastrocnemius muscle were preserved in the colitis model as compared with the control. The data demonstrated that DSS-induced chronic colitis had no significant impact on femoral artery endothelial function or ischemic limb recovery in female mice.

Keywords: inflammatory bowel disease; limb ischemia; endothelial function; ROS

Citation: Wu, H.; Zhu, Q.; Liu, X.; Hao, H.; Sun, Z.; Wang, M.; Hill, M.A.; Xu, C.; Liu, Z. Recovery of Ischemic Limb and Femoral Artery Endothelial Function Are Preserved in Mice with Dextran Sodium Sulfate-Induced Chronic Colitis. *Biology* **2022**, *11*, 1169. https://doi.org/10.3390/biology11081169

Academic Editor: Gaetano Santulli

Received: 4 July 2022
Accepted: 2 August 2022
Published: 4 August 2022

1. Introduction

Inflammatory bowel disease (IBD), encompassing ulcerative colitis (UC) and Crohn's disease (CD), is characterized by chronic and recurrent intestinal inflammation. IBD not only alters the function of the gastrointestinal system, but can also affect many other organ systems, which are referred to as extraintestinal manifestations (EIMs) [1,2]. Several population-based studies have previously shown that patients with IBD are at an increased risk of developing cardiovascular diseases (CVDs), including peripheral artery disease (PAD), ischemic heart disease, and cerebrovascular disease [3,4]. However, the underlying mechanisms for links between IBD and CVDs remain poorly defined.

Systemic inflammation and endothelial cell dysfunction are considered among the important factors for the development and progression of CVDs [5,6]. Several inflammatory biomarkers, including C-reactive protein, calprotectin, TNF-α, IL-6, and IL-17, are significantly increased in patients with IBD [7]. Studies have demonstrated that endothelial cell dysfunction, as reflected by decreases in flow-mediated dilatation and pulse arterial tonometry indices, is present in patients with IBD [8,9]. Consistent with this, our recent meta-analysis showed that patients with IBD were significantly associated with endothelial cell dysfunction, increased arterial stiffness, and carotid intima-media thickness [10].

Dextran sodium sulfate (DSS) induces significant damages to the intestinal epithelium. The epithelial damage compromises the barrier integrity and subsequently exposes the mucosal and submucosal immune cells to bacterial antigens, resulting in significant local and systemic inflammation [11]. Our recent study showed that abdominal aortic endothelial cell function was impaired selectively in female mice with chronic colitis [12]. The French population-based cohort study demonstrated that the risk of developing PAD was significantly increased 1.27-fold in patients with IBD compared with the general population [4]. To our knowledge, there have been no animal studies to evaluate limb ischemia recovery and femoral artery endothelial function in DSS-induced colitis. The present study was aimed to test the hypothesis that experimental colitis could lead to femoral artery endothelial cell dysfunction and impaired hind limb ischemia recovery.

2. Materials and Methods

2.1. Animals

All mice procedures were conducted in accordance with the Guide for the Care and Use of Laboratory Animals of National Institutes of Health (NIH). The protocols were approved by the Institutional Animal Care and Use Committee of the University of Missouri, MO, USA. Eight-week-old C57BL/6 female mice were ordered from Jackson Lab (Bar Harbor, ME, USA). Mice were fed standard laboratory chow and maintained under controlled conditions (22 $\pm$ 1 °C, 12 h light/12 h dark cycle) until euthanasia.

2.2. Chronic DSS-Induced Colitis and Critical Limb Ischemia Mouse Model

To induce colitis, mice were administered with 2.5% (*wt/vol*) DSS solution (MP Biomedicals, Santa Ana, CA, USA) as described [11]. Mice for the chronic model were randomly assigned to either control (n = 8) or DSS (n = 8) treatment groups. Mice were treated with 3 cycles of DSS (seven days of DSS treatment with fourteen days of drinking water between each cycle) and subsequently sacrificed on day 51 as shown in Figure 1A. DSS solution was freshly made every other day. On day 33, unilateral hind limb ischemia was produced by femoral artery ligation as previously described [13]. In brief, mice were anesthetized with 2% isoflurane/O_2, and placed in supine position over a heated pad to maintain body temperature. The left femoral artery was exposed through a 1 cm incision, and then separated from the femoral vein. The proximal femoral artery was occluded and the distal femoral artery proximal to the popliteal artery was ligated with 7–0 suture. The segment of femoral artery between the proximal and distal knots was transected, and the incision was closed using 5–0 Vicryl sutures.

2.3. Analysis of Colitis and Ischemic Limbs

All mice were weighed every three days and evaluated for fecal occult blood using Compliance Gold (Germaine Laboratories, San Antonio, TX, USA) and stool consistency. The colitis severity was determined using disease activity index (DAI) for each mouse as described [14]. At the day of euthanasia, colon length was measured from the anal verge to the ileocecal junction. Gastrocnemius muscle and colon tissues were collected and embedded in paraffin and cut into 5 µm-thick tissue sections. The colon tissue and gastrocnemius muscle sections were stained with hematoxylin and eosin (H&E) and examined using an automated upright microscope system (Leica DM5500B, Wetzlar, Germany). The level of inflammation in colon tissue was determined using a scoring system for inflammation-associated histological changes as described [11]. The degeneration of gastrocnemius muscle was examined in the H&E-stained sections morphologically.

2.4. Analyses of Plasma Cytokines

While under isoflurane anesthesia, blood was collected from each mouse into K3 EDTA micro tubes (Sarstedt, Nümbrecht, Germany). Each blood sample was centrifuged at 3000× g for 10 min at 4 °C to obtain the plasma. The levels of plasma cytokines were measured using a mouse cytokine/chemokine discovery assay (Eve Technologies, Calgary, AB, Canada) [15].

2.5. Laser Doppler Perfusion Imaging and Function Assessment of Ischemic Limb

Limb blood flow was measured noninvasively using a laser doppler perfusion imaging system (Moor Instruments, Wilmington, DE, USA) as described [16]. Briefly, the evaluation of limb perfusion was conducted by acquiring flow images of the ischemic and contralateral limb before the creation of limb ischemia (baseline), and on days 0, 3, 7, 14, and 21 post-ischemia. After removal of excessive hair from the limbs, mice were placed on a heating pad at 37 °C to minimize temperature variation. The flux ratio of the total blood perfusion (ischemic limb blood flow/control limb blood flow) was used to determine the recovery of limb circulation. Assessments of limb function and ischemia were conducted at the same time points as blood flow measurements. Limb function was evaluated using the Tarlov scale and limb ischemia was assessed using the modified ischemia scale, as described [17,18].

2.6. Analysis of Vascular Density and ROS Production

Gastrocnemius muscle tissues of the ischemic limb were collected and embedded in optimal cutting temperature compound (Sakura Finetek, Torrance, CA, USA), and frozen in liquid nitrogen immediately. Tissue vascular density was evaluated using CD31 immunofluorescent staining as described [19]. Briefly, the frozen cross sections (5 µm in thickness) of gastrocnemius muscle tissue were sequentially fixed with 4% paraformaldehyde for 10 min, washed with PBS (x3), and blocked with 5% bovine serum albumin in PBS for 30 min at room temperature. After washing with PBS, the sections were incubated with Alexa Fluor 594 anti-mouse CD31 immunofluorescent antibody (1:500, Biolegend, San Diego, CA, USA) at 4 °C overnight. After rinsing in PBS (x3), the sections were incubated with 4′,6-diamidino-2-phenylindole (DAPI) for 10 min at room temperature, and then washed in PBS (x3), and examined using a fluorescence microscope (Olympus CKX53, Tokyo, Japan). Vascular density was determined using Image J software (NIH, Bethesda, MD, USA) with VesselJ plugin [19]. Four independent fields were examined for each section to calculate the mean vascular density. Tissue ROS levels were measured using dihydroethidium assay, as described [20]. Briefly, the frozen sections (5 µm in thickness) of gastrocnemius muscle tissue were prepared and incubated with 5 µM dihydroethidium (DHE; Molecular Probes, Eugene, OR, USA) for 15 min. Images were captured using a fluorescence microscope (Olympus CKX53, Tokyo, Japan), and analyzed with Image J software to calculate the mean gray value.

2.7. Evaluation of Femoral Artery Endothelial Function Ex Vivo

The femoral artery of the contralateral limb was carefully excised and prepared in ice-cold physiological saline solution (PSS) containing (in mM) 130 NaCl, 4.7 KCl, 14.9 $NaHCO_3$, 1.6 $CaCl_2$, 1.18 KH_2PO_4, 1.17 $MgSO_4$, 5.5 glucose, and 0.026 EDTA, pH 7.4, gassed with 5% CO_2 and 95% O_2. The femoral artery section was then carefully cut into rings (2 mm in length) under a stereo microscope. The arterial rings were mounted in a wire myograph system (620M; Danish Myo Technology, Hinnerup, Denmark). The organ chambers were filled with 5 mL PSS at 37 °C and aerated continuously with carbogen (95% O_2 + 5% CO_2). Based on vessel length/tension relationships, a preload tension of 3 mN/mm was applied to the rings. The vascular preparations were allowed to equilibrate for one hour with replacement of PSS every 20 min. To determine vascular contractility, the rings were incubated with high-potassium PSS containing (in mM): 74.7 NaCl, 60 KCl, 14.9 $NaHCO_3$, 1.6 $CaCl_2$, 1.18 KH_2PO_4, 1.17 $MgSO_4$, 5.5 glucose, and 0.026 EDTA (pH 7.4). After adequate washout with PSS (x3), a concentration–response curve was made to the endothelium-dependent vasodilator acetylcholine (Ach, 10^{-9} to 10^{-5} M, accumulative) and a concentration-response curve to the endothelium-independent vasodilator nitroglycerin (NTG, 10^{-9} to 10^{-5} M) for each ring after submaximal precontraction with phenylephrine (PE, 10^{-6} M) as described [12].

2.8. Statistical Analysis

All data were presented as mean ± SEM, and analyzed using an unpaired, two-tailed Student's t-test or two-way ANOVA followed by the Bonferroni correction. All statistical analyses were performed using Prism 8.0 software (GraphPad Software Inc., La Jolla, CA, USA). A two-sided $p < 0.05$ was considered statistically significant.

3. Results

3.1. Evaluation of Chronic DSS-Induced Colitis Models

Chronically DSS-treated mice developed significant chronic colitis, as demonstrated by fluctuating body weight (Figure 1B) together with a fluctuating DAI scoring (Figure 1C), and significant decreases in colon length (Figure 1D). Marked tissue damage and lamina propria inflammatory cell infiltration were evident in colon tissues of the colitis model, compared to the controls (Figure 1E). In addition, the plasma levels of granulocyte colony-stimulating factor (G-CSF), IL-6, IL-17, TNF-α, and chemokine ligand 1 (CXCL1) were significantly increased in the colitis model (Figure 2).

3.2. Blood Flow and Function Recoveries of the Ischemic Limb Were Preserved in Mice with Chronic Colitis

To determine if experimental chronic colitis could affect the recovery of ischemic limb, a unilateral hind limb ischemia model was produced by femoral artery ligation (Figure 3A), and laser doppler perfusion imaging of ischemic limb was conducted at baseline (before surgery at day 33) and day 33, 36, 40, 47, 54 after DSS administration (Figure 3B). There were no differences in blood flow recovery (ischemic limb blood perfusion/normal limb blood perfusion) (Figure 3C), functional score (Figure 3D), and ischemia score (Figure 3E) in mice with and without colitis. H&E staining showed marked fatty degeneration and atrophy in gastrocnemius muscle of the ischemic limb in mice with or without colitis (Figure 3F).

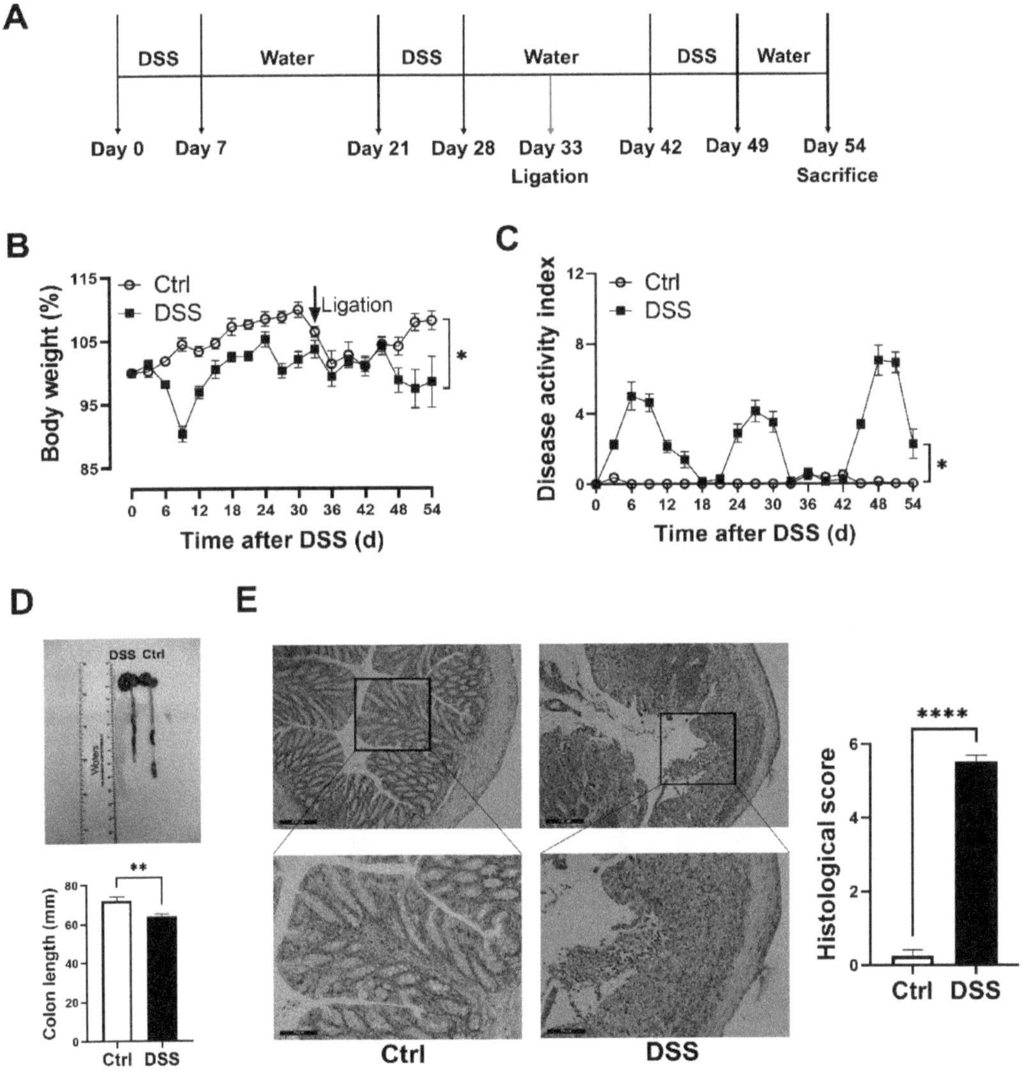

Figure 1. Female mice with chronic colitis displayed significant inflammation in the colon. (**A**) Experimental scheme illustrating DSS treatment and femoral ligation protocol for chronic colitis and limb ischemia models. Changes in (**B**) body weight and (**C**) DAI during DSS treatment. (**D**) Colon length and representative photographs for colon tissue in colitis and control mice. (**E**) Representative images of H&E staining of colon tissue and summary of histological score (scale bar, 100 μm and 50 μm). Results are expressed as mean ± SEM. * $p < 0.05$, ** $p < 0.01$, **** $p < 0.0001$, by unpaired 2-tailed Student's t-test, $n = 8$ mice in each group. DAI, disease activity index.

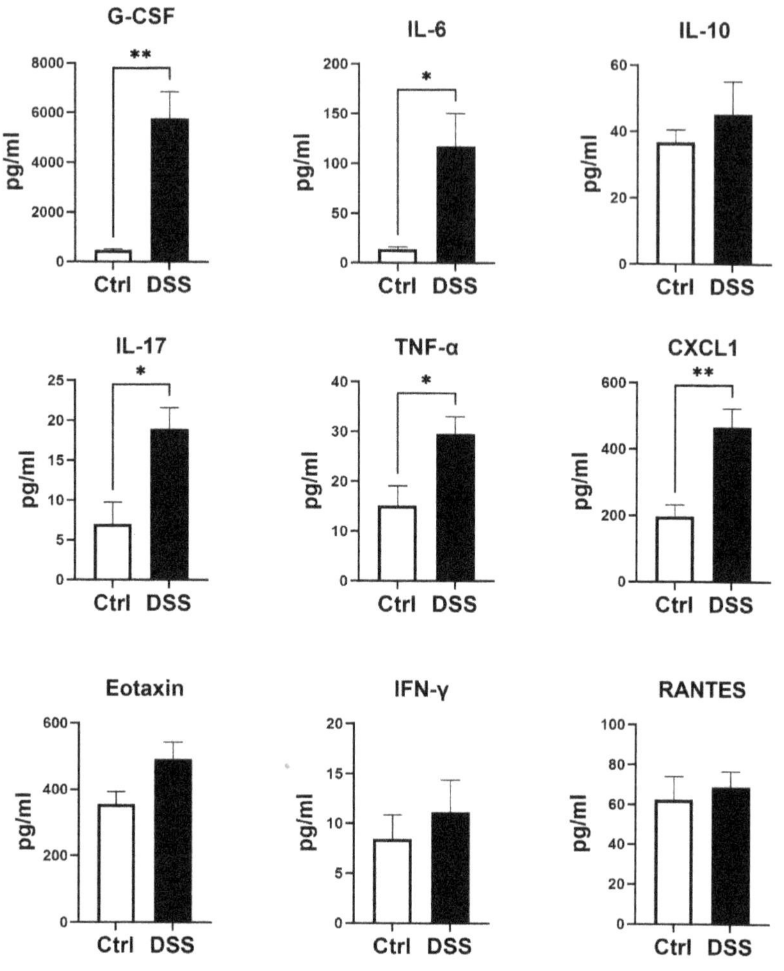

Figure 2. Levels of plasma cytokines in colitis model. Results are expressed as mean ± SEM. * $p < 0.05$, ** $p < 0.01$, by unpaired 2-tailed Student's t-test, $n = 5$ mice in each group. G-CSF, granulocyte colony-stimulating factor; TNF-α, tumor necrosis factor alpha; IFN-γ, interferon gamma; IL, interleukin; CXCL1, chemokine ligand 1; RANTES, regulated upon activation, normal T cell expressed and presumably secreted.

3.3. No Differences in Tissue ROS Level and Vascular Density in Mice with Chronic Colitis

Chronic colitis triggers a systemic inflammation with a potential increase in ROS production. Thus, frozen gastrocnemius muscle sections were prepared for ROS measurement using a DHE assay. To our surprise, there was no significant difference in ROS level in the freshly prepared ischemic gastrocnemius muscle cross-sections in mice with or without colitis (Figure 3G). To assess vascular density in the gastrocnemius muscle of the ischemic limb with or without colitis, gastrocnemius muscle frozen sections were incubated with anti-mouse CD31 immunofluorescent antibody. No difference in vascular density was observed in the gastrocnemius muscle of the ischemic limb in mice with or without colitis (Figure 3H).

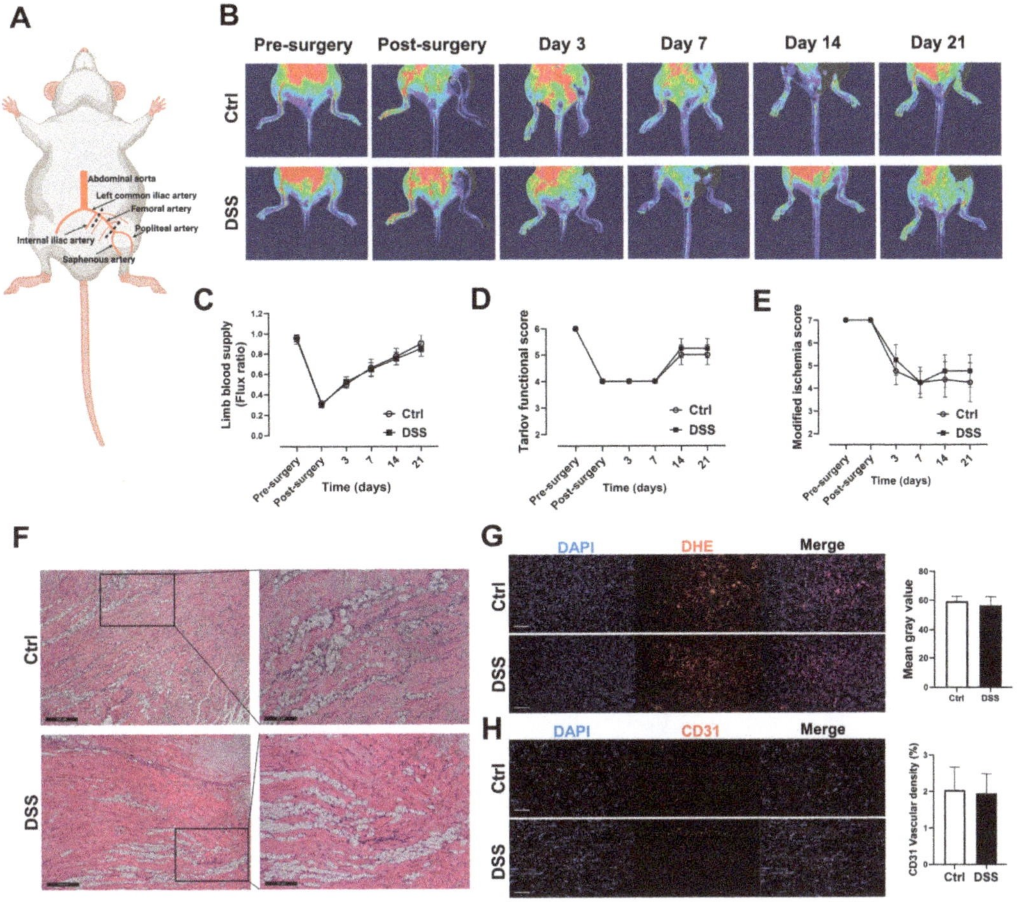

Figure 3. No significant changes in limb ischemia recovery, vascular density, and ROS levels in chronic colitis model. (**A**) Anatomy schematic of unilateral femoral artery ligation. (**B**,**C**) Analysis of laser doppler perfusion images for limb blood supply. (**D**,**E**) Functional score and ischemia score to assess limb function. (**F**) Representative images of H&E staining of gastrocnemius muscle from ischemic limb (scale bar, 100 μm and 50 μm). (**G**) Representative images of ROS in gastrocnemius muscle from ischemic limb using DHE staining (scale bar, 50 μm) and graph for quantification of ROS levels in muscle. (**H**) Representative images of vascular density in gastrocnemius muscle from ischemic limb using CD31 immunofluorescent antibody (scale bar, 50 μm) and graph for quantification of vascular density in muscle. Results are expressed as mean ± SEM. $n = 7$ mice in each group.

3.4. Femoral Artery Endothelium-Dependent and -Independent Vasodilation Were Preserved in Mice with Chronic Colitis

To determine if experimental colitis could negatively affect femoral artery endothelial function of the contralateral limb in mice, endothelium-intact rings of the femoral artery were assessed for endothelium-dependent and endothelium-independent relaxation after submaximal contraction with PE. There were no differences in PE-induced submaximal precontraction (Figure 4A), Ach-induced endothelium-dependent relaxation (Figure 4B), or NTG-induced endothelium-independent relaxation in mice with or without colitis (Figure 4C).

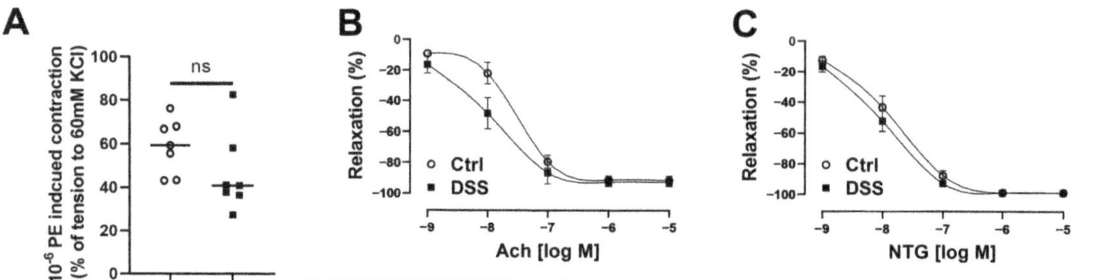

Variable	Maximal relaxation (%)	EC_{50} value (-log[M])
Control	91.44 ± 2.41	7.50 ± 0.12
DSS	92.71 ± 2.47	7.97 ± 0.21

EC50, effective concentration causing 50% of the maximal relaxation.

Variable	Maximal relaxation (%)	EC_{50} value (-log[M])
Control	98.57 ± 1.00	7.89 ± 0.12
DSS	98.67 ± 0.60	8.12 ± 0.09

EC50, effective concentration causing 50% of the maximal relaxation.

Figure 4. Femoral artery contractility and endothelial function were preserved in the chronic colitis model. (**A**) PE-induced contractility of femoral artery. (**B**) Ach-induced endothelium-dependent relaxation of femoral artery, maximal relaxation and EC50 for Ach were calculated in the table. (**C**) NTG-induced endothelium-independent relaxation of femoral artery, maximal relaxation and EC50 for NTG were calculated in the table. Results are expressed as mean ± SEM. By unpaired 2-tailed Student's *t*-test (**A**), in two-way ANOVA followed by Bonferroni correction(**B**,**C**), $n = 7$ mice in each group. PE, phenylephrine; Ach, acetylcholine; NTG, nitroglycerin.

4. Discussion

In this study, we demonstrated that: (1) DSS treatment successfully induced chronic colitis in mice with significant increases in plasma proinflammatory cytokines; (2) no significant difference in blood flow recovery of ischemic limb in mice with and without chronic colitis; (3) no significant changes in ROS levels and vascular density in gastrocnemius muscle of the ischemic limb in mice with chronic colitis; and (4) no significant changes in endothelium-dependent and endothelium-independent relaxation of contralateral limb femoral artery in mice with chronic colitis.

CVDs remain the major cause of morbidity and mortality despite aggressive treatment of traditional risk factors [21]. Chronic inflammation plays a key role in the initiation and progression of CVDs. Indeed, a variety of systemic inflammatory diseases, such as systemic lupus erythematosus, rheumatoid arthritis, and psoriasis, are associated with an increased risk of CVDs and premature CVDs [22]. Similarly, patients with IBD also have an increased risk of CVDs over controls [3,23]. PAD, a form of CVD with reduced blood flow to peripheral arteries, is associated with atherosclerosis and vascular inflammation [24]. In a Chinese population-based cohort study, it was reported that the risk of PAD was increased in patients with IBD [25]. A French cohort study also showed that the risk of PAD was significantly increased in patients with IBD compared with the general population with the highest relative risk in those younger than 35 years [4].

Several inflammatory biomarkers, including C-reactive protein, TNF-α, IL-6, IL-17, and calprotectin, are significantly increased in patients with IBD [7]. Systemic inflammation, as indicated by TNF-α gene expression levels in peripheral blood monocytes, is associated with impairment in walking time in patients with PAD [26]. In addition, circulating Th17-associated cytokines (IL-6, IL-17, and TNF-α) are correlated to the severity and progression of carotid artery plaques [27]. The data from the present study also showed that the plasma levels of proinflammatory cytokines, G-CSF, IL-6, IL-17, TNF-α, and CXCL1, were significantly increased in mice with chronic colitis, suggesting the presence of significant systematic inflammation in mice with chronic colitis. However, there were no differences in tissue ROS levels in the ischemic limbs of mice with and without colitis.

Inflammation and angiogenesis have long been coupled together in IBD. Microvascular density assessed by CD31 staining was increased in IBD mucosa, and IBD mucosal extracts induced a potent angiogenic response in corneal angiogenesis assay [28]. TNF-α is a major inflammatory mediator that is involved in angiogenesis [29]. Studies have demonstrated

TNF receptor-deficient mice exhibited an enhanced hind limb angiogenesis, while a reduced angiogenesis was observed in TNFR2-knockout mice [30]. Notably, it has been reported that sarcopenia is commonly present in the IBD population [31]. Increased serum levels of TNF-α are associated with muscle impairment by promoting protein degradation through the muscle-specific ubiquitin ligases [32]. The data from the present work demonstrated no difference in vascular density in the gastrocnemius muscle of the ischemic limbs in mice with or without colitis.

Endothelial dysfunction and arterial stiffness are considered among the important factors for the initiation and progression of CVDs. Chronic inflammation has been known to be associated with endothelial dysfunction [33]. Our recent meta-analysis demonstrated that patients with IBD were significantly associated with endothelial cell dysfunction, increased arterial stiffness, and carotid intima-media thickness [10]. In the present study, we observed that plasma TNF-α, IL-6, and IL-17 were significantly increased in mice with colitis. IL-6 and TNF-α are considered potent oxidative stress-inducing agents, thus leading to endothelial dysfunction [34,35]. IL-17 may induce endothelial dysfunction through activating RhoA/Rho-kinase [36]. Significant decreases in retinal blood flow have been reported in mice with T-cell transfer colitis and DSS-induced colitis [37,38]. Recent animal studies showed that the perivascular sensory neurotransmitter function of mesenteric arteries is significantly impaired in an IL-10 knockout mouse colitis model [39]. Recently, we demonstrated that abdominal aortic endothelial function was impaired selectively in female mice with chronic colitis [12]. To our surprise, the data from the present work showed that no significant changes in endothelium-dependent and endothelium-independent relaxation of the contralateral limb femoral artery were observed in the colitis model.

There were several limitations in the present study. First, only female mice were used based on the data from our recent study. Although abdominal aortic endothelial dysfunction occurs selectively in female mice, it is possible that endothelial dysfunction and/or impaired ischemic limb recovery may also occur in male mice with chronic colitis. Second, disease duration of the animal model could be an important factor for endothelial dysfunction of the femoral artery and ischemic limb recovery. Thus, a longer time point may be needed to demonstrate the effect of IBD on endothelial function of femoral artery and ischemic limb in the mouse model. Third, no studies were conducted to evaluate the potential protective effect of estrogen on limb ischemia recovery in the present study. Finally, there are several colitis models, such as the IL-10 knockout mouse model with spontaneous colitis, trinitrobenzene sulfonic acid-induced colitis, polyinosinic-polycytidylic acid-induced colitis, and T-cell transfer-induced colitis. These models could be used to determine if endothelial dysfunction and impaired recovery of ischemic limb could be demonstrated in IBD in the future.

In conclusion, despite demonstrating a successful implementation of the colitis model and the existence of a systemic inflammatory environment, DSS-induced chronic colitis had no significant impact on femoral artery endothelial cell function or the recovery of ischemic limbs in female mice.

Author Contributions: Conceptualization, H.W., C.X. and Z.L.; methodology, Q.Z.; software, H.W.; formal analysis, X.L.; resources, Z.S.; data curation, H.H.; writing—original draft preparation, H.W.; writing—review and editing, M.A.H. and Z.L.; visualization, M.W.; supervision, C.X. and Z.L.; project administration, H.H.; funding acquisition, Z.L. All authors have read and agreed to the published version of the manuscript.

Funding: The present work was partially supported by US NIH grants ES026200 and HL148196 (ZL).

Institutional Review Board Statement: The experimental protocols were reviewed and approved by the Institutional Animal Care and Use Committee of the University of Missouri, Columbia, MO, USA (protocol code 9227).

Informed Consent Statement: Not applicable.

Data Availability Statement: Not applicable.

Conflicts of Interest: The authors declare no conflict of interest.

References

1. Rogler, G.; Singh, A.; Kavanaugh, A.; Rubin, D.T. Extraintestinal Manifestations of Inflammatory Bowel Disease: Current Concepts, Treatment, and Implications for Disease Management. *Gastroenterology* **2021**, *161*, 1118–1132. [CrossRef]
2. Wu, H.; Hu, T.; Hao, H.; Hill, M.A.; Xu, C.; Liu, Z. Inflammatory bowel disease and cardiovascular diseases: A concise review. *Eur. Heart J. Open* **2022**, *2*, oeab029. [CrossRef]
3. Rungoe, C.; Basit, S.; Ranthe, M.F.; Wohlfahrt, J.; Langholz, E.; Jess, T. Risk of ischaemic heart disease in patients with inflammatory bowel disease: A nationwide Danish cohort study. *Gut* **2013**, *62*, 689–694. [CrossRef] [PubMed]
4. Kirchgesner, J.; Beaugerie, L.; Carrat, F.; Andersen, N.N.; Jess, T.; Schwarzinger, M. Increased risk of acute arterial events in young patients and severely active IBD: A nationwide French cohort study. *Gut* **2018**, *67*, 1261–1268. [CrossRef] [PubMed]
5. Lerman, A.; Zeiher, A.M. Endothelial function: Cardiac events. *Circulation* **2005**, *111*, 363–368. [CrossRef] [PubMed]
6. Ferrucci, L.; Fabbri, E. Inflammageing: Chronic inflammation in ageing, cardiovascular disease, and frailty. *Nat. Rev. Cardiol.* **2018**, *15*, 505–522. [CrossRef] [PubMed]
7. Neurath, M.F. Cytokines in inflammatory bowel disease. *Nat. Rev. Immunol.* **2014**, *14*, 329–342. [CrossRef]
8. Ozturk, K.; Guler, A.K.; Cakir, M.; Ozen, A.; Demirci, H.; Turker, T.; Demirbas, S.; Uygun, A.; Gulsen, M.; Bagci, S. Pulse Wave Velocity, Intima Media Thickness, and Flow-mediated Dilatation in Patients with Normotensive Normoglycemic Inflammatory Bowel Disease. *Inflamm. Bowel Dis.* **2015**, *21*, 1314–1320. [CrossRef]
9. Roifman, I.; Sun, Y.C.; Fedwick, J.P.; Panaccione, R.; Buret, A.G.; Liu, H.; Rostom, A.; Anderson, T.J.; Beck, P.L. Evidence of endothelial dysfunction in patients with inflammatory bowel disease. *Clin. Gastroenterol. Hepatol.* **2009**, *7*, 175–182. [CrossRef]
10. Wu, H.; Xu, M.; Hao, H.; Hill, M.A.; Xu, C.; Liu, Z. Endothelial Dysfunction and Arterial Stiffness in Patients with Inflammatory Bowel Disease: A Systematic Review and Meta-Analysis. *J. Clin. Med.* **2022**, *11*, 3179. [CrossRef]
11. Wirtz, S.; Popp, V.; Kindermann, M.; Gerlach, K.; Weigmann, B.; Fichtner-Feigl, S.; Neurath, M.F. Chemically induced mouse models of acute and chronic intestinal inflammation. *Nat. Protoc.* **2017**, *12*, 1295–1309. [CrossRef]
12. Wu, H.; Hu, T.; Zhang, L.; Xia, X.; Liu, X.; Zhu, Q.; Wang, M.; Sun, Z.; Hao, H.; Cui, Y.; et al. Abdominal Aortic Endothelial Dysfunction Occurs in Female Mice With Dextran Sodium Sulfate-Induced Chronic Colitis Independently of Reactive Oxygen Species Formation. *Front. Cardiovasc. Med.* **2022**, *9*, 871335. [CrossRef] [PubMed]
13. Niiyama, H.; Huang, N.F.; Rollins, M.D.; Cooke, J.P. Murine model of hindlimb ischemia. *J. Vis. Exp.* **2009**, *23*, e1035. [CrossRef]
14. Cooper, H.S.; Murthy, S.N.; Shah, R.S.; Sedergran, D.J. Clinicopathologic study of dextran sulfate sodium experimental murine colitis. *Lab. Investig.* **1993**, *69*, 238–249. [PubMed]
15. Ogrodnik, M.; Zhu, Y.; Langhi, L.G.P.; Tchkonia, T.; Krüger, P.; Fielder, E.; Victorelli, S.; Ruswhandi, R.A.; Giorgadze, N.; Pirtskhalava, T.; et al. Obesity-Induced Cellular Senescence Drives Anxiety and Impairs Neurogenesis. *Cell Metab.* **2019**, *29*, 1061–1077.e8. [CrossRef]
16. Wang, H.; Agarwal, P.; Xiao, Y.; Peng, H.; Zhao, S.; Liu, X.; Zhou, S.; Li, J.; Liu, Z.; He, X. A Nano-In-Micro System for Enhanced Stem Cell Therapy of Ischemic Diseases. *ACS Cent. Sci.* **2017**, *3*, 875–885. [CrossRef]
17. Tarlov, I.M. Spinal cord compression studies. III. Time limits for recovery after gradual compression in dogs. *AMA Arch. Neurol. Psychiatry* **1954**, *71*, 588–597. [CrossRef] [PubMed]
18. Westvik, T.S.; Fitzgerald, T.N.; Muto, A.; Maloney, S.P.; Pimiento, J.M.; Fancher, T.T.; Magri, D.; Westvik, H.H.; Nishibe, T.; Velazquez, O.C.; et al. Limb ischemia after iliac ligation in aged mice stimulates angiogenesis without arteriogenesis. *J. Vasc. Surg.* **2009**, *49*, 464–473. [CrossRef] [PubMed]
19. Rabiolo, A.; Bignami, F.; Rama, P.; Ferrari, G. VesselJ: A New Tool for Semiautomatic Measurement of Corneal Neovascularization. *Investig. Ophthalmol. Vis. Sci.* **2015**, *56*, 8199–8206. [CrossRef] [PubMed]
20. Wang, Q.; Zou, M.H. Measurement of Reactive Oxygen Species (ROS) and Mitochondrial ROS in AMPK Knockout Mice Blood Vessels. *Methods Mol. Biol.* **2018**, *1732*, 507–517. [PubMed]
21. Roth, G.A.; Mensah, G.A.; Johnson, C.O.; Addolorato, G.; Ammirati, E.; Baddour, L.M.; Barengo, N.C.; Beaton, A.Z.; Benjamin, E.J.; Benziger, C.P.; et al. Global Burden of Cardiovascular Diseases and Risk Factors, 1990–2019: Update From the GBD 2019 Study. *J. Am. Coll. Cardiol.* **2020**, *76*, 2982–3021. [CrossRef]
22. Mason, J.C.; Libby, P. Cardiovascular disease in patients with chronic inflammation: Mechanisms underlying premature cardiovascular events in rheumatologic conditions. *Eur. Heart J.* **2015**, *36*, 482–489c. [CrossRef]
23. Choi, Y.J.; Lee, D.H.; Shin, D.W.; Han, K.D.; Yoon, H.; Shin, C.M.; Park, Y.S.; Kim, N. Patients with inflammatory bowel disease have an increased risk of myocardial infarction: A nationwide study. *Aliment. Pharmacol. Ther.* **2019**, *50*, 769–779. [CrossRef]
24. Criqui, M.H.; Matsushita, K.; Aboyans, V.; Hess, C.N.; Hicks, C.W.; Kwan, T.W.; McDermott, M.M.; Misra, S.; Ujueta, F.; on behalf of the American Heart Association Council on Epidemiology and Prevention; et al. Lower Extremity Peripheral Artery Disease: Contemporary Epidemiology, Management Gaps, and Future Directions: A Scientific Statement From the American Heart Association. *Circulation* **2021**, *144*, e171–e191. [CrossRef]

25. Lin, T.Y.; Chen, Y.G.; Lin, C.L.; Huang, W.S.; Kao, C.H. Inflammatory Bowel Disease Increases the Risk of Peripheral Arterial Disease: A Nationwide Cohort Study. *Medicine* **2015**, *94*, e2381. [CrossRef]
26. Pande, R.L.; Brown, J.; Buck, S.; Redline, W.; Doyle, J.; Plutzky, J.; Creager, M.A. Association of monocyte tumor necrosis factor α expression and serum inflammatory biomarkers with walking impairment in peripheral artery disease. *J. Vasc. Surg.* **2015**, *61*, 155–161. [CrossRef]
27. Liu, Z.; Lu, F.; Pan, H.; Zhao, Y.; Wang, S.; Sun, S.; Li, J.; Hu, X.; Wang, L. Correlation of peripheral Th17 cells and Th17-associated cytokines to the severity of carotid artery plaque and its clinical implication. *Atherosclerosis* **2012**, *221*, 232–241. [CrossRef]
28. Danese, S.; Sans, M.; de la Motte, C.; Graziani, C.; West, G.; Phillips, M.H.; Pola, R.; Rutella, S.; Willis, J.; Gasbarrini, A.; et al. Angiogenesis as a novel component of inflammatory bowel disease pathogenesis. *Gastroenterology* **2006**, *130*, 2060–2073. [CrossRef]
29. Sainson, R.C.; Johnston, D.A.; Chu, H.C.; Holderfield, M.T.; Nakatsu, M.N.; Crampton, S.P.; Davis, J.; Conn, E.; Hughes, C.C.W. TNF primes endothelial cells for angiogenic sprouting by inducing a tip cell phenotype. *Blood* **2008**, *111*, 4997–5007. [CrossRef]
30. Luo, D.; Luo, Y.; He, Y.; Zhang, H.; Zhang, R.; Li, X.; Dobrucki, W.; Sinusas, A.J.; Sessa, W.; Min, W. Differential functions of tumor necrosis factor receptor 1 and 2 signaling in ischemia-mediated arteriogenesis and angiogenesis. *Am. J. Pathol.* **2006**, *169*, 1886–1898. [CrossRef]
31. Ryan, E.; McNicholas, D.; Creavin, B.; Kelly, M.E.; Walsh, T.; Beddy, D. Sarcopenia and Inflammatory Bowel Disease: A Systematic Review. *Inflamm. Bowel Dis.* **2019**, *25*, 67–73. [CrossRef]
32. Cohen, S.; Nathan, J.A.; Goldberg, A.L. Muscle wasting in disease: Molecular mechanisms and promising therapies. *Nat. Rev. Drug Discov.* **2015**, *14*, 58–74. [CrossRef] [PubMed]
33. Xu, S.; Ilyas, I.; Little, P.J.; Li, H.; Kamato, D.; Zheng, X.; Luo, S.; Li, Z.; Liu, P.; Han, J.; et al. Endothelial Dysfunction in Atherosclerotic Cardiovascular Diseases and Beyond: From Mechanism to Pharmacotherapies. *Pharmacol. Rev.* **2021**, *73*, 924–967. [CrossRef]
34. Wassmann, S.; Stumpf, M.; Strehlow, K.; Schmid, A.; Schieffer, B.; Böhm, M.; Nickenig, G. Interleukin-6 induces oxidative stress and endothelial dysfunction by overexpression of the angiotensin II type 1 receptor. *Circ. Res.* **2004**, *94*, 534–541. [CrossRef] [PubMed]
35. Gao, X.; Belmadani, S.; Picchi, A.; Xu, X.; Potter, B.J.; Tewari-Singh, N.; Capobianco, S.; Chilian, W.M.; Zhang, C. Tumor necrosis factor-alpha induces endothelial dysfunction in Lepr(db) mice. *Circulation* **2007**, *115*, 245–254. [CrossRef]
36. Nguyen, H.; Chiasson, V.L.; Chatterjee, P.; Kopriva, S.E.; Young, K.J.; Mitchell, B.M. Interleukin-17 causes Rho-kinase-mediated endothelial dysfunction and hypertension. *Cardiovasc. Res.* **2013**, *97*, 696–704. [CrossRef]
37. Watts, M.N.; Leskova, W.; Carter, P.R.; Zhang, S.; Kosloski-Davidson, M.; Grisham, M.B.; Harris, N.R. Ocular dysfunction in a mouse model of chronic gut inflammation. *Inflamm. Bowel Dis.* **2013**, *19*, 2091–2097. [CrossRef]
38. Watts, M.N.; Eshaq, R.S.; Carter, P.R.; Harris, N.R. Decreased retinal blood flow in experimental colitis; improvement by eye drop administration of losartan. *Exp. Eye Res.* **2013**, *115*, 22–26. [CrossRef]
39. Norton, C.E.; Grunz-Borgmann, E.A.; Hart, M.L.; Jones, B.W.; Franklin, C.L.; Boerman, E.M. Role of perivascular nerve and sensory neurotransmitter dysfunction in inflammatory bowel disease. *Am. J. Physiol. Heart Circ. Physiol.* **2021**, *320*, H1887–H1902. [CrossRef] [PubMed]

biology MDPI

Article

Neuregulin (NRG-1β) Is Pro-Myogenic and Anti-Cachectic in Respiratory Muscles of Post-Myocardial Infarcted Swine

Cristi L. Galindo [1,*], Van Thuan Nguyen [1], Braxton Hill [1], Ethan Easterday [1], John H. Cleator [2] and Douglas B. Sawyer [3,*]

1 Department of Biology, Ogden College of Science & Engineering, Western Kentucky University, Bowling Green, KY 42101, USA; vanthuan.nguyen@wku.edu (V.T.N.); braxton.hill843@topper.wku.edu (B.H.); ethan.easterday493@topper.wku.edu (E.E.)
2 Centennial Heart at Skyline, 3443 Dickerson Pike, Suite 430, Nashville, TN 37207, USA; john.cleator@hcahealthcare.com
3 Department of Cardiac Services, Maine Medical Center, Scarborough, ME 04074, USA
* Correspondence: cristi.galindo@wku.edu (C.L.G.); douglas.sawyer@mainehealth.org (D.B.S.)

Simple Summary: Neuregulin is a growth factor that has been shown to prevent adverse remodeling in the heart and may represent a therapeutic for patients with systolic heart failure. A common symptom in heart failure is shortness of breath, which has been related in part to impaired skeletal muscle function. Since neuregulin directly activates skeletal muscle, in addition to heart tissue, we hypothesized that neuregulin might directly affect intercostal muscle gene expression changes in heart disease. We tested this hypothesis by performing global gene expression analysis of intercostal muscle tissue collected from pigs treated with recombinant neuregulin after the induction of myocardial infarction, an experimental model clinically similar to a human heart attack. We found that neuregulin-treated pigs had massive changes in global gene expression consistent with new muscle cell formation, as compared to untreated pigs. These data suggest that neuregulin is an important mediator of muscle function that can potentially be used to study heart disease-associated muscle dysfunction and the development of new therapeutics aimed at muscle repair in heart failure, as well as other diseases associated with muscle dysfunction and weakness.

Abstract: Neuregulin-1β (NRG-1β) is a growth and differentiation factor with pleiotropic systemic effects. Because NRG-1β has therapeutic potential for heart failure and has known growth effects in skeletal muscle, we hypothesized that it might affect heart failure-associated cachexia, a severe co-morbidity characterized by a loss of muscle mass. We therefore assessed NRG-1β's effect on intercostal skeletal muscle gene expression in a swine model of heart failure using recombinant glial growth factor 2 (USAN-cimaglermin alfa), a version of NRG-1β that has been tested in humans with systolic heart failure. Animals received one of two intravenous doses (0.67 or 2 mg/kg) of NRG-1β bi-weekly for 4 weeks, beginning one week after infarct. Based on paired-end RNA sequencing, NRG-1β treatment altered the intercostal muscle gene expression of 581 transcripts, including genes required for myofiber growth, maintenance and survival, such as MYH3, MYHC, MYL6B, KY and HES1. Importantly, NRG-1β altered the directionality of at least 85 genes associated with cachexia, including myostatin, which negatively regulates myoblast differentiation by down-regulating MyoD expression. Consistent with this, MyoD was increased in NRG-1β-treated animals. In vitro experiments with myoblast cell lines confirmed that NRG-1β induces ERBB-dependent differentiation. These findings suggest a NRG-1β-mediated anti-atrophic, anti-cachexia effect that may provide additional benefits to this potential therapy in heart failure.

Keywords: myocardial infarction; neuregulin; glial growth factor 2; pre-clinical therapy; skeletal muscle; gene expression; RNA sequencing

Citation: Galindo, C.L.; Nguyen, V.T.; Hill, B.; Easterday, E.; Cleator, J.H.; Sawyer, D.B. Neuregulin (NRG-1β) Is Pro-Myogenic and Anti-Cachectic in Respiratory Muscles of Post-Myocardial Infarcted Swine. *Biology* **2022**, *11*, 682. https://doi.org/10.3390/biology11050682

Academic Editor: Gaetano Santulli

Received: 26 February 2022
Accepted: 1 April 2022
Published: 29 April 2022

1. Introduction

Chronic heart failure (CHF) is a complex clinical syndrome resulting from any disorder that impairs cardiac function such as ventricular filling or the ejection of blood into the systemic circulation. Impaired muscle formation can be a cause of several muscle-related [1] diseases as well as age-related muscle deterioration, also known as sarcopenia [2,3]. Although the precise mechanisms are not fully understood, some studies recently showed that systemic oxidative stress [4], exercise intolerance [5], a low grade of inflammation, abnormal energy metabolism, transition of myofibers, mitochondrial dysfunction, a reduction in muscular strength, and muscle atrophy play an important role in skeletal muscle dysfunction/abnormalities in the setting of CHF [6,7]. However, abnormalities in skeletal muscle metabolism, total muscle mass, and peak functional capacity in patients with CHF cannot be fully explained by lowered cardiac output [8] or inadequate skeletal muscle oxygenation [9].

The neuregulin (NRG) family is a large class of neuronal growth factors that induce signaling via type I epidermal growth factor receptor (EGFR) tyrosine kinases (ErbB2, ErbB3, and ErbB4). NRG-1 is essential for the development of the sympathetic nervous system [10] and is required for the formation and maintenance of neuromuscular synapses. In postembryonic skeletal muscles, local production of NRG-1 by motor neurons increases acetylcholine receptor expression and accumulation at motor nerve terminals [11]. Several NRG-1 isoforms, including NRG-1α, NRG-1β Type I, and NRG-1β type II (also called glial growth factor 2, GGF2), were shown to stimulate myoblast differentiation in immortalized and primary myoblasts [12–14]. Additionally, ErbB3 is induced during muscle cell differentiation [15] and regulates muscle progenitor cell fate and thus the balance between stem cell renewal and differentiation [16].

We previously reported results from a pre-clinical trial of the effects of a pharmaceutical grade of NRG-1/GGF2 (cimaglermin alfa) using a swine model of heart failure. In this previously published study [17], a balloon was inserted into the descending coronary artery of pigs to induce myocardial infarction. GGF2 was administrated intravenously twice weekly, starting at 1 week after infarction. Tissues were collected 35 days thereafter, and left ventricular tissue remote from the infarct was interrogated for global gene expression using microarrays. GGF2-treated animals exhibited better cardiac function, including less maladaptive myocardial remodeling, reduced fibrosis, and gene expression changes consistent with better cardiomyocyte survival but less myofibroblast-driven extracellular matrix remodeling. In the study presented here, we performed deep sequencing on intercostal muscles of these same pigs to determine the effects of GGF2 treatment on post-MI skeletal muscle function. We also compared these results with GGF2-mediated expression changes in cardiac muscle and myopathic genes obtained by analyzing repository expression data. As detailed herein, our results support a role for exogenously delivered GGF2 in skeletal muscle remodeling and repair after cardiac injury.

2. Materials and Methods

Animals: Intercostal muscle samples were acquired from swine that were used to test the efficacy of NRG treatment on swine after myocardial infarction, the details of which were previously published [17]. Briefly, eight pigs underwent balloon occlusion surgery to introduce a left ventricular infarct. At one week post-MI, five of the animals received biweekly intravenous injections of GGF2 (three at a dose of 0.67 mg/kg and two at 2 mg/kg). The remaining swine served as untreated controls. As reported previously [17–19], high-dose treatment was associated with symptomatic hypoglycemia. In consultation with veterinary staff and IACUC, dose adjustment was made for the remaining swine in the treatment group.

Sequencing: Total RNA was isolated using RNeasy Mini kit (Qiagen), following the manufacturer's instructions, RNA integrity was confirmed using Agilent Bioanalyzer, and RNA sequencing was performed by the Vanderbilt Technologies for Advanced Genomics core on a total of eight animals, all of which received myocardial infarction as previously described [17]. RNA libraries were constructed using the Illumina TruSeq Stranded Total RNA kit with Ribo-Zero Gold. Samples were sequenced on the Illumina HiSeq 2500 using v3 SBS chemistry. Libraries were sequenced on paired-end-50 flow cell runs at ~35 M PF reads per sample.

Data analysis: Sequences were aligned to the susScr3 assembly of the pig genome and analyzed using the Tophat2 + CuffDiff platform provided by DNAnexus, Inc. (Mountain View, CA, USA). Genes were annotated using the Ensemble, UCSC, and NCBI iGenomes datasets. Partek Genomics Suite 6.6 was also used to perform statistical analyses of aligned sequences, as well as hierarchical clustering and principal components analysis (PCA). Differential sequences that were determined to be statistically significant (multiple hypothesis-corrected p-value < 0.05 and fold-difference > 2.0) among the three different groups of animals were compiled into a single list and the annotations were manually verified using information from UCSC, Ensemble, and NCBI databases. For uncharacterized transcripts, the differential Sus scrofa sequences were queried against the human reference genome using the BLAT tool provided on the UCSC Genome Browser website to identify orthologous human genes. Uncharacterized transcripts were included in the final list if they were at least 75% similar to an annotated human gene.

Functional enrichment analysis with Bonferroni multiple hypothesis adjustment was performed using GSEA, available online from the BROAD Institute. Individual gene functions were identified using public databases and literature reports. Functional subgroups of genes that were detected by exhaustive, manual annotation could generally be recapitulated using statistical functional analysis tools but were missed due to lower significance relative to hundreds of other, largely redundant ontological categories (e.g., "tissue development"). To better identify these potentially relevant functional groups while minimizing user bias, we also organized genes hierarchically according to their biological processes within the Gene Ontology Consortium classification system. The top three non-redundant biological processes were chosen based on the numbers of differential genes included and functional specificity, whereby non-informative categories, such as "Regulation of Cell Process", were ignored. A divisive hierarchical clustering approach was applied for consecutive subgrouping. All ontologies generated from the gene lists, irrespective of statistical significance, were examined to gain a complete functional overview; however, a statistically significant overlap (Bonferroni-adjusted p-value < 0.05) was required for inclusion in the subsequently generated functional hierarchy reported herein.

Cell Culture: C2C12 cells and L6 myoblasts were cultured in growth media (GM) consisting of Dulbecco's modified Eagle's medium (DMEM) (25 mM glucose) supplemented with 10% fetal bovine serum (FBS) and 1% antibiotics (penicillin + streptomycin). GM was changed every alternate day until cells were 90% confluent. GM was then changed to differentiation media (DM) consisting of DMEM (25 mM glucose) supplemented with 2% horse serum and 1% antibiotics. Compounds or vehicle controls were added at Day 0 (i.e., the time at which the cells were switched from GM to DM). The additional methodology is provided in Supplementary Figures S1–S7.

Quantification of myoblast fusion and myotube numbers. C2C12 or L6 cells were differentiated according to the conditions of described experiments and fixed in 4% paraformaldehyde for 25 min. Next, the cells were stained with Jenner–Giemsa dyes as described in (Velica). The multiple microphotographs of stained cells were taken using an inverted microscope, and at least 4 different fields were used to estimate the total number of cells and the number of myotubes. The cells with more than 3 nuclei were counted as the number of myotubes. The fusion index was determined as the number of cells with more than 3 nuclei per cell divided by the total number of cells. In some experiments, the myotube numbers were estimated without the cell staining based on microphotographs of growing cells.

Western blot and immunoprecipitation analysis. The cells were solubilized by scraping into the cold NP-40 lysis buffer (10 mM Tris-HCl, pH 7.5, 150 mM NaCl, 1% Nonidet P-40 containing protease inhibitors, and 1 mM Na3VO4). The lysates were then clarified by centrifugation (14,000× g, 10 min). Aliquots containing equal amounts of protein were subjected to 4–20% gradient SDS-PAGE. Subsequently, proteins were transferred to nitrocellulose membranes, and the membrane was blocked by incubation with Odyssey blocking buffer (LI-COR) for 1 h at room temperature. The membrane was then incubated overnight at 4 °C with the indicated antibody in the same blocking solution washed 3 times in the TBS buffer containing 0.05% Tween 20 (TBSTw) and incubated for 1 h with HRP-conjugated secondary antibody in SEA BLOCK blocking buffer (Thermo Fisher, Waltham, MA, USA). The membranes were then washed three times with the TBSTw buffer and visualized using the Fujifilm LAS-4000 imaging system. For immunoprecipitation, the cells were grown and differentiated in 10 cm tissue culture dishes and lysed in 0.5 mL of cold NP-40 lysis buffer. The lysates were precleared by incubation with 100 µL of protein A agarose beads (Santa-Cruz, Dallas, TX, USA) for 2 h at 4 °C and then incubated with 5 µg of C-18 antibodies against ErbB-2 and 50 µL of protein A agarose overnight at 4 °C. The beads were washed 3 times in TBSTw, then boiled in 1× Laemmli sample buffer for 2 min, and subjected to Western blot.

3. Results

3.1. GGF2 Alters Intercostal Muscle Gene Expression

Comparison of GGF2-treated and untreated pigs yielded a total of 581 differentially expressed transcripts including 500 annotated swine genes and 81 uncharacterized transcripts with at least 75% similarity to an annotated human gene (Supplementary Table S1). Not surprisingly, there were nearly twice as many transcriptional changes in animals treated with the high dose of GGF2 (512 genes) compared to the post-MI swine that received the lower dose (311 genes) versus untreated controls. These dose-dependent differences resulted in a clear separation of the two groups from one another, secondarily to untreated samples, upon hierarchical clustering of standardized hybridization values (Figure 1a). Of the 242 genes differentially expressed between treated and untreated swine at both doses (Figure 1b), only one transcript was altered in disparate directions compared to controls (MYLIP, down-regulated 1.7-fold in low dose-treated and up-regulated two-fold in high dose-treated animals, Supplementary Table S1). Of the remaining 241 similarly altered genes, 91 and 150 were up and down-regulated in treated animals, respectively (Supplementary Table S1).

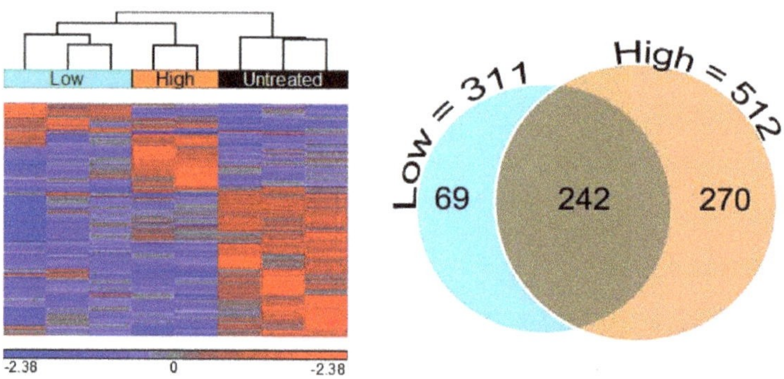

(a) (b)

Figure 1. Low and high doses of GGF2 alter intercostal skeletal muscle transcription. (**a**) Hierarchical clustering of 581 genes significantly differentially expressed between GGF2-treated and untreated post-MI swine. Columns represent individual samples for each group, and rows represent individual probes/genes. Red, blue, and gray represent the highest, lowest, and medial fluorescent signal values, respectively. (**b**) Venn diagram showing numbers of genes differentially expressed in post-MI swine treated with GGF2 at low (teal circle, 311 genes) and high doses (orange circle, 512 genes) compared to untreated control animals.

3.2. Dose-Dependent Effects of GGF2

More than half of the genes that were altered in both low dose and high dose GGF2-treated swine (133 out of 242) exhibited a "dose-like" expression pattern (61 up-regulated and 72 down-regulated transcripts). Notable examples of up-regulated genes were those important for muscle cell development and contraction, as well as genes involved in amino acid and carbohydrate metabolism. Down-regulated genes displaying a "dose-like" response included those involved in lipid metabolism and insulin signaling, ECM structure and maintenance, and regulation of vascular functions. Notable dose-specific genes included those important for myofiber growth and muscle growth maintenance, such as slow-twitch MYL6B (down-regulated) and myosin heavy chain genes MYH3 and MYHC (up-regulated). The gene encoding ladybird homeobox 1 (LBX1), which is a key regulator of migratory muscle-specific stem cell precursors required for the development of forelimb muscles, including the diaphragm, was also up-regulated (Figure 2).

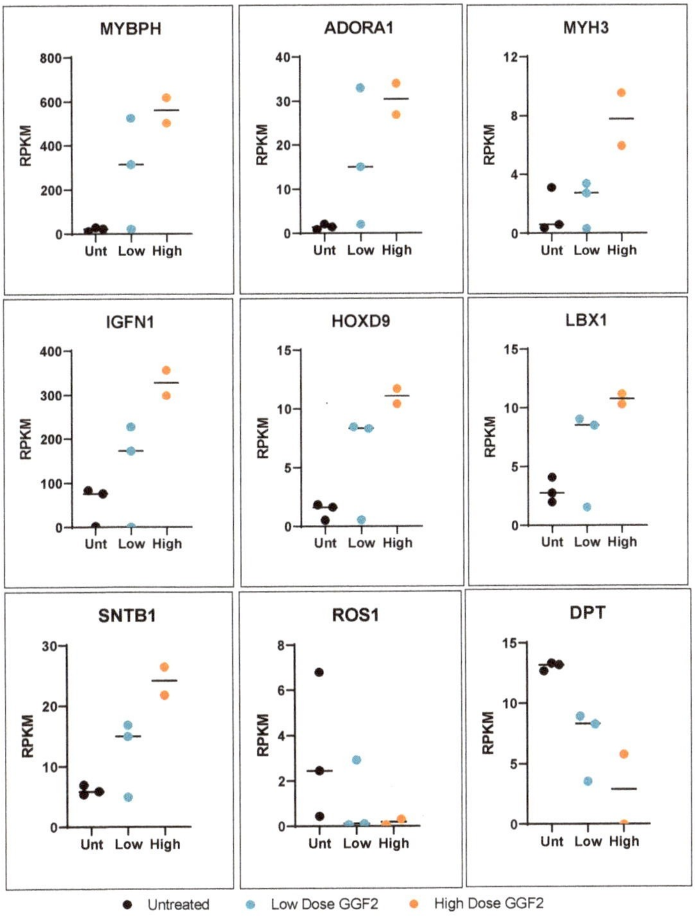

Figure 2. Dose-like response genes altered in GGF-2 treated pigs post-MI. Dot plots show relative counts as reads per kilobase per transcript, per million mapped reads (RPKM) for genes exhibiting fold-differences for GGF2 versus untreated pigs (y-axis). Individual samples are indicated on the x-axis and by color. Black represents untreated (Unt), teal represents low-dose (0.67 mg/kg/day), and high dose (2 mg/kg/day) is shown in orange.

3.3. Functional Enrichment of GGF2 Altered Genes

Although many more genes were altered in post-MI pigs that received the higher dose than those treated with low dose GGF2, enriched functions were similar. The most significantly over-represented biological processes for both gene lists were tissue development, cellular proliferation and differentiation, organ morphogenesis, and ECM organization (Figure 3a). Functional analysis of the larger gene set (500+ genes altered in either of the two doses) resulted in nearly identical results. The gene lists were therefore combined for further analyses and the average fold-change and p-values were used where applicable.

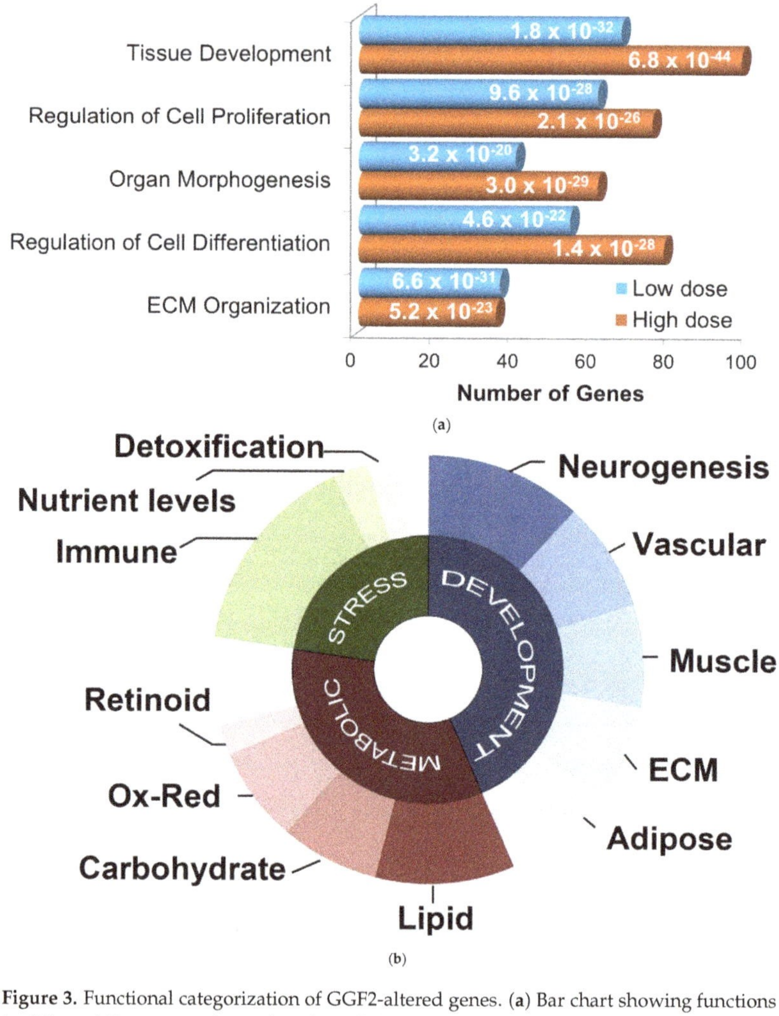

Figure 3. Functional categorization of GGF2-altered genes. (**a**) Bar chart showing functions enriched in differentially expressed gene lists from GGF-2 treated versus untreated LV tissues, identified using Ingenuity Pathway Analysis software. Functions are listed on the y-axis, and the x-axis indicates the number of differentially expressed genes in each category. Teal and orange represent low and high doses of GGF2, respectively. Enrichment p-value is indicated on each bar. (**b**) Pie chart showing top gene ontology categories for GGF-2 altered genes manually grouped into three generalized associative categories: development (**blue**), metabolic (**red**), and stress response (**green**). Uncharacterized and poorly characterized transcripts were excluded.

To identify relevant functional subgroups of GGF2-altered genes and thereby obtain more specific mechanistic information, genes were hierarchically ordered based on their official Gene Ontology Consortium classifications. Based on this heuristic approach, the top non-redundant biological processes encompassing 445 out of all 581 differentially expressed genes were (1) developmental process (267 genes, $p = 1 \times 10^{-21}$, enrichment score (ES) = 1.8), (2) regulation of primary metabolic process (208 genes, $p = 1.8 \times 10^{-3}$, ES = 1.4), and (3) response to stress (139 genes, $p = 1 \times 10^{-2}$, ES = 1.5). Some genes were classified in more than one of these categories and thus were "counted" more than once for subsequent functional subdivisions (i.e., $267 + 208 + 139 = 614 \neq 445$ genes). Developmental genes were further parsed by tissue type (neurogenesis, vascular, muscular, ECM, and adipose), as were metabolic and stress response genes (Figure 3b).

Notably, more than two dozen GGF2-altered genes have known roles in fat cell metabolism or signaling. For example, adiponectin (ADIPOQ), predicted gene with 89.7% similarity to human adiponectin receptor 1 (ADIPOR1), predicted gene with 83.7% similarity to human adipogenesis regulatory factor (ADIRF), CCAAT/enhancer binding protein (C/EBP), alpha (CEBPA), perilipin (PLIN1) and peroxisome proliferator-activated receptor gamma (PPARG), which were all down-regulated (2.9–28.2-fold, Supplementary Table S1) encode markers of adipogenesis. Consistent with lowered adipogenesis, genes encoding lipid sensors and fatty acid metabolism were altered, as were genes associated with carbohydrate metabolism and insulin signaling. ECM genes were also mainly down-regulated, whereas muscle-associated transcripts were generally up-regulated (Supplementary Table S1).

3.4. Genes Altered in Skeletal and Cardiac Muscles of GGF2-Treated Post-MI Pigs

We previously analyzed gene expression in left ventricular tissues remote from infarct in these same animals using *Sus scrofa* microarrays [17]. This prior study indicated that GGF2 treatments reduced profibrotic transcripts, lowered percentages of activated fibroblasts, and inhibited cardiac fibrosis [17]. We thus compared our prior ventricular data to transcriptional changes in intercostal skeletal muscle tissues to identify potentially "global" GGF2-mediated alterations. As shown in Table 1, there were 32 genes that were similarly altered in skeletal and cardiac muscle of GGF2-treated post-MI pigs, 18 of which were previously shown to be functionally relevant specifically in skeletal muscle and/or the cardiovascular system [20–53]. To determine whether these commonly altered genes are functionally associated, we performed protein–protein interaction network functional enrichment analysis, using the publicly available online tool, STRING. For context, we also included NRG1 and relevant ErbB receptors (ErbB2-4). The resulting network included 24 of the input proteins, with three subgroups based on K means clustering (Figure 4).

Table 1. Genes differentially expressed in cardiac and skeletal muscle of GGF2-treated post-MI pigs.

Gene	Name	SM	LV	Function
NR4A3	Nuclear Receptor Subfamily 4 Group A Member 3	10.4	12.8	β-adrenergic inducible, Regulates transcription of fatty acid and muscle mass genes [20,21]
FOS	Fos Proto-Oncogene, AP-1 Transcription Factor Subunit	4.9	5.0	Activate phospholipid synthesis, Regulators of cell proliferation, muscle cell differentiation, and Transformation [22–24]
RND	Rho Family GTPase	3.6	2.6	Regulate the organization of the actin cytoskeleton in response to extracellular growth factors [54]
BTG2	BTG Anti-Proliferation Factor 2	3.3	4.1	Cell cycle regulation, cell proliferation in skeletal muscle [25]
PPP1R15A	Protein Phosphatase 1 Regulatory Subunit 15A	2.8	1.7	Down-regulates the TGF-beta, growth arrest and DNA damage-inducible protein, promoting cell death [55]

Table 1. *Cont.*

Gene	Name	SM	LV	Function
IER5	Immediate Early Response 5	2.6	4.3	Cell regulation, proliferation, and resistance to thermal stress. Dephosphorylates HSF1, and ribosomal protein S6 [56]
PAAF1	Proteasomal ATPase Associated Factor 1	2.4	1.5	Associated with Heart Conduction Disease, regulation of association of proteasome components [26,57]
ATP1B4	ATPase Na+/K+ Transporting Family Member Beta 4	2.0	1.5	Transporting protein, Transcriptional coregulator during muscle development [27,28]
COL1A2	Collagen Type 1 Alpha 2 Chain	−1.8	−2.7	Fibrillar forming collagen, putative down-regulated c-Myc target gene, or upregulate let-7b in skin fibroblasts [58].
EFEMP1	EGF Containing Fibulin Extracellular Matrix Protein 1	−1.8	−1.7	Binds EGFR receptor, autophosphorylation and the activation of downstream signaling pathways, Negative regulator of chondrocyte differentiation [59].
LAMB1	Laminin Subunit Beta 1	−1.8	−1.5	Cell adhesion, Differentiation, Encoding laminin subunit beta-1, are associated with COB with variable muscular or ocular abnormalities, Expressed in skeletal muscle [29,30]
LUM	Lumican	−2.0	−1.9	A collagen binding proteoglycan with increased expression in hearts, Regulate tissue repair, collagen fibril organization [31,32]
THY1	Thy-1 Cell Surface Antigen	−2.0	−2.6	Cell surface glycoprotein and A pathogenic CF fraction in cardiac fibrosis [33,34]
ECM1	Extracellular Matrix Protein 1	−2.1	−2.2	Response to elevated platelet cytosolic Ca2+ and ERK Signaling, Upregulated in cardiac aging and myocardial infarction [35,36]
FAP	Fibroblast Activation Protein Alpha	−2.2	−1.8	Expressed in stromal fibroblasts of epithelial cancers, tissue remodeling, healing wounds, Correlate coronary heart disease [37–39]
DPT	Dermatopontin	−2.3	−2.1	Extracellular matrix proteins involved in cell-matrix interaction, Postulated to modify the behavior of TGF-beta [40]
FBN1	Fibrillin 1	−2.3	−2.6	Calcium ion binding and extracellular matrix structural constituent, Differential regulation in smooth muscle cells [41–43]
PRPS2	Phosphoribosyl Pyrophosphate Synthetase 2	−2.3	−1.7	Phosphoribosyl pyrophosphate synthetase, protein homodimerization activity, and carbohydrate catabolic process-related genes [60,61]
NID1	Nidogen 1	−2.4	−1.8	Basement membrane glycoproteins, Laminin interactions in the heart [44–46]
PLSCR4	Phospholipid Scramblase 4	−2.5	−2.3	Protein coding gene; upregulated in hypertrophic mouse hearts [47,62]
WNT5A	Wnt Family Member 5A	−2.5	−1.5	Canonical and non-canonical wnt pathways, Regulating developmental pathways during embryogenesis
FSCN1	Fascin Actin-Bundling Protein 1	−2.5	−1.5	Organize F-actin; Involved in cell migration, motility, adhesion, and cellular interactions
COL4A1	Collagen Type IV Alpha 1 Chain	−2.6	−1.6	Alpha protein of Type IV collagen, components of basement membranes, Upregulated in the skeletal muscle response [48,49]
ACER3	Alkaline Ceramidase 3	−2.6	−1.5	Protein coding gene

Table 1. *Cont.*

Gene	Name	SM	LV	Function
COL5A2	Nidogen 1	−2.6	−2.5	Alpha chain for fibrillar collagen; cardiac repair and involved in Muscle-Invasive Bladder [48,50,51]
PRTFDC1	Phosphoribosyl Transferase Domain Containing 1	−2.7	−2.3	Protein Coding gene, protein homodimerization activity, and magnesium ion binding
CD55	CD55 molecule (Cromer Blood Group)	Reduced	−2.2	Glycoprotein; Regulates cell decay dysferlin is expressed in skeletal and cardiac muscles [52]
COL3A1	Collagen Type III Alpha 1 Chain	−2.8	−1.9	Fibrillar collagen found in extensible connective tissues and the vascular system [53]
HBB	Hemoglobin Subunit Beta	−3.0	−5.9	Oxygen transport from the lung, Endogenous inhibitor of enkephalin-degrading enzymes such as DPP3, and as a selective antagonist of the P2RX3 receptor which is involved in pain signaling [63]
HBA	Hemoglobin Subunit Alpha	−3.6	−5.9	Iron ion binding and oxygen transport from the lung to the various peripheral tissue [64]
ARMCX2	Armadillo Repeat Containing X-Linked 2	−3.7	−1.8	Regulate the dynamics and distribution of mitochondria in neural cells; involved in tissue development and tumorigenesis [65]

Yellow highlighted rows indicate those with relevant muscle and/or heart functions. SM = skeletal (intercostal) muscle, LV = left ventricular (cardiac muscle).

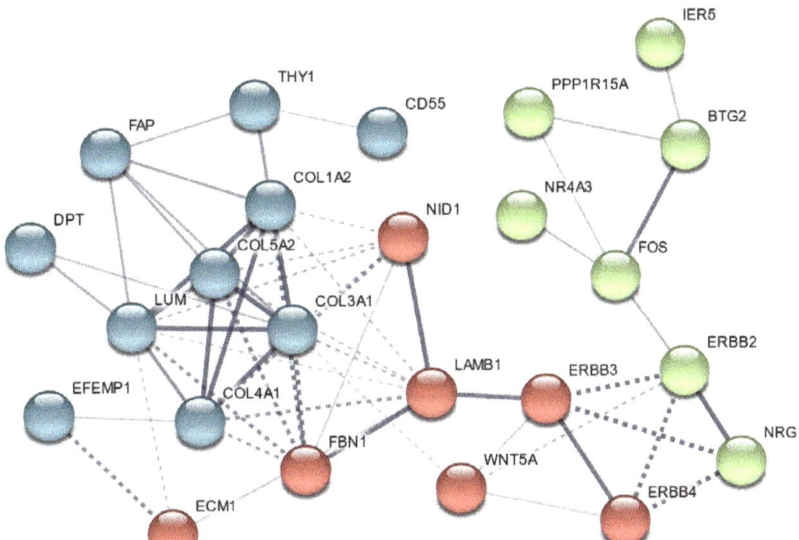

Figure 4. Genes commonly altered in cardiac and intercostal skeletal muscle encode functionally associated proteins. A vector graphic network created using STRING network version 11.5 with K means clustering of proteins encoded by differentially expressed genes commonly altered in both skeletal muscle and cardiac tissues of GGF2-treated post-MI pigs is shown. Full gene names with functions are provided in Table 1, except for NRG1 (neuregulin 1) and its receptors (ERBB2-4). Bubbles represent individual proteins, and lines represent associations between proteins. Line thickness indicates edge confidence, low (0.150), medium (0.400), high (0.700) and highest (0.900). Line shape indicates the predicted mode of action. Each color represents an individual cluster.

3.5. Comparison to Skeletal Muscle Myopathies

GGF2-altered genes important for muscle development included those that have been previously implicated in myocyte cell dysfunction, gross alterations in muscle mass, and heart disease (Table 1). We therefore compared GGF2-altered genes to four animal models of skeletal myopathy and 11 human myopathies (Table 2) and found 230 cachexia transcripts reversibly altered in GGF2-treated pigs (e.g., up-regulated in myopathic skeletal muscle and down-regulated in GGF2-treated pigs). The greatest overlap among experimental animal studies was the mouse denervation model of myopathy (70 genes), followed by cardiac cachexia (46 genes) and 24 h starvation (31 genes) (Figure 5a). Conversely, only five of the genes differentially expressed in rats with experimental Type I diabetes, compared to non-diabetic controls, were altered in the reverse direction in GGF2-treated versus untreated pigs. A similar situation was observed for Type II diabetes in humans (only 8 overlapping genes). Interestingly, the greatest overlap observed for human diseases was for Pompe disease (63 genes), followed closely by tibial muscular dystrophy (55 genes).

Table 2. Gene Expression Omnibus (GEO) myopathy studies included in meta-analysis.

GEO Study ID	Description	Tissue	Species
GSE1557 [66]	Cardiac cachexia (n = 4)	Left ventricle	Rat
GSE52676 [67]	Starvation (n = 6) Denervation (n = 9) Type 1 Diabetes (n = 3)	Soleus	Mouse
GSE45331 [68]	Myotonic dystrophy type 2 (n = 6) Control (n = 4)	Vastus lat.	Human
GSE48574 [69]	ISCU (n = 3) vs. Control (n = 5)	Vastus lat.	Human
GSE38680 [70]	Pompe (n = 9) vs. Control (n = 10) Pompe (n = 11) vs. Control (n = 7)	Biceps Quad	Human
GSE34111 [71]	Cancer cachexia (n = 12) vs. Control (n = 6)	Quad	Human
GSE42806 [72]	Tibial muscular dystrophy (n = 7) vs. Healthy (n = 5)	Extensor digitorum longus	Human
GSE25941 [73]	Female: old (78 ± 1 years, n = 11) vs. young (25 ± 1 years, n = 8) Male: old (78 ± 1 years, n = 10) vs. young (25 ± 1 years, n = 7)	Vastus lat.	Human
GSE9103 [74]	Old (n = 65.1 ±1.5, n = 10) vs. young (22.7 ± 0.7, n = 10)	Vastus lat.	Human
GSE5110 [75]	48 h immobilization vs. control: male subjects (n = 5)	Vastus lat.	Human
GSE21496 [76]	48 h suspension vs. control: sedentary male subjects (n = 7)	Vastus lat.	Human
GSE43760 [77]	Metabolic syndrome (n = 6) vs. healthy (n = 6)	Vastus lat.	Human
GSE27536	COPD low BMI (n = 6) vs. healthy (n = 12) COPD normal BMI (n = 8) vs. healthy (n = 12)	Vastus lat.	Human
GSE6798 [78]	Obese + PCOS (n = 16) vs. Control (n = 13)	Vastus lat.	Human
GSE8157 [79]	Obese + PCOS (n = 10) vs. Control (n = 13)	Vastus lat.	Human
GSE19420	Type 2 diabetes (n = 10) vs. normoglycemic subjects (n = 12)	Vastus lat.	Human
GSE25462 [80]	Type 2 diabetes (n = 10) vs. normoglycemic + no family history of diabetes (n = 15) Type 2 diabetes (n = 10) vs. normoglycemic + family history of type 2 diabetes (n = 25)	Quad	Human

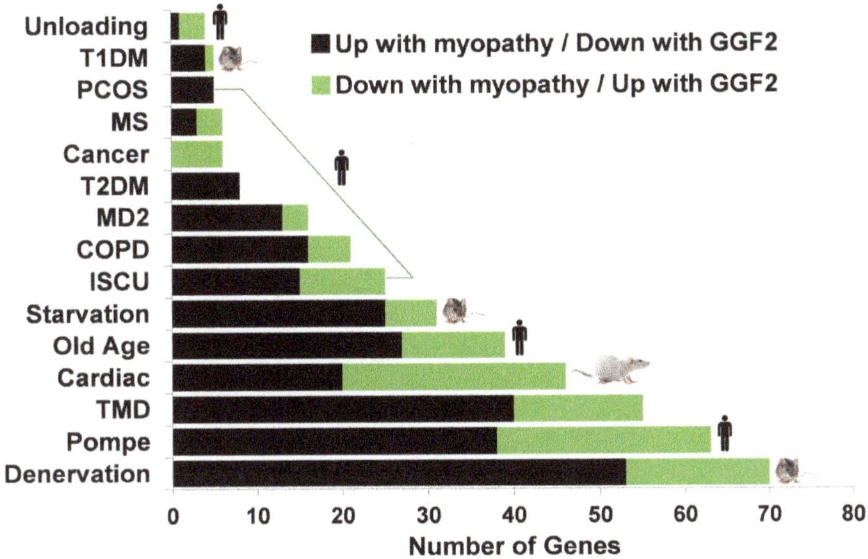

(a)

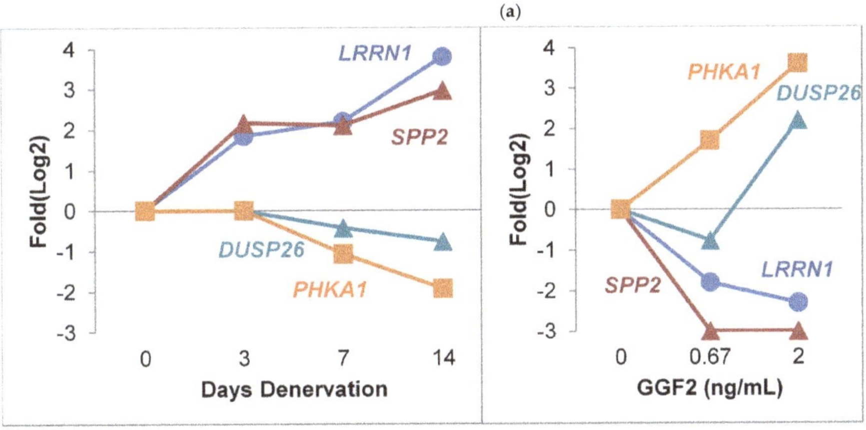

(b)

Figure 5. GGF2 reversibly alters myopathic genes identified through meta-analysis of repository gene expression data. (**a**) Bar chart showing numbers of genes (x-axis) significantly altered in various myopathies (study details listed in Table 2) that were reversibly expressed in GGF2-treated post-MI pigs. Species indicated by an icon at end of bar. Color of bar indicates directionality in listed experimental conditions relative to respective controls. Abbreviations are T1DM = (Type I Diabetes Model), PCOS = polycystic ovary syndrome, MS, T2DM = Type II diabetes mellitus, MD2 = myotonic dystrophy type II, COPD = chronic obstructive pulmonary disease, ISCU = iron-sulfur cluster scaffold homolog myopathy, and TMD = tibial muscular dystrophy. (**b**) Line graphs comparing alterations in four genes after denervation in an animal model of cachexia (left panel, GSE52676) and the same genes reversibly altered in GGF2-treated pigs (right panel). Log2 fold difference (Experiment vs. Control) is shown on the y axis, and time (in days) or dose (in ng/mL) is labeled on the x-axis. Colored lines correspond to genes, as labeled. LRRN1 = Leucine Rich Repeat Neuronal 1, SPP2 = Secreted Phosphoprotein 2, DUSP26 = Dual Specificity Phosphatase 26, PHKA1 = Phosphorylase Kinase Regulatory Subunit Alpha 1.

Notable examples of overlapping transcripts included the gene encoding thrombospondin 4 (THBS4), which was down-regulated 2.6-fold in GGF2-treated swine ($p = 0.001$) and up-regulated in denervated animal muscles (1.6-fold, $p = 0.048$) and in humans with Pompe (2-fold, $p = 0.046$), tibial muscular dystrophy (3.1-fold, $p = 0.011$), iron-sulfur cluster myopathy (2-fold, $p = 0.046$) and old age (1.6-fold, $p = 0.039$). Other myotrophic genes that were reversibly down-regulated in response to GGF2 treatment were CHRNA1, which encodes the alpha subunit of the nicotinic cholinergic receptor, syndecan 4 (SDC4), dermatopontin (DPT), the stem cell marker KIT, the cardiac muscle myofibrillar stretch-sensor ankyrin repeat domain 1 (ANKRD1), matrix metalloproteinase 2 (MMP2) and a negative regulator of insulin secretion (GPR137B). GGF2 also induced the expression of genes that are down-regulated in humans and/or animal experimental models of myopathy, including KLF10 (encodes Kruppel-like factor 10), purinergic receptor P2Y (P2RY1), the gene encoding the lactate transporter MCT4 (SLC16A3), retinoid receptor γ (RXRG), triadin (TRDN), myosin heavy chain (MYH1), epidermal growth factor (EGF), and the skeletal muscle isoform of phosphorylase kinase (PHKA1) that when mutated causes muscle glycogenosis (Figure 5b).

3.6. GGF2/NRG-1β Stimulates ErbB2-Dependent Myoblast Differentiation

To validate NRG-induced myogenesis, immature L6 or C2C12 myoblasts were cultured in differentiation media and treated with GGF2 or recombinant human NRG-1β for 5–6 days. Consistent with previous studies [12,13,81], NRG-1β and GGF2 enhanced myoblast fusion and myotube formation (Figure 6a and Figure S1). GGF2/NRG-1β-induced muscle cell differentiation was accompanied by increased protein levels of differentiation markers, including myosin heavy chain (MHC) and myotubule (M)-cadherin, as well as phosphorylated AKT, glycogen synthase kinase 3-α (GSK-3α), and focal adhesion kinase (FAK) (Figure 6b and Figure S1). ErbB2 expression and phosphorylation were increased during differentiation (Figure S2), and differentiation was abrogated by the addition of the ErbB2 inhibitor TAK165 (Figure 6a), as well as by other pharmacological inhibitors or siRNA (Figures S3–S7).

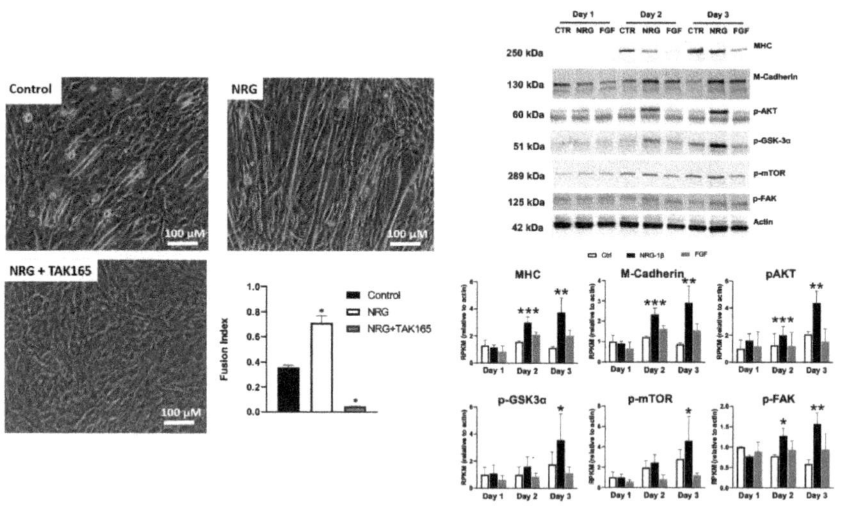

(a) (b)

Figure 6. NRG-1β treatment of cultured myoblasts enhances differentiation. (a) C2C12 cells were cultured in differentiation media (DMEM + 2% horse serum) for 5 days in the absence (control) or presence of 10 ng/mL of recombinant neuregulin-1β (NRG, 10 ng/mL) or with both NRG and 0.2 μM of the ErbB2 receptor inhibitor TAK164. (**b**) Western blots of lysates from C2C12 cells grown in differentiation media for 5 days in the absence (control, CTR) or presence of recombinant NRG-1β (NRG) or fibroblast growth factor (FGF). Full blots provided as Supplementary Figures S8–S14. Bar graphs, grouped by day, show results of quantification by densitometry for the indicated proteins (n = 2, n = 3 for p-FAK). Bar colors indicate treatment type: CTRL = white, NRG-1β = black, and FGF = gray. Asterisks indicate statistical significance, * p value < 0.05), ** p value < 0.005), *** $p < 0.0005$).

3.7. A Putative GGF2-Induced Skeletal Muscle Signaling Pathway

Based on gene set enrichment analysis (GSEA), top transcription factors for GGF2-altered genes included the pro-adipogenic factor CCAAT enhancer binding protein β (C/EBP-β, 101 genes, FDR q value = 3.8×10^{-33}), as well as the myogenic differentiation factors myoblast determination protein 1 (MYOD, 19 genes, FDR q value = 5.4×10^{-9}) and myocyte enhancer factor-2 (MEF2, 21 genes, FDR q value = 1.1×10^{-8}). A more comprehensive functional analysis using Ingenuity Pathway Analysis (IPA) software likewise identified these same three factors, as well as all three relevant ERBB receptors (Table 3), suggesting that more than one skeletal muscle cell type contributed to the observed transcriptional changes. Similarly, 45 genes were altered downstream of fibroblast growth factor (FGF2, $p = 4.8 \times 10^{-20}$, Table 3), which is a known regulator of skeletal muscle proliferation and differentiation and additionally influences intramuscular adiposity by regulating trans-differentiation of fibro-adipogenic progenitors [82].

In addition to providing significance (p-value), IPA functional analysis additionally predicts directionality for some upstream factors (Table 3), based on known regulatory information derived from the scientific literature and other knowledge repositories. This information, along with manual searches of differentially expressed genes using PubMed and various online gene, protein, and signaling pathway databases, was used to construct a potentially meaningful functional network (Figure 7). This heuristic approach resulted in a clear pattern of down-regulated versus up-regulated genes, proteins, and predicted regulatory factors which generally correlated to cell types. Using these results as a guide, cell-type-specific proliferation and differentiation was inferred beginning with mesenchymal stem cells (MSC), which normally proliferate in response to platelet derived growth factor-BB (PDGF BB, 36 downstream genes altered, $p = 3.7 \times 10^{-22}$, activation Z score = 1.6, Table 3) and epidermal growth factor (EGF, transcriptionally up-regulated 3.5-fold, Supplementary Table S1 and 49 downstream genes affected, Table 3) and differentiate into fibro-adipogenic progenitors or myoblasts in response to platelet-derived growth factor receptor alpha (PDGFRA, transcriptionally downregulated -2.1-fold, Supplementary Table S1) and ladybird homeobox 1 (LBX1, transcriptionally upregulated 3.4-fold, Supplementary Table S1), respectively.

Concomitantly, myogenesis inhibitor Twist1 was also predicted as inhibited ($Z = -2.4$, Table 3). Twist 1 maintains skeletal muscle progenitors, in part by inhibiting transactivation of myocyte enhancer factor 2C (MEF2C), which was predicted as being activated in GGF2-treated post-MI pig skeletal muscle ($Z = 2.9$, Table 3). Continuing this process revealed generalized decreases in adipocyte and fibroblast lineages (blue boxes, Figure 7), along with increased MSCs and myoblast lineages (red boxes, Figure 7).

Table 3. Predicted upstream regulators for GGF2-altered Genes.

Upstream Regulator	Gene Expression	No. Downstream Targets Altered (p-Value)	Z Score (Predicted State)
Transforming growth factor β1 (TGF-β1)	-	114 (1.6×10^{-23})	Z = −2.297 (Inhibited)
Platelet derived growth factor-BB (PDGF BB)	-	36 (3.7×10^{-22})	Z = 1.567
Angiotensinogen (AGT)	−3.8	67 (8.5×10^{-20})	Z = −3.032 (Activated)
Fibroblast growth factor 2 (FGF2)	-	45 (4.8×10^{-20})	nd
CAMP Responsive Element Binding Protein 1 (CREB1)	-	46 (3.5×10^{-19})	Z = 2.424 (Activated)
Erb-B2 Receptor Tyrosine Kinase 2 (ERBB2)	-	60 (5.7×10^{-19})	nd
PPARG coactivator 1 α (PCG-1α)	-	37 (4.4×10^{-18})	Z = 1.877
Aryl hydrocarbon receptor (AHR)	-	45 (4.5×10^{-17})	Z = 3.63 (Activated)
Twist family BHLH transcription factor 1 (TWIST1)	-	23 (6.7×10^{-17})	Z = −2.449 (Inhibited)
Epidermal growth factor (EGF)	3.5	49 (5.3×10^{-16})	nd
Cadherin associated protein α1 (α-catenin)	-	19 (3.0×10^{-15})	Z = 3.118 (Activated)
Erb-B3 Receptor Tyrosine Kinase 3 (ERRB3)	-	18 (2.5×10^{-13})	nd
Mothers against DPP homolog 7 (SMAD7)	2.4	18 (3.3×10^{-12})	Z = 3.11 (Activated)
CCAAT enhancer binding protein β (C/EBP-β)	-	37 (1.1×10^{-11})	Z = −2.091 (Inhibited)
AKT serine/threonine kinase 1 (AKT1)	-	26 (1.9×10^{-11})	Z = −1.855
Enalapril (Hypertension medication)	-	13 (9.3×10^{-11})	Z = 3.073 (Activated)
CCAAT enhancer binding protein α (C/EBP-α)	−9.3	28 (2.5×10^{-11})	Z = −1.812
Brain-derived neurotrophic factor (BDNF)		30 (6.7×10^{-11})	nd
Transforming growth factor β3 (TGF-β3)	-	17 (7.6×10^{-11})	Z = −1.937
Myocilin (MYOC)	−2.5	13 (1.7×10^{-10})	nd
Myocyte enhancer factor 2C (MEF2C)	-	15 (3.7×10^{-10})	Z = 2.912 (Activated)
Peroxisome proliferator activated receptor γ (PPAR-γ)	−5.4	36 (8.6×10^{-10})	Z = −1.987
Sterol regulatory element-binding transcription factor 1 (SREBF1)	-	20 (2.4×10^{-9})	Z = −1.937
Erb-B3 Receptor Tyrosine Kinase 3 (ERRB4)	-	12 (3.4×10^{-8})	Z = 2.388 (Activated)
Transforming growth factor β1 (TGF-β1)	-	15 (4.3×10^{-8})	Z = −1.634
microRNA-29b-3p (miR-29b-3p)	-	13 (4.8×10^{-8})	Z = 3.097 (Activated)
Thapsigargin (Calcium reuptake inhibitor)	-	20 (6.5×10^{-8})	Z = 3.450 (Activated)
microRNA-335-3p (miR-335-3p)	-	8 (6.7×10^{-8})	Z = 2.828 (Activated)
CCAAT enhancer binding protein δ (C/EBP-δ)	-	13 (1.3×10^{-7})	Z = −1.813
Peroxisome proliferator activated receptor α (PPAR-α)	-	30 (2.5×10^{-7})	Z = −1.363
Clopidogrel (antiplatelet blood-thinning medication)		11 (4.6×10^{-7})	Z = 3.302 (Activated)
microRNA lethal 7a-5p (Let-7a-5p)	-	17 (1.4×10^{-6})	Z = 3.682 (Activated)
Myocardin (MYOCD)	-	9 (4.4×10^{-6})	Z = 2.759 (Activated)
Cell death inducing DFFA like effector C (CIDEC)	−9.4	6 (1.1×10^{-5})	nd
Msh homeobox 1 (MSX1)	−1.7	5 (1.8×10^{-5})	Z = −1.982
microRNA-30c-5p (miR-30c-5p)	-	17 (2.1×10^{-5})	Z = 3.117 (Activated)
sterol regulatory element binding transcription factor 2 (SREBF2)	-	9 (5.8×10^{-5})	Z = −2.394 (Inhibited)
microRNA-21 (miR-21)	-	17 (3.3×10^{-5})	Z = −2.668 (Inhibited)
Myogenic differentiation 1 (MYOD1)	2.2	16 (3.7×10^{-5})	nd
Wnt family member 5a (WNT5a)	−2.5	10 (8.8×10^{-5})	Z = −2.394 (Inhibited)
Peroxisome proliferator activated receptor δ (PPAR-δ)	-	15 (8.5×10^{-4})	Z = −1.214
26s Proteosome (protein complex)	-	11 (5.5×10^{-4})	Z = −2.035 (Inhibited)
Transcription factor 4 (TCF4)	-	18 (8.7×10^{-4})	Z = −1.554
PPARG coactivator 1 β (PCG-1β)	-	5 (1.6×10^{-2})	Z = −2.186 (Inhibited)

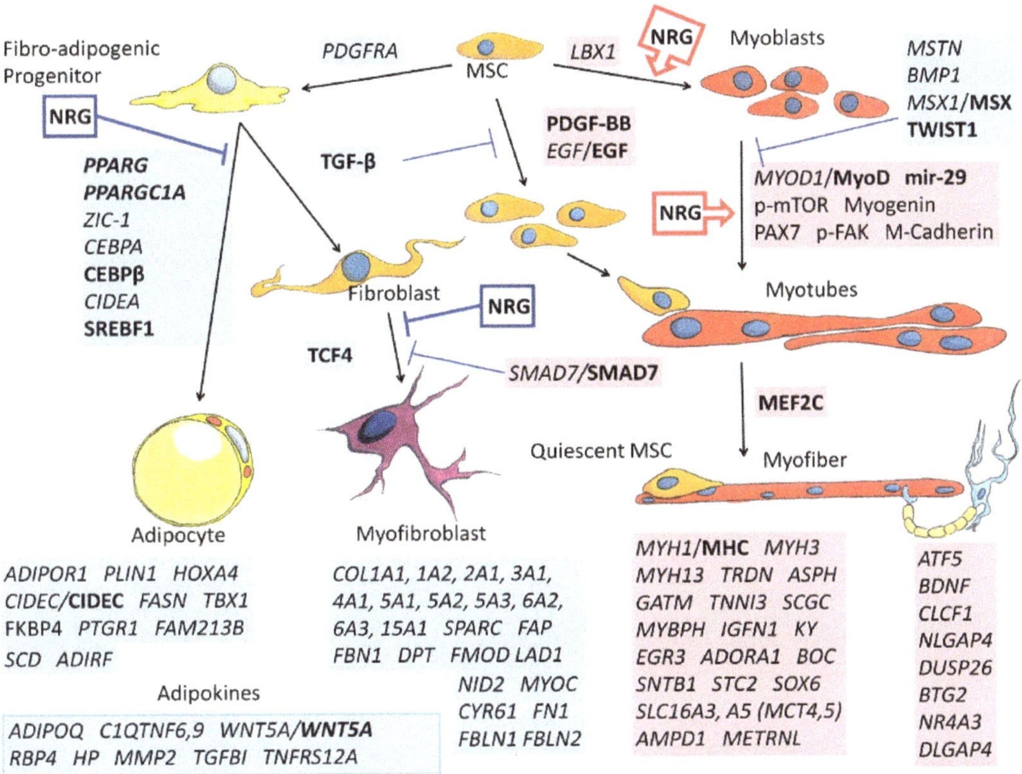

Figure 7. Depiction of inferred relationships of altered genes in GGF2-treated post-MI pigs. Transcripts that were up-regulated (highlighted in red) or down-regulated (blue) in intercostal muscle of neuregulin-treated pigs are italicized. Proteins identified as altered or activated based on Western blot analyses are also indicated. Bold indicates protein level inference based on functional analyses as described in the text. Known effects of neuregulin (NRG) on cell transitions based on published in vitro studies are indicated in red boxes with arrows (for promoting) or in blue (for inhibiting). Transcript abbreviations are official gene (*italic*) or protein symbols, and p- indicates phosphorylation (determined by Western blotting of lysates from NRG-treated myoblasts). MSC = mesenchymal stem cell; mir-29 = microRNA 29. Commissioned illustrations by Andrew Celso Gutierrez.

4. Discussion

The results presented herein show that exogenous delivery of a pharmaceutical-grade version of NRG/GGF2 not only improves post-MI cardiac function but may also promote better intercostal skeletal muscle function via shifting the balance of stem-cell derived differentiated cell types. The global expression profiles exhibited in muscles of treated post-MI pigs were overwhelmingly consistent with fewer adipocytes and myofibroblasts and conversely with higher numbers of myofibers. The combined methodologies comparing transcript levels with basic functional analyses, a meta-analysis of previously published muscle studies, and manual review of public databases and scientific literature provided a surprisingly complete and plausible molecular signature relevant to NRG-mediated multi-organ responses to cardiac injury.

In addition to predictive alterations in the activity levels of local transcription factors and intracellular signaling molecules, the functional analysis identified multiple circulating factors that could account for some of the observed transcriptional changes in skeletal muscle. Inhibition of TGF-β1, for example, might reflect systemic changes given that

TGF-β is generally increased with chronic inflammation and has been proposed as a biomarker of coronary artery disease in the setting of acute heart failure [83]. Local skeletal muscle responses to lower circulating levels of AGT might likewise result from GGF2-mediated alterations in blood volume and blood pressure within the cardiovascular system. In support of this, GGF2 treatment altered several genes consistent with effects of the nonsteroidal anti-inflammatory drug clopidogrel (11 genes, $p = 4.6 \times 10^{-7}$, Z = 3.302, Table 3) and the hypertension drug enalapril (13 genes, $p = 9.3 \times 10^{-11}$, Z = 3.073, Table 3). The SERCA2 inhibitor thapsigargin was similarly identified based on the directionality and expression of a cluster of 20 genes (6.5×10^{-8}, Z = 3.450), suggesting that in skeletal muscle GGF2 mimics some of the cardioprotective actions associated with commonly used heart medications, possibly involving prolonged calcium-induced contractions and increased blood flow.

Although this study did not include sequencing of small RNAs, several micro-RNAs were nonetheless predicted as being present at the local tissue level or in the circulation. For instance, miR-29b-3p was predicted as being activated based on functional analysis (Z = 3.1, Table 3), which is notable, because miR-29b-3p promotes muscle differentiation by reducing myoblast proliferation and inducing myotube formation [84,85]. Similarly predicted as activated (Z = 2.8, Table 3), miR-335-3p was reported to play a role in muscle regeneration, influencing both myoblast differentiation and fiber type transformation [86,87]. On the other hand, sarcopenia-associated miR-21 [88,89] was predicted to be inhibited, based on the alteration of 17 mRNA targets in post-MI pigs (Table), which is of interest because mir-21 was recently reported to be a circulating biomarker for accelerated sarcopenia in patients with chronic heart failure [88]. GGF2 might also influence heart failure-associated muscle dysfunction via increased mir-30c-5p, which might represent a therapeutic target based on protection against myocardial ischemia-reperfusion injury in rats [90] and decreased plasma levels in patients with coronary heart disease [91].

Considered together, the data presented here strongly support the role of NRG/GGF2 in skeletal muscle maintenance and turnover in general and especially in the setting of cardiac injury and heart failure. Systemic treatments with circulating growth facts, such as NRG, may be problematic due to diverse effects, given the importance of NRG in multiple interrelated systems. On the other hand, these types of growth factor-based treatment have the benefit of potentially correcting abnormalities in multiple organs. In this study, we provide insights into the underlying molecular mechanisms of NRG-mediated effects inducible by cardiac injury-mediated stress in an "unrelated" tissue that is affected secondarily by heart dysfunction. Future studies will be aimed at dissecting cell-level effects of NRG/GGF2 and further investigating those identified downstream factors that might represent putative and highly specific therapeutic candidates for other muscle-involved disorders. In addition, future investigation examining respiratory muscle function will be needed to understand the potential clinical benefits of NRG/GGF2's actions in skeletal muscle in this context.

5. Patents

There are no patents associated with this study.

Supplementary Materials: The following are available online at https://www.mdpi.com/article/10.3390/biology11050682/s1, Table S1: Differentially expressed genes in GGF2 treated post-MI pigs. Figures S1–S7: Supplementary methods for cell culture experiments. Figures S8–S14: Western blot membranes of (a) MHC (~250 kDa) protein detected with anti-MHC (MF20; 1:1000; R&D, Minneapolis, MN, USA) antibody, (b) M-Cadherin (~130 kDa) protein detected with anti-M Cadherin (D4B9L; 1:1000, Cell Signaling Technology, Danvers, MA, USA) antibody, (c) p-AKT (~60 kDa) protein detected with anti-phospho-AKT (Ser473, D9E; 1:1000, Cell Signaling Technology, Danvers, MA, USA) antibody (d) p-GSK3α (~51 kDa) protein detected with anti-GSK-3α (Ser21; 1:1000, Cell Signaling Technology, Danvers, MA, USA) antibody (e) p-mTOR (~289 kDa) protein detected with anti-phosho-mTOR (Ser2448; 1:1000, Cell Signaling Technology, Danvers, MA, USA) antibody (f) p-FAK (~125 kDa) protein detected with anti-phospho-FAK (Tyr576/577; 1:1000, Cell Signaling

Technology, Danvers, MA, USA) antibody (g) actin (~kDa) protein detected with anti-actin (C-11; 1:1000, Santa Cruz Biotechnology, Inc., Dallas, TX, USA) antibody.

Author Contributions: Conceptualization, D.B.S.; methodology, D.B.S. and C.L.G.; software, C.L.G.; validation, V.T.N., B.H. and E.E.; formal analysis, C.L.G.; investigation, J.H.C.; resources, D.B.S.; data curation, C.L.G.; writing—original draft preparation, D.B.S. and C.L.G.; writing—review and editing, D.B.S., C.L.G., V.T.N., B.H. and E.E.; visualization, D.B.S. and C.L.G.; supervision, D.B.S.; project administration, D.B.S.; funding acquisition, D.B.S. and C.L.G. All authors have read and agreed to the published version of the manuscript.

Funding: This research was funded by the Ogden College of Science at WKU, National Institutes of Health National Heart Lung and Blood Institute (P20 HL101425, U01 HL100398, K01 HL121045-01) and of General Medical Sciences (P20GM103436), the National Science Foundation (EPSCoR) and Acorda Therapeutics, Inc. The APC was waived by the journal.

Institutional Review Board Statement: This study was approved by the Vanderbilt Institutional Animal Care and Use Committee (IACUC, protocol number M/10/117) and conducted according to Association for the Accreditation of Laboratory Animal Care (AAALAC) International standards.

Informed Consent Statement: Not applicable.

Data Availability Statement: Original study data are available in the Gene Expreassion Omnibus under Accession series number GSE48255 (https://www.ncbi.nlm.nih.gov/bioproject/PRJNA209383). Data were accessed June 2017 and February 2022.

Acknowledgments: The authors would like to thank Andrew Celso Gutierrez for providing the custom, hand-drawn cell illustrations displayed in Figure 7.

Conflicts of Interest: The authors declare no conflict of interest. The funders had no role in the design of the study; in the collection, analyses, or interpretation of data; in the writing of the manuscript, or in the decision to publish the results.

References

1. Kharraz, Y.; Guerra, J.; Pessina, P.; Serrano, A.L.; Muñoz-Cánoves, P. Understanding the Process of Fibrosis in Duchenne Muscular Dystrophy. *BioMed Res. Int.* **2014**, *2014*, 965631. [CrossRef] [PubMed]
2. Larsson, L.; Degens, H.; Li, M.; Salviati, L.; Lee, Y.I.; Thompson, W.; Kirkland, J.L.; Sandri, M. Sarcopenia: Aging-Related Loss of Muscle Mass and Function. *Physiol. Rev.* **2019**, *99*, 427–511. [CrossRef] [PubMed]
3. Seene, T.; Kaasik, P. Muscle weakness in the elderly: Role of sarcopenia, dynapenia, and possibilities for rehabilitation. *Eur. Rev. Aging Phys. Act.* **2012**, *9*, 109–117. [CrossRef]
4. Yokota, T.; Kinugawa, S.; Hirabayashi, K.; Yamato, M.; Takada, S.; Suga, T.; Nakano, I.; Fukushima, A.; Matsushima, S.; Okita, K.; et al. Systemic oxidative stress is associated with lower aerobic capacity and impaired skeletal muscle energy metabolism in heart failure patients. *Sci. Rep.* **2021**, *11*, 2272. [CrossRef] [PubMed]
5. Zizola, C.; Schulze, P.C. Metabolic and structural impairment of skeletal muscle in heart failure. *Heart Fail. Rev.* **2013**, *18*, 623–630. [CrossRef]
6. Keller-Ross, M.L.; Larson, M.; Johnson, B.D. Skeletal Muscle Fatigability in Heart Failure. *Front. Physiol.* **2019**, *10*, 129. [CrossRef]
7. Kinugawa, S.; Takada, S.; Matsushima, S.; Okita, K.; Tsutsui, H. Skeletal Muscle Abnormalities in Heart Failure. *Int. Heart J.* **2015**, *56*, 475–484. [CrossRef]
8. Maskin, C.S.; Forman, R.; Sonnenblick, E.H.; Frishman, W.H.; LeJemtel, T.H. Failure of dobutamine to increase exercise capacity despite hemodynamic improvement in severe chronic heart failure. *Am. J. Cardiol.* **1983**, *51*, 177–182. [CrossRef]
9. Mancini, D.M.; Coyle, E.; Coggan, A.; Beltz, J.; Ferraro, N.; Montain, S.; Wilson, J.R. Contribution of intrinsic skeletal muscle changes to 31P NMR skeletal muscle metabolic abnormalities in patients with chronic heart failure. *Circulation* **1989**, *80*, 1338–1346. [CrossRef]
10. Britsch, S.; Li, L.; Kirchhoff, S.; Theuring, F.; Brinkmann, V.; Birchmeier, C.; Riethmacher, D. The ErbB2 and ErbB3 receptors and their ligand, neuregulin-1, are essential for development of the sympathetic nervous system. *Genes Dev.* **1998**, *12*, 1825–1836. [CrossRef]
11. Sandrock, A.W., Jr.; Goodearl, A.D.; Yin, Q.W.; Chang, D.; Fischbach, G.D. ARIA is concentrated in nerve terminals at neuromuscular junctions and at other synapses. *J. Neurosci.* **1995**, *15*, 6124–6136. [CrossRef] [PubMed]
12. Florini, J.R.; Samuel, D.S.; Ewton, D.Z.; Kirk, C.; Sklar, R.M. Stimulation of myogenic differentiation by a neuregulin, glial growth factor 2. Are neuregulins the long-sought muscle trophic factors secreted by nerves? *J. Biol. Chem.* **1996**, *271*, 12699–12702. [CrossRef] [PubMed]
13. Kim, D.; Chi, S.; Lee, K.H.; Rhee, S.; Kwon, Y.K.; Chung, C.H.; Kwon, H.; Kang, M.S. Neuregulin stimulates myogenic differentiation in an autocrine manner. *J. Biol. Chem.* **1999**, *274*, 15395–15400. [CrossRef]

14. Ford, B.D.; Han, B.; Fischbach, G.D. Differentiation-dependent regulation of skeletal myogenesis by neuregulin-1. *Biochem. Biophys. Res. Commun.* **2003**, *306*, 276–281. [CrossRef]
15. Jo, S.A.; Zhu, X.; Marchionni, M.A.; Burden, S.J. Neuregulins are concentrated at nerve-muscle synapses and activate ACh–receptor gene expression. *Nature* **1995**, *373*, 158–161. [CrossRef] [PubMed]
16. Van Ho, A.T.; Hayashi, S.; Brohl, D.; Aurade, F.; Rattenbach, R.; Relaix, F. Neural crest cell lineage restricts skeletal muscle progenitor cell differentiation through Neuregulin1-ErbB3 signaling. *Dev. Cell* **2011**, *21*, 273–287. [CrossRef]
17. Galindo, C.L.; Kasasbeh, E.; Murphy, A.; Ryzhov, S.; Lenihan, S.; Ahmad, F.A.; Williams, P.; Nunnally, A.; Adcock, J.; Song, Y.; et al. Anti-remodeling and anti-fibrotic effects of the neuregulin-1beta glial growth factor 2 in a large animal model of heart failure. *J. Am. Heart Assoc.* **2014**, *3*, e000773. [CrossRef]
18. Huang, Z.; Sawyer, D.B.; Troy, E.L.; McEwen, C.; Cleator, J.H.; Murphy, A.; Caggiano, A.O.; Eisen, A.; Parry, T.J. Species-specific effects of neuregulin-1beta (cimaglermin alfa) on glucose handling in animal models and humans with heart failure. *Toxicol. Appl. Pharmacol.* **2017**, *332*, 92–99. [CrossRef]
19. Parry, T.J.; Ganguly, A.; Troy, E.L.; Luis Guerrero, J.; Iaci, J.F.; Srinivas, M.; Vecchione, A.M.; Button, D.C.; Hackett, C.S.; Zolty, R.; et al. Effects of neuregulin GGF2 (cimaglermin alfa) dose and treatment frequency on left ventricular function in rats following myocardial infarction. *Eur. J. Pharmacol.* **2017**, *796*, 76–89. [CrossRef]
20. Liu, Q.; Zhu, X.; Xu, L.; Fu, Y.; Garvey, W.T. 6-Mercaptopurine augments glucose transport activity in skeletal muscle cells in part via a mechanism dependent upon orphan nuclear receptor NR4A3. *Am. J. Physiol.-Endocrinol. Metab.* **2013**, *305*, E1081–E1092. [CrossRef]
21. Pearen, M.A.; Ryall, J.G.; Maxwell, M.A.; Ohkura, N.; Lynch, G.S.; Muscat, G.E. The orphan nuclear receptor, NOR-1, is a target of beta-adrenergic signaling in skeletal muscle. *Endocrinology* **2006**, *147*, 5217–5227. [CrossRef] [PubMed]
22. Guo, Z.; Luo, C.; Zhu, T.; Li, L.; Zhang, W. Elevated c-fos expression is correlated with phenotypic switching of human vascular smooth muscle cells derived from lower limb venous varicosities. *J. Vasc. Surg. Venous Lymphat. Disord.* **2021**, *9*, 242–251. [CrossRef] [PubMed]
23. Trouche, D.; Grigoriev, M.; Lenormand, J.-L.; Robin, P.; Alexandre Leibovitch, S.; Sassone-Corsi, P.; Harel-Bellan, A. Repression of c-fos promoter by MyoD on muscle cell differentiation. *Nature* **1993**, *363*, 79–82. [CrossRef] [PubMed]
24. Zhao, L.; Ouyang, Y.; Bai, Y.; Gong, J.; Liao, H. miR-155-5p inhibits the viability of vascular smooth muscle cell via targeting FOS and ZIC3 to promote aneurysm formation. *Eur. J. Pharmacol.* **2019**, *853*, 145–152. [CrossRef]
25. Shah, V.O.; Dominic, E.A.; Moseley, P.; Pickett, G.; Fleet, M.; Ness, S.; Raj, D.S.C. Hemodialysis Modulates Gene Expression Profile in Skeletal Muscle. *Am. J. Kidney Dis.* **2006**, *48*, 616–628. [CrossRef]
26. Tamboli, R.A.; Hajri, T.; Jiang, A.; Marks-Shulman, P.A.; Williams, D.B.; Clements, R.H.; Melvin, W.; Bowen, B.P.; Shyr, Y.; Abumrad, N.N.; et al. Reduction in inflammatory gene expression in skeletal muscle from Roux-en-Y gastric bypass patients randomized to omentectomy. *PLoS ONE* **2011**, *6*, e28577. [CrossRef]
27. Pestov, N.B.; Adams, G.; Shakhparonov, M.I.; Modyanov, N.N. Identification of a novel gene of the X,K-ATPase beta-subunit family that is predominantly expressed in skeletal and heart muscles. *FEBS Lett.* **1999**, *456*, 243–248. [CrossRef]
28. Pestov, N.B.; Zhao, H.; Basrur, V.; Modyanov, N.N. Isolation and characterization of BetaM protein encoded by ATP1B4—A unique member of the Na,K-ATPase β-subunit gene family. *Biochem. Biophys. Res. Commun.* **2011**, *412*, 543–548. [CrossRef]
29. Roediger, M.; Miosge, N.; Gersdorff, N. Tissue distribution of the laminin β1 and β2 chain during embryonic and fetal human development. *J. Mol. Histol.* **2010**, *41*, 177–184. [CrossRef]
30. Radmanesh, F.; Caglayan, A.O.; Silhavy, J.L.; Yilmaz, C.; Cantagrel, V.; Omar, T.; Rosti, B.; Kaymakcalan, H.; Gabriel, S.; Li, M.; et al. Mutations in LAMB1 cause cobblestone brain malformation without muscular or ocular abnormalities. *Am. J. Hum. Genet.* **2013**, *92*, 468–474. [CrossRef]
31. Engebretsen, K.V.; Lunde, I.G.; Strand, M.E.; Waehre, A.; Sjaastad, I.; Marstein, H.S.; Skrbic, B.; Dahl, C.P.; Askevold, E.T.; Christensen, G.; et al. Lumican is increased in experimental and clinical heart failure, and its production by cardiac fibroblasts is induced by mechanical and proinflammatory stimuli. *FEBS J.* **2013**, *280*, 2382–2398. [CrossRef] [PubMed]
32. Mohammadzadeh, N.; Lunde, I.G.; Andenæs, K.; Strand, M.E.; Aronsen, J.M.; Skrbic, B.; Marstein, H.S.; Bandlien, C.; Nygård, S.; Gorham, J.; et al. The extracellular matrix proteoglycan lumican improves survival and counteracts cardiac dilatation and failure in mice subjected to pressure overload. *Sci. Rep.* **2019**, *9*, 9206. [CrossRef] [PubMed]
33. Li, Y.; Song, D.; Mao, L.; Abraham, D.M.; Bursac, N. Lack of Thy1 defines a pathogenic fraction of cardiac fibroblasts in heart failure. *Biomaterials* **2020**, *236*, 119824. [CrossRef] [PubMed]
34. Cheng, K.; Ibrahim, A.; Hensley, M.T.; Shen, D.; Sun, B.; Middleton, R.; Liu, W.; Smith, R.R.; Marbán, E. Relative roles of CD90 and c-kit to the regenerative efficacy of cardiosphere-derived cells in humans and in a mouse model of myocardial infarction. *J. Am. Heart Assoc.* **2014**, *3*, e001260. [CrossRef] [PubMed]
35. Hamada, T.; Wessagowit, V.; South, A.P.; Ashton, G.H.; Chan, I.; Oyama, N.; Siriwattana, A.; Jewhasuchin, P.; Charuwichitratana, S.; Thappa, D.M.; et al. Extracellular matrix protein 1 gene (ECM1) mutations in lipoid proteinosis and genotype-phenotype correlation. *J. Investig. Dermatol.* **2003**, *120*, 345–350. [CrossRef]
36. Hardy, S.A.; Mabotuwana, N.S.; Murtha, L.A.; Coulter, B.; Sanchez-Bezanilla, S.; Al-Omary, M.S.; Senanayake, T.; Loering, S.; Starkey, M.; Lee, R.J.; et al. Novel role of extracellular matrix protein 1 (ECM1) in cardiac aging and myocardial infarction. *PLoS ONE* **2019**, *14*, e0212230. [CrossRef]

37. Uitte de Willige, S.; Keane, F.M.; Bowen, D.G.; Malfliet, J.; Zhang, H.E.; Maneck, B.; McCaughan, G.W.; Leebeek, F.W.G.; Rijken, D.C.; Gorrell, M.D. Circulating fibroblast activation protein activity and antigen levels correlate strongly when measured in liver disease and coronary heart disease. *PLoS ONE* **2017**, *12*, e0178987. [CrossRef]
38. Li, M.; Cheng, X.; Rong, R.; Gao, Y.; Tang, X.; Chen, Y. High expression of fibroblast activation protein (FAP) predicts poor outcome in high-grade serous ovarian cancer. *BMC Cancer* **2020**, *20*, 1032. [CrossRef]
39. Solano-Iturri, J.D.; Beitia, M.; Errarte, P.; Calvete-Candenas, J.; Etxezarraga, M.C.; Loizate, A.; Echevarria, E.; Badiola, I.; Larrinaga, G. Altered expression of fibroblast activation protein-α (FAP) in colorectal adenoma-carcinoma sequence and in lymph node and liver metastases. *Aging* **2020**, *12*, 10337–10358. [CrossRef]
40. Kim, T.; Ahmad, K.; Shaikh, S.; Jan, A.T.; Seo, M.-G.; Lee, E.J.; Choi, I. Dermatopontin in Skeletal Muscle Extracellular Matrix Regulates Myogenesis. *Cells* **2019**, *8*, 332. [CrossRef]
41. Bastos, A.N.; Alves, M.M.; Monte-Alto-Costa, A.; Machado, D.G.; Cavalcante, G.J.; Panico, M.; Porto, L.C. α-smooth muscle actin, fibrillin-1, apoptosis and proliferation detection in primary varicose lower limb veins of women. *Int. Angiol.* **2011**, *30*, 262–271. [PubMed]
42. Chen, L.; Ge, Q.; Black, J.L.; Deng, L.; Burgess, J.K.; Oliver, B.G. Differential regulation of extracellular matrix and soluble fibulin-1 levels by TGF-β_1 in airway smooth muscle cells. *PLoS ONE* **2013**, *8*, e65544. [CrossRef] [PubMed]
43. Franken, R.; Teixido-Tura, G.; Brion, M.; Forteza, A.; Rodriguez-Palomares, J.; Gutierrez, L.; Garcia Dorado, D.; Pals, G.; Mulder, B.J.; Evangelista, A. Relationship between fibrillin-1 genotype and severity of cardiovascular involvement in Marfan syndrome. *Heart* **2017**, *103*, 1795–1799. [CrossRef] [PubMed]
44. McNiven, V.; Ito, Y.A.; Hartley, T.; Kernohan, K.; Miller, E.; Armour, C.M. NID1 variant associated with occipital cephaloceles in a family expressing a spectrum of phenotypes. *Am. J. Med. Genet. Part A* **2019**, *179*, 837–841. [CrossRef] [PubMed]
45. Zimmermann, K.; Hoischen, S.; Hafner, M.; Nischt, R. Genomic sequences and structural organization of the human nidogen gene (NID). *Genomics* **1995**, *27*, 245–250. [CrossRef]
46. Kim, E.H.; Galchev, V.I.; Kim, J.Y.; Misek, S.A.; Stevenson, T.K.; Campbell, M.D.; Pagani, F.D.; Day, S.M.; Johnson, T.C.; Washburn, J.G.; et al. Differential protein expression and basal lamina remodeling in human heart failure. *Proteom. Clin. Appl.* **2016**, *10*, 585–596. [CrossRef]
47. Wadley, G.D.; Lamon, S.; Alexander, S.E.; McMullen, J.R.; Bernardo, B.C. Noncoding RNAs regulating cardiac muscle mass. *J. Appl. Physiol.* **2018**, *127*, 633–644. [CrossRef]
48. Chen, L.; Bai, J.; Li, Y. miR-29 mediates exercise-induced skeletal muscle angiogenesis by targeting VEGFA, COL4A1 and COL4A2 via the PI3K/Akt signaling pathway. *Mol. Med. Rep.* **2020**, *22*, 661–670. [CrossRef]
49. Labelle-Dumais, C.; Schuitema, V.; Hayashi, G.; Hoff, K.; Gong, W.; Dao, D.Q.; Ullian, E.M.; Oishi, P.; Margeta, M.; Gould, D.B. COL4A1 Mutations Cause Neuromuscular Disease with Tissue-Specific Mechanistic Heterogeneity. *Am. J. Hum. Genet.* **2019**, *104*, 847–860. [CrossRef]
50. Meienberg, J.; Rohrbach, M.; Neuenschwander, S.; Spanaus, K.; Giunta, C.; Alonso, S.; Arnold, E.; Henggeler, C.; Regenass, S.; Patrignani, A.; et al. Hemizygous deletion of COL3A1, COL5A2, and MSTN causes a complex phenotype with aortic dissection: A lesson for and from true haploinsufficiency. *Eur. J. Hum. Genet.* **2010**, *18*, 1315–1321. [CrossRef]
51. Meng, X.Y.; Shi, M.J.; Zeng, Z.H.; Chen, C.; Liu, T.Z.; Wu, Q.J.; Li, S.; Li, S. The Role of COL5A2 in Patients with Muscle-Invasive Bladder Cancer: A Bioinformatics Analysis of Public Datasets Involving 787 Subjects and 29 Cell Lines. *Front. Oncol.* **2018**, *8*, 659. [CrossRef] [PubMed]
52. Wenzel, K.; Zabojszcza, J.; Carl, M.; Taubert, S.; Lass, A.; Harris, C.L.; Ho, M.; Schulz, H.; Hummel, O.; Hubner, N.; et al. Increased susceptibility to complement attack due to down-regulation of decay-accelerating factor/CD55 in dysferlin-deficient muscular dystrophy. *J. Immunol.* **2005**, *175*, 6219–6225. [CrossRef] [PubMed]
53. Kuivaniemi, H.; Tromp, G. Type III collagen (COL3A1): Gene and protein structure, tissue distribution, and associated diseases. *Gene* **2019**, *707*, 151–171. [CrossRef] [PubMed]
54. Li, Y.H.; Ghavampur, S.; Bondallaz, P.; Will, L.; Grenningloh, G.; Pu Schel, A.W. Rnd1 regulates axon extension by enhancing the microtubule destabilizing activity of SCG10. *J. Biol. Chem.* **2009**, *284*, 363–371. [CrossRef] [PubMed]
55. Song, Y.; Liu, S.; Zhao, Z.; Zhang, Y.; Yang, Y.; Luo, B. Relationship between PPP1R15A gene polymorphism (rs611251) and Epstein-Barr virus-associated tumors. *Acta Virol.* **2017**, *61*, 445–452. [CrossRef] [PubMed]
56. Ishikawa, Y.; Sakurai, H. Heat-induced expression of the immediate-early gene IER5 and its involvement in the proliferation of heat-shocked cells. *FEBS J.* **2015**, *282*, 332–340. [CrossRef]
57. Nakamura, M.; Basavarajaiah, P.; Rousset, E.; Beraud, C.; Latreille, D.; Henaoui, I.S.; Lassot, I.; Mari, B.; Kiernan, R. Spt6 levels are modulated by PAAF1 and proteasome to regulate the HIV-1 LTR. *Retrovirology* **2012**, *9*, 13. [CrossRef]
58. Liu, J.; Luo, C.; Yin, Z.; Li, P.; Wang, S.; Chen, J.; He, Q.; Zhou, J. Downregulation of let-7b promotes COL1A1 and COL1A2 expression in dermis and skin fibroblasts during heat wound repair. *Mol. Med. Rep.* **2016**, *13*, 2683–2688. [CrossRef]
59. Camaj, P.; Seeliger, H.; Ischenko, I.; Krebs, S.; Blum, H.; De Toni, E.N.; Faktorova, D.; Jauch, K.W.; Bruns, C.J. EFEMP1 binds the EGF receptor and activates MAPK and Akt pathways in pancreatic carcinoma cells. *Biol. Chem.* **2009**, *390*, 1293–1302. [CrossRef]
60. Lei, B.; Wan, B.; Peng, J.; Yang, Y.; Lv, D.; Zhou, X.; Shu, F.; Li, F.; Zhong, L.; Wu, H.; et al. PRPS2 Expression Correlates with Sertoli-Cell Only Syndrome and Inhibits the Apoptosis of TM4 Sertoli Cells. *J. Urol.* **2015**, *194*, 1491–1497. [CrossRef]
61. Miao, W.; Wang, Y. Targeted Quantitative Kinome Analysis Identifies PRPS2 as a Promoter for Colorectal Cancer Metastasis. *J. Proteome Res.* **2019**, *18*, 2279–2286. [CrossRef] [PubMed]

62. Kim, S.; Song, J.; Ernst, P.; Latimer, M.N.; Ha, C.-M.; Goh, K.Y.; Ma, W.; Rajasekaran, N.-S.; Zhang, J.; Liu, X.; et al. MitoQ regulates redox-related noncoding RNAs to preserve mitochondrial network integrity in pressure-overload heart failure. *Am. J. Physiol. Heart Circ. Physiol.* **2020**, *318*, H682–H695. [CrossRef] [PubMed]
63. Son, G.H.; Park, S.H.; Kim, Y.; Kim, J.Y.; Kim, J.W.; Chung, S.; Kim, Y.H.; Kim, H.; Hwang, J.J.; Seo, J.S. Postmortem mRNA expression patterns in left ventricular myocardial tissues and their implications for forensic diagnosis of sudden cardiac death. *Mol. Cells* **2014**, *37*, 241–247. [CrossRef] [PubMed]
64. Nauck, M.A.; Quast, D.R.; Wefers, J.; Meier, J.J. GLP-1 receptor agonists in the treatment of type 2 diabetes–state-of-the-art. *Mol. Metab.* **2021**, *46*, 101102. [CrossRef] [PubMed]
65. Wang, T.; Zhong, H.; Qin, Y.; Wei, W.; Li, Z.; Huang, M.; Luo, X. ARMCX Family Gene Expression Analysis and Potential Prognostic Biomarkers for Prediction of Clinical Outcome in Patients with Gastric Carcinoma. *BioMed Res. Int.* **2020**, *2020*, 3575038. [CrossRef]
66. Wellner, M.; Dechend, R.; Park, J.K.; Shagdarsuren, E.; Al-Saadi, N.; Kirsch, T.; Gratze, P.; Schneider, W.; Meiners, S.; Fiebeler, A.; et al. Cardiac gene expression profile in rats with terminal heart failure and cachexia. *Physiol. Genom.* **2005**, *20*, 256–267. [CrossRef]
67. Soares, R.J.; Cagnin, S.; Chemello, F.; Silvestrin, M.; Musaro, A.; De Pitta, C.; Lanfranchi, G.; Sandri, M. Involvement of microRNAs in the regulation of muscle wasting during catabolic conditions. *J. Biol. Chem.* **2014**, *289*, 21909–21925. [CrossRef]
68. Screen, M.; Jonson, P.H.; Raheem, O.; Palmio, J.; Laaksonen, R.; Lehtimaki, T.; Sirito, M.; Krahe, R.; Hackman, P.; Udd, B. Abnormal splicing of NEDD4 in myotonic dystrophy type 2: Possible link to statin adverse reactions. *Am. J. Pathol.* **2014**, *184*, 2322–2332. [CrossRef]
69. Crooks, D.R.; Natarajan, T.G.; Jeong, S.Y.; Chen, C.; Park, S.Y.; Huang, H.; Ghosh, M.C.; Tong, W.H.; Haller, R.G.; Wu, C.; et al. Elevated FGF21 secretion, PGC-1alpha and ketogenic enzyme expression are hallmarks of iron-sulfur cluster depletion in human skeletal muscle. *Hum. Mol. Genet.* **2014**, *23*, 24–39. [CrossRef]
70. Palermo, A.T.; Palmer, R.E.; So, K.S.; Oba-Shinjo, S.M.; Zhang, M.; Richards, B.; Madhiwalla, S.T.; Finn, P.F.; Hasegawa, A.; Ciociola, K.M.; et al. Transcriptional response to GAA deficiency (Pompe disease) in infantile-onset patients. *Mol. Genet. Metab.* **2012**, *106*, 287–300. [CrossRef]
71. Gallagher, I.J.; Stephens, N.A.; MacDonald, A.J.; Skipworth, R.J.; Husi, H.; Greig, C.A.; Ross, J.A.; Timmons, J.A.; Fearon, K.C. Suppression of skeletal muscle turnover in cancer cachexia: Evidence from the transcriptome in sequential human muscle biopsies. *Clin. Cancer Res. Off. J. Am. Assoc. Cancer Res.* **2012**, *18*, 2817–2827. [CrossRef] [PubMed]
72. Screen, M.; Raheem, O.; Holmlund-Hampf, J.; Jonson, P.H.; Huovinen, S.; Hackman, P.; Udd, B. Gene expression profiling in tibial muscular dystrophy reveals unfolded protein response and altered autophagy. *PLoS ONE* **2014**, *9*, e90819. [CrossRef] [PubMed]
73. Raue, U.; Trappe, T.A.; Estrem, S.T.; Qian, H.R.; Helvering, L.M.; Smith, R.C.; Trappe, S. Transcriptome signature of resistance exercise adaptations: Mixed muscle and fiber type specific profiles in young and old adults. *J. Appl. Physiol.* **2012**, *112*, 1625–1636. [CrossRef] [PubMed]
74. Lanza, I.R.; Short, D.K.; Short, K.R.; Raghavakaimal, S.; Basu, R.; Joyner, M.J.; McConnell, J.P.; Nair, K.S. Endurance exercise as a countermeasure for aging. *Diabetes* **2008**, *57*, 2933–2942. [CrossRef]
75. Urso, M.L.; Scrimgeour, A.G.; Chen, Y.W.; Thompson, P.D.; Clarkson, P.M. Analysis of human skeletal muscle after 48 h immobilization reveals alterations in mRNA and protein for extracellular matrix components. *J. Appl. Physiol.* **2006**, *101*, 1136–1148. [CrossRef]
76. Reich, K.A.; Chen, Y.W.; Thompson, P.D.; Hoffman, E.P.; Clarkson, P.M. Forty-eight hours of unloading and 24 h of reloading lead to changes in global gene expression patterns related to ubiquitination and oxidative stress in humans. *J. Appl. Physiol.* **2010**, *109*, 1404–1415. [CrossRef]
77. Poelkens, F.; Lammers, G.; Pardoel, E.M.; Tack, C.J.; Hopman, M.T. Upregulation of skeletal muscle inflammatory genes links inflammation with insulin resistance in women with the metabolic syndrome. *Exp. Physiol.* **2013**, *98*, 1485–1494. [CrossRef]
78. Skov, V.; Glintborg, D.; Knudsen, S.; Jensen, T.; Kruse, T.A.; Tan, Q.; Brusgaard, K.; Beck-Nielsen, H.; Hojlund, K. Reduced expression of nuclear-encoded genes involved in mitochondrial oxidative metabolism in skeletal muscle of insulin-resistant women with polycystic ovary syndrome. *Diabetes* **2007**, *56*, 2349–2355. [CrossRef]
79. Skov, V.; Glintborg, D.; Knudsen, S.; Tan, Q.; Jensen, T.; Kruse, T.A.; Beck-Nielsen, H.; Hojlund, K. Pioglitazone enhances mitochondrial biogenesis and ribosomal protein biosynthesis in skeletal muscle in polycystic ovary syndrome. *PLoS ONE* **2008**, *3*, e2466. [CrossRef]
80. Jin, W.; Goldfine, A.B.; Boes, T.; Henry, R.R.; Ciaraldi, T.P.; Kim, E.Y.; Emecan, M.; Fitzpatrick, C.; Sen, A.; Shah, A.; et al. Increased SRF transcriptional activity in human and mouse skeletal muscle is a signature of insulin resistance. *J. Clin. Investig.* **2011**, *121*, 918–929. [CrossRef]
81. Suarez, E.; Bach, D.; Cadefau, J.; Palacin, M.; Zorzano, A.; Guma, A. A novel role of neuregulin in skeletal muscle. Neuregulin stimulates glucose uptake, glucose transporter translocation, and transporter expression in muscle cells. *J. Biol. Chem.* **2001**, *276*, 18257–18264. [CrossRef] [PubMed]
82. Mathes, S.; Fahrner, A.; Ghoshdastider, U.; Rudiger, H.A.; Leunig, M.; Wolfrum, C.; Krutzfeldt, J. FGF-2-dependent signaling activated in aged human skeletal muscle promotes intramuscular adipogenesis. *Proc. Natl. Acad. Sci. USA* **2021**, *118*. [CrossRef] [PubMed]

83. Qiu, X.; Ma, F.; Zhang, H. Circulating Levels of IL-13, TGF-beta1, and Periostin as Potential Biomarker for Coronary Artery Disease with Acute Heart Failure. *Evid.-Based Complement. Altern. Med.* **2021**, *2021*, 1690421. [CrossRef] [PubMed]
84. Qin, C.Y.; Cai, H.; Qing, H.R.; Li, L.; Zhang, H.P. Recent advances on the role of long non-coding RNA H19 in regulating mammalian muscle growth and development. *Yi chuan = Hereditas* **2017**, *39*, 1150–1157. [CrossRef]
85. Wei, W.; He, H.B.; Zhang, W.Y.; Zhang, H.X.; Bai, J.B.; Liu, H.Z.; Cao, J.H.; Chang, K.C.; Li, X.Y.; Zhao, S.H. miR-29 targets Akt3 to reduce proliferation and facilitate differentiation of myoblasts in skeletal muscle development. *Cell Death Dis.* **2013**, *4*, e668. [CrossRef]
86. Greco, S.; De Simone, M.; Colussi, C.; Zaccagnini, G.; Fasanaro, P.; Pescatori, M.; Cardani, R.; Perbellini, R.; Isaia, E.; Sale, P.; et al. Common micro-RNA signature in skeletal muscle damage and regeneration induced by Duchenne muscular dystrophy and acute ischemia. *FASEB J.* **2009**, *23*, 3335–3346. [CrossRef]
87. Huang, B.; Jiao, Y.; Zhu, Y.; Ning, Z.; Ye, Z.; Li, Q.X.; Hu, C.; Wang, C. Putative MicroRNA-mRNA Networks Upon Mdfi Overexpression in C2C12 Cell Differentiation and Muscle Fiber Type Transformation. *Front. Mol. Biosci.* **2021**, *8*, 675993. [CrossRef]
88. Qaisar, R.; Karim, A.; Muhammad, T.; Shah, I.; Khan, J. Circulating MicroRNAs as Biomarkers of Accelerated Sarcopenia in Chronic Heart Failure. *Glob. Heart* **2021**, *16*, 56. [CrossRef]
89. Yanai, K.; Kaneko, S.; Ishii, H.; Aomatsu, A.; Ito, K.; Hirai, K.; Ookawara, S.; Ishibashi, K.; Morishita, Y. MicroRNAs in Sarcopenia: A Systematic Review. *Front. Med.* **2020**, *7*, 180. [CrossRef]
90. Sun, M.; Guo, M.; Ma, G.; Zhang, N.; Pan, F.; Fan, X.; Wang, R. MicroRNA-30c-5p protects against myocardial ischemia/reperfusion injury via regulation of Bach1/Nrf2. *Toxicol. Appl. Pharmacol.* **2021**, *426*, 115637. [CrossRef]
91. Meng, S.; Hu, Y.; Zhu, J.; Feng, T.; Quan, X. miR-30c-5p acts as a therapeutic target for ameliorating myocardial ischemia-reperfusion injury. *Am. J. Transl. Res.* **2021**, *13*, 2198–2212. [PubMed]

biology MDPI

Review

The Interplay of Hypoxia Signaling on Mitochondrial Dysfunction and Inflammation in Cardiovascular Diseases and Cancer: From Molecular Mechanisms to Therapeutic Approaches

Esmaa Bouhamida [1,2], Giampaolo Morciano [1,2], Mariasole Perrone [1], Asrat E. Kahsay [1], Mario Della Sala [1], Mariusz R. Wieckowski [3], Francesco Fiorica [4], Paolo Pinton [1,2], Carlotta Giorgi [1,*] and Simone Patergnani [1,2,*]

[1] Department of Medical Sciences and Laboratory for Technologies of Advanced Therapies (LTTA), University of Ferrara, 44121 Ferrara, Italy; bhmsme@unife.it (E.B.); mrcgpl@unife.it (G.M.); prrmsl@unife.it (M.P.); khsstn@unife.it (A.E.K.); mario.dellasala@unife.it (M.D.S.); paolo.pinton@unife.it (P.P.)
[2] Translational Research Center, Maria Cecilia Hospital GVM Care & Research, 48022 Cotignola, Italy
[3] Laboratory of Mitochondrial Biology and Metabolism, Nencki Institute of Experimental Biology of the Polish Academy of Sciences, 02-093 Warsaw, Poland; m.wieckowski@nencki.edu.pl
[4] Department of Radiation Oncology and Nuclear Medicine, AULSS 9 Scaligera, Ospedale Mater Salutis di Legnago, 37045 Verona, Italy; francesco.fiorica@aulss9.veneto.it
* Correspondence: carlotta.giorgi@unife.it (C.G.); simone.patergnani@unife.it (S.P.)

Citation: Bouhamida, E.; Morciano, G.; Perrone, M.; Kahsay, A.E.; Della Sala, M.; Wieckowski, M.R.; Fiorica, F.; Pinton, P.; Giorgi, C.; Patergnani, S. The Interplay of Hypoxia Signaling on Mitochondrial Dysfunction and Inflammation in Cardiovascular Diseases and Cancer: From Molecular Mechanisms to Therapeutic Approaches. *Biology* **2022**, *11*, 300. https://doi.org/10.3390/biology11020300

Academic Editor: Nathan Lanning

Received: 4 January 2022
Accepted: 9 February 2022
Published: 12 February 2022

Simple Summary: The regulation of hypoxia has recently emerged as having a central impact in mitochondrial function and dysfunction in various diseases, including the major disorders threatening worldwide: cardiovascular diseases and cancer. Despite the studies in this matter, its effective role in protection and disease progression even though its direct molecular mechanism in both disorders is still to be elucidated. This review aims to cover the current knowledge about the effect of hypoxia on mitochondrial function and dysfunction, and inflammation, in cardiovascular diseases and cancer, and reports further therapeutic strategies based on the modulation of hypoxic pathways.

Abstract: Cardiovascular diseases (CVDs) and cancer continue to be the primary cause of mortality worldwide and their pathomechanisms are a complex and multifactorial process. Insufficient oxygen availability (hypoxia) plays critical roles in the pathogenesis of both CVDs and cancer diseases, and hypoxia-inducible factor 1 (HIF-1), the main sensor of hypoxia, acts as a central regulator of multiple target genes in the human body. Accumulating evidence demonstrates that mitochondria are the major target of hypoxic injury, the most common source of reactive oxygen species during hypoxia and key elements for inflammation regulation during the development of both CVDs and cancer. Taken together, observations propose that hypoxia, mitochondrial abnormality, oxidative stress, inflammation in CVDs, and cancer are closely linked. Based upon these facts, this review aims to deeply discuss these intimate relationships and to summarize current significant findings corroborating the molecular mechanisms and potential therapies involved in hypoxia and mitochondrial dysfunction in CVDs and cancer.

Keywords: hypoxia; HIF-1α; mitochondria; oxidative stress; inflammation; cardiovascular diseases; cancer; therapeutic target

1. Introduction

Cardiovascular disease (CVD) and cancer are the major issues threatening public health globally and relatively widespread with enhanced morbidity and mortality. New strategies to improve their prevention and treatment are priorities worldwide. Besides, accumulating insights reported risk factors potentially linking both disorders such as environmental factors (obesity, tobacco, sedentary lifestyle, and diet), genetics, cellular, and signaling mechanisms [1]. Furthermore, a growing number of studies demonstrate that the incidence of CVDs risk is higher in patients with cancer disorders [2] and various ancillary processes correlated to CVDs have been identified to have a role in the pathogenesis of cancer [3,4]. Therefore, understanding the mechanism overlapping in these disorders paves the way to improve and elucidate putative therapeutic targets and preventative approaches, ultimately developing emerged cardio-oncology research.

Oxygen (O_2) delivery is an effective prerequisite to ensure the normal function of the cell and is fundamental for a wide range of physiological responses including, cell metabolism, and growth. O_2 serves as an important for aerobic respiration that yields the primary cellular energy, the adenosine triphosphate (ATP) [5]. This process normally occurs at the powerhouse of O_2 consumption in the cell, the mitochondria, mainly via oxidative phosphorylation (OXPHOS) and tricarboxylic acid (TCA). However, when O_2 supply is insufficient to meet cellular energy demand, cells undergo hypoxia and are forced to use anaerobic respiration, which generates less than a tenth of the amount of aerobic respiration. Subsequently, mitochondria are severely affected by hypoxia, they sense the O_2 levels and initiate cellular responses to hypoxia. Along with those lines, mitochondria are considered the key source of hypoxic damage in the human body [6]. Therefore, hypoxia introduces dysfunctional feedback resulting in mitochondrial damage that exacerbates oxidative stress and inflammatory signaling, correlating to mitochondrial metabolism upon hypoxia [7]. It is well known that hypoxia is a hallmark of various diseases. Indeed, at cellular levels, emerging evidence reported the pathophysiological of both CVDs and cancer disorder involve complicated and coordinated signaling pathways triggered during a decline of tissues or cells O_2 stress (hypoxic milieu). We have selected to discuss both CVDs and cancer as major disorders characterized by the involvement of hypoxia in which most importantly has a crucial impact in promoting several processes, such as oxidative stress, inflammation, as well as cell death, and it is considered as one of the key common features with dual roles in both mentioned disorders [8].

In response to hypoxia, each cell displays numerous types of signals at the transcriptional and translational levels, consisting of the activation of a group of genes termed as hypoxic-inducible genes, which are involved in different biological processes, including cellular metabolism (lactate dehydrogenase-A (LDH-A) or pyruvate dehydrogenase kinase isoform 1 (PDK)) [9,10], angiogenesis (vascular endothelial growth factor-A (VGF-A)) [11], erythropoiesis (erythropoietin (EPO)) [12], and inflammation (inducible nitric oxide synthase (iNOS)) [13].

Hypoxic-inducible factor-1 (HIF-1) is a central regulator mediating the cellular response in hypoxic conditions. It is composed of a heterodimer of constitutively expressed subunits, which are the O_2-regulated subunits HIF-α and HIF-β [14,15]. Research of Gregg Semenza's laboratory at Johns Hopkins Medical Institutions led to HIF-1 exploration after discerning hypoxia response element (HRE), putative sequences in the 3′-flanking region of the human EPO gene. Further research found that the transcriptional activation of several regulatory genes is triggered by the binding of a particular protein to the HRE, which is induced by hypoxia. This protein was later identified as HIF-1 [16].

Over the last few decades, multiple pioneers of research on HIF-1α have strikingly revolutionized the comprehension of the O_2 balance. In fact, HIF-1α has drawn much attention in many research fields, as it is outlined as the master O_2 regulator within the cells, a hallmark transcriptional factor in the cellular response to a hypoxic environment, and a primary element for the regulation of several genes during hypoxic condition [17]. It is foreseeable that the disruption in hypoxia-related pathways contributes for several pathological states including CVDs as well as cancer, pointing out the key role of HIF-1 as a breaking point and a major cause specifically between both mentioned disorders [18].

This review aims to cover the actual comprehensions about the effect of hypoxia-mediating signaling pathways on mitochondrial function, and inflammation in key human diseases such as CVDs and cancer. The complexity of these impacts will be examined in the light of recent studies that shall help us to better dissecting the molecular mechanism and designing potential therapeutic approaches against both disorders.

2. Molecular Characteristic and Regulation of HIF-1

In mammalian cells, HIF-1 has been demonstrated to play a fundamental impact in cellular and systemic O_2 homeostasis [19], which mediates adaptation to hypoxia through activation of a multitude of genes encoding proteins needed for improving tissue O_2 homeostasis, energy metabolism, and efficient management of hypoxia-induced toxic stress [20]. HIF-1 is a heterodimeric trans-acting DNA-binding transcription factor that constitutionally comprises expressed subunit HIF-1β (aryl hydrocarbon receptor nuclear translocator, ARNT) and an O_2-sensitive expressed HIF-1α subunit (or its analogues HIF-2α and HIF-3α) a master transcriptional regulator in response to hypoxia and a key modulator for the induction of genes that facilitate adaptation and survival of cells [21]. Both subunits, α and β, exhibit basic helix–loop–helix (bHLH) motifs and belong to the bHLH-Per-ARNT-Sim (PAS) homology protein family. The bHLH domain is a DNA-binding domain that can bind HREs to target specific genes [22,23].

In normal conditions, the HIF-1α subunit is hydroxylated by HIF prolyl-4-hydroxylases at proline 402 and 564 in the O_2-dependent degradation domain (ODDD) of the α-subunits, causing its ubiquitination and proteasomal destruction via the ubiquitin-proteasome (26S) mechanism, which is able to incite constantly the proteasomal degradation. Von Hippel–Lindau (pVHL), which also acts as a tumor suppressor, binds the ubiquitin ligase complex E3, that targets HIF-1α subunit destruction in O_2-dependent degradation domain. Because of this, during normoxia HIF-1α protein has a short half-life [24,25]. In contrast, under hypoxia, the repression of O_2-dependent propyl-hydroxylase-1, -2, and -3 enzyme activity (PHD1, -2, -3) results in suppression of HIF-1α hydroxylation. HIF-1α protein is then stabilized, accumulates in the cytosol, and transferred into the nucleus, where it creates a heterodimer complex with HIF-1β and binds to HREs with a consensus sequence (5′-RCGTG-3′) in promoter or enhancer of target genes to activate a concerted transcriptional response (Figure 1). The nuclear translocation of HIF-1α is not enough to stimulate the target genes transcription [23]. The HIF-1α/HIF-1β (HIF-1) transcription factor recruits various cofactors that are fundamental for full transcription activity, including CREB-binding protein (CBP)/p300 and transcription intermediary factor 2 steroid-receptor activator that ultimately binds to CTAD domain. Another cofactor recognized is the M2 isoform of pyruvate kinase (PKM2), which enhances the binding of the complex HIF-1 to HRE [26].

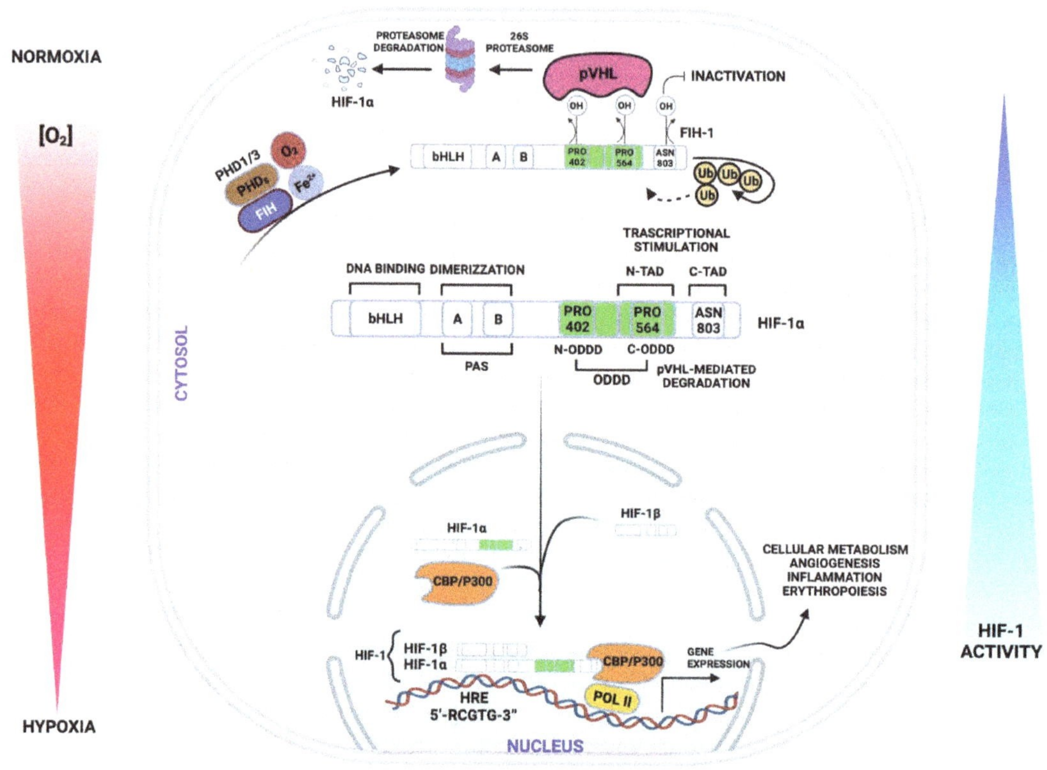

Figure 1. Schematic illustration representing the regulation of Hypoxic-inducible factor-1 α (HIF-1α) protein in response to normoxia and hypoxia. During normoxia, HIF-1α protein is hydroxylated by propyl-hydroxylases (PHDs) and factor inhibiting HIF (FIH). Both oxygen-dependent proteins are stimulated in normal condition and suppress HIF-1α activity. The hydroxylated prolyl residues permit the binding of HIF-1α by the von Hippel–Lindau protein (pVHL), resulting in ubiquitination and ultimate proteasomal destruction. During hypoxia or PHD inhibition, HIF-1α moves to the nucleus, heterodimerizes with HIF-1β, and subsequently binds to hypoxia response element (HRE) in the putative region of target gene to enhance their transcription.

3. Role of Hypoxia Signaling and Mitochondria in Cardiovascular Diseases (CVDs) and Cancer

3.1. Hypoxia Signaling and Mitochondria in CVDs

Hypoxia stimulates multiple processes to adapt to insufficient levels of O_2 in the environment. Therefore, it has mostly negative consequences for cardiovascular functions, ultimately manifesting in pathology. HIF-1α is an O_2-sensitive transcription factor that regulates adaptive metabolic responses to hypoxia and elicits a crucial impact in various CVDs, such as ischemic heart disease (IHD) and heart failure (HF) [27]. Recent insights demonstrated the effect of HIF-1α signaling in the progression of heart disorders [28,29], or its cardioprotective role after I/R in animal model deficient of PHD3 and HL-1 cardiomyocytes [30,31] (Figure 2). Furthermore, HIF-1α overexpression caused mitigation of ischemia/reperfusion (I/R)-enhanced cardiomyocytes loss, suggesting that HIF-1α may drastically influence cardiomyocytes endurance [32]. Recently, studies documented that the HIF signaling pathway not only stimulates disease progression but has a cardioprotective effect as well as the potential for cell recovery from cellular stress in various disorders. This seems to be linked to the duration of hypoxic exposure as well as the stabilization of HIF-1. In fact, chronic exposure to hypoxia is found to increase ischemic ventricular arrhythmias and further cell death [33]. While intermittent exposure to hypoxia reduces arrhythmia during I/R, as also stimulates protective effects against myocardial infarction in rodents [34–37]. The prolonged HIF-1α upregulation was shown to promote dilated cardiomyopathy in transgenic mice with PHD2 depletion [38]. HIF-1α can also influence mitochondrial function and alleviate the severity of ischemic heart. HIF-1α can also influence mitochondrial function and alleviate the severity of ischemic heart. HIF-1α can upregulate mitophagy, mitochondrial autophagy, in cardiac cells through HIF-1α/(BNIP3) BCL2 and adenovirus E1B 19-kD-interacting protein 3 pathways, thereby stimulating their survival following myocardial ischemia-reperfusion. This is only applied to the role of HIF-1α-mediated mitophagy at an early phase of ischemia, which may result in cardiac protection, while prolonged autophagy may activate cell death in H9C2 cardiomyoblasts and Sprague Dawley rat models [39]. In contrast, other studies revealed that HIF-1 activation enhances BNIP3 expression, resulting in (H9C2) cardiomyocyte death, which is a hallmark of ischemia and HF [40,41] (Table 1). Major interplays have been identified between hypoxia-mediated mitochondrial function and mitophagy in cardiomyocytes. In brief, upregulation of the inner mitochondrial membrane protein (IMM), Optic atrophy 1 (Opa1) stimulates mitophagy and mitochondrial function in response to hypoxia in mouse cardiomyocytes [42]. FUN14 Domain Containing 1 (FUNDC1) is an OMM protein that accumulates on the mitochondrial associated membranes (MAMs). Several recent studies reported its effective role to mediate mitophagy during ischemic conditions in cardiomyocytes, and it thus conveys cardioprotection [43]. Although the cardioprotective effect of mitophagy in the ischemic heart is widely demonstrated, during the reperfusion stage, mitophagy has a defective impact on cardiac function, and this may be due to the repression of FUNDC1-dependent mitophagy and necrosis upregulation [44]. The contribution of FUNDC1 in response to hypoxia may provide new insight in favor of therapeutic target approaches in CVDs, and further research focused on the FUNDC1-HIF-1 axis may be beneficial. Furthermore, a novel protein, WD Repeat Domain 26 (WDR26), has been detected to localize into the mitochondria, promoting mitophagy in H9C2 cells during hypoxia, suggesting its pivotal effect in hypoxia-enhanced mitophagy [45]. In another line of evidence, HIF-1α accumulation directs mitophagy and promotes the differentiation of H9C2 cells [46]. Therefore, HIF-1 modulates hundreds of genes in diverse biological pathways, and most of them influence mitochondrial function. HIF-1 activates mitochondria-specific genes crucial to a metabolic shift away from OXPHOS to glycolysis, including LDH-A, phosphoglycerate kinase-1 (PGK1), to adapt to hypoxic stress [47]. HIF-1 elevates glycolysis by upregulating glycolysis enzyme production, increasing glucose transporters, and repressing the mitochondrial energy metabolism [18]. Moreover, HIF-1α promotes PDK-1 activation, which phosphorylates and inhibits pyruvate dehydrogenase (PDH), from converting pyruvate to acetyl CoA to fuel the mitochondrial TCA cycle and preventing the formation of iron-sulfur (Fe/S) clusters, thereby attenuating complex I activity [48]. It also inhibits the expression of mitochondrial encoded subunits in OXPHOS complexes by blocking the nuclear–mitochondrial interaction [49]. Recent pieces of evidence have also reported HIF-1α in improving mitochondrial function, reducing cellular oxidative stress, and stimulating the cardioprotection [39].

Furthermore, HIF-1α improves mitochondrial respiratory function by triggering various cardioprotective signaling pathways, including the phosphoinositide-3-kinase/Akt (PI3K/AKT) and Janus kinase (JAK) 2/signal transducer and activator of transcription (STAT) 3, to protect the heart during I/R injury [50]. Indeed, Nanayakkara and colleagues reported the transcriptional role of HIF-1α during hypoxia in regulating frataxin expression levels, a highly conserved nuclear-encoded mitochondrial protein, expressed in tissues such as the heart, neurons, and liver with a high metabolic rate [51], which served as a cardioprotective element against ischemic injury. Ultimately, enhanced frataxin levels can alleviate mitochondrial iron overload, thereby preserving mitochondrial membrane integrity and the cardiomyocyte's viability [52]. HIF-1α stabilization permits cells and tissues to adapt to the hypoxic response in I/R, thus protecting cardiomyocytes against IHD and improving patient prognosis [20]. On the other hand, several lines of evidence outlined the impact of HIF-1α localization dependent on the mitochondrial function regulation, which has been thought to be contingent on its trafficking to the nucleus. Nevertheless, studies documented that HIF-1α not only localizes to the nucleus after exposure to hypoxia or preconditioning but surprisingly it localizes also to the mitochondria [53,54]. For instance, Mylonis and collaborators reported that HIF-1α at the outer mitochondrial membrane (OMM) attenuates hypoxia-induced apoptosis [55]. In alignment with this regard, during the elevation of oxidative stress, HIF-1α translocates to the mitochondria to reduce mitochondrial reactive oxygen species (mtROS) in the human umbilical vein endothelial (HUVEC) during hypoxia [56].

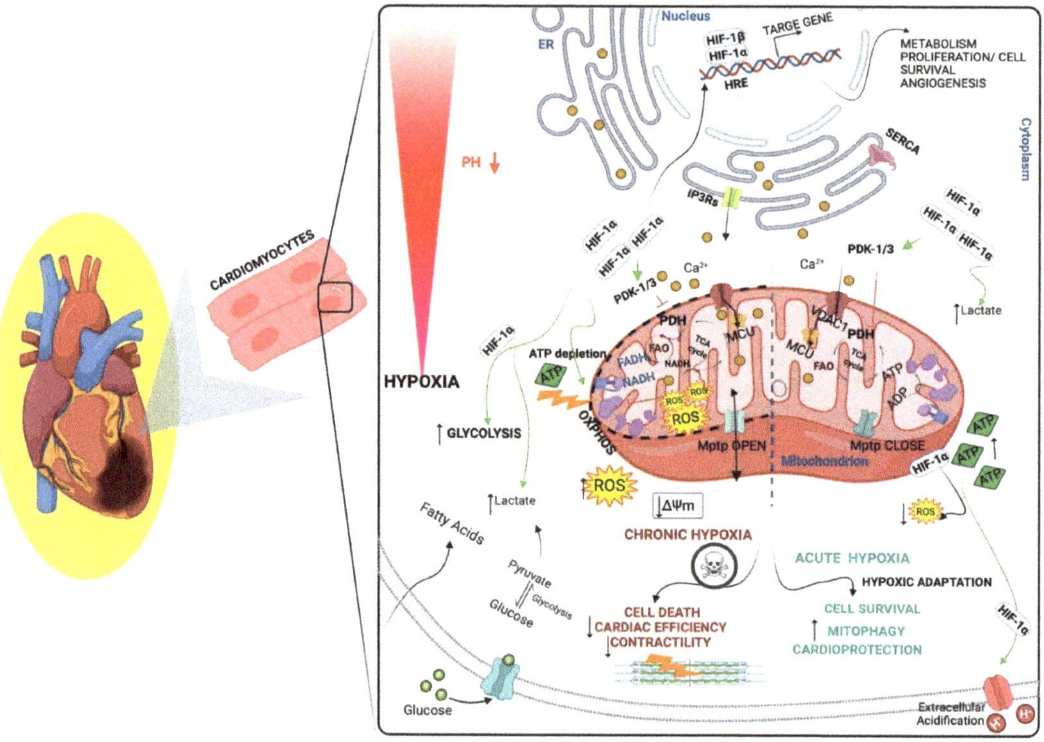

Figure 2. Hypoxic-inducible factor-1 α (HIF-1α) signaling and mitochondria in cardiovascular diseases (CVDs). Sudden decrease in oxygen (O_2) levels results in abrupt biochemical and metabolic changes. Hypoxia causes the accumulation of HIF-1α that moves to the nucleus to activate genes crucial to a metabolic switch away from the mitochondrial oxidative phosphorylation system (OXPHOS) to glycolysis, the cardiomyocyte relies on anaerobic respiration instead of oxidative phosphorylation, which in turn causes disruption of the mitochondrial membrane potential ($\Delta\Psi_m$) and adenosine triphosphate (ATP) depletion, affecting the mitochondrial Permeability Transition Pore (mPTP) opening and subsequently inhibiting contractile function. Hypoxia triggers a switch in cellular metabolism to anaerobic glycolysis, causing acidification of the cell as protons (H^+) accumulates. Cardiomyocyte damage and mitochondrial deficiency are relatively linked to the degree of hypoxia exposure and due to the dual effect of HIF-1α; in acute hypoxia (right), HIF-1α acts as cardioprotective against oxidative damage by alleviating ROS generation and stimulating the removal of unwanted mitochondria through mitophagy. While (left) HIF-1α enhances ROS levels and increases cell death, ultimately, reduces cardiac efficiency and contractility.

Table 1. Representative list summarizing the effect of HIF-1α in cardiovascular diseases. HIF-1α^{tg} mice: HIF-1α^{tg} transgenic mice models. CH rodent models: chronic hypoxic rodent models. LPS: lipopolysaccharide.

Cardiovascular Disorders	In Vivo/In Vitro	Animal Models	Cell Lines	HIF-1α Effect	References
Ischemia-reperfusion injury (I/R injury)	In vitro	-	Rat neonatal ventricular cardiomyocytes cells	Cardioprotective effect, overexpression of HIF-1α elevated target genes (iNOS, VEGF, HSP70, and GLUT1-4)	[32]
	In Vivo/in vitro	PHD3$^{-/-}$ mice	HL-1 cardiomyocytes	Cardioprotection, PHD3 deletion increased HIF-1α, resulted in cardiomyocytes death suppression	[30,31]
	In Vivo/in vitro	Sprague Dawley (SD)/rat model	H9C2 cardiomyoblasts	Cardioprotection, BNIP3-mediated autophagy modulation	[39]
Myocardial infraction (MI)	In Vivo	Post-MI mice	-	Cardioprotection, upregulated angiogenesis	[57]
	In vivo/in vitro	MI-mice	Rat neonatal cardiomyocytes	Detrimental, stimulated apoptosis through p53 following MI	[58]
Heart failure (HF)	In vivo	HIF-1α^{tg}mice	-	Detrimental, prolonged HIF-1α accumulation increased disease development	[28]
Myocarditis	In vitro	-	H9C2 cardiomyoblasts	Detrimental, repression of HIF-1α improved cardiomyocytes at odds with LPS-stimulated cell death	[29]
Dilated cardiomyopathy	In vivo	PHD2$^{-/-}$ mice	-	Detrimental, prolonged HIF-1α upregulation promoted dilated cardiomyopathy	[38]
Cyanotic congenital heart disease (CCHD)	In Vivo	CH rodent models	-	Cardioprotection, HIF-1α overexpression alleviated maladapted metabolic	[59]
Cardiac hypertrophy	In Vivo	HIF-1α KO mice	-	Cardioprotection by controlling negatively TGF-β	[60]

3.2. *Hypoxia Signaling and Mitochondria in Cancer*

Multicellular organisms have adopted several mechanisms to rapidly adjust to hypoxia, prolonging survival in the absence of adequate resource [61]. A common observation of most tumors is an insufficient amount of O_2, the severity of which varies between tumor types. In proliferating and expanding tumor tissue, the adaptation of tumor cells to exhausted oxygen supply is mainly mediated by HIF-1. Such metabolic adjustment is pivotal for cancer cell survival and proliferation in response to environmental stimuli [62].

In this section, we will discuss how hypoxia affects the mitochondrial function through TCA citric acid cycle, electron transport chain, and its dual role of hypoxia in ROS production and mitigation and finally briefly consider hypoxia-induced mitochondrial distribution and morphology in cancer.

3.2.1. Hypoxia-Induced Modulation of Krebs Cycle and Oxidative Respiration

TCA cycle represents the metabolic engine within cells. It is found to be inhibited by HIF-1α through the induction of the enzyme PDK1. PDK1 inactivates the TCA cycle enzymes by phosphorylating PDH, which converts pyruvate to acetyl-CoA. PDK1 overexpression in HIF-1α-silenced cells led to an increased ATP production, reduced ROS generation and prevented the hypoxia-mediated apoptosis [9]. Indeed, the accumulation of 2-hydroxyglutarate (2-HG), an essential epigenetic regulator in cancer cells, has been documented to enhance the stabilization of HIF-1α [63]. Importantly, HIF-1α was found to act as a metabolic switch from glycolysis to OXPHOS for regulatory T cells glioblastoma. Specific ablation of HIF-1α in regulatory T cells resulted in enhanced pyruvate import into mitochondria [64]. Taken together, TCA cycle metabolites resulted to be affected by hypoxia to limit substrate availability for phosphorylation and epigenetic modifications to change cell function and fate.

In addition, the effect of HIF-1α-mediated metabolic reprograming also alters OXPHOS respiration [65]. Evidently, MCF-7 carcinomas cells exposed to 24h hypoxia showed reduced OXPHOS flux and decrease 2-OG dehydrogenase as well as glutaminase activities, without functional alteration of respiratory complexes I and IV [66] (Table 2). Other possible ways by which hypoxic cells reduce oxidative metabolism involve small RNAs. HIFs increased transcription of genes encoding microRNAs (miRs), small RNAs that link to mRNAs in sequential mode to either suppress their translation or promote their degeneration. Among the list of targets of miR-210, there are the iron-sulfur cluster enzyme (ISCU) genes, required for mitochondrial complex I function, which are particularly found downregulated under hypoxia [67]. Moreover, a bioinformatics survey and PCR real-time experiments demonstrated the involvement of both nuclear respiratory factors (NRF-1) and HIF-1α in modulating voltage-dependent anion-selective channel 1 (VDAC1) promoter during nutrients deprivation or hypoxia [68]. Taken together, cancer cells activate a metabolic switch that involve HIF-1α and impaired Krebs cycle and OXPHOS function to sustain glycolysis metabolism.

Table 2. Representative list summarizing the effect of HIF-1α on mitochondria in cancer.

Cancer Disorders	In Vivo/In Vitro	Animal Models	Cell Lines	HIF-1α Effect	References
Human Breast ductal carcinoma	In vitro	-	MCF-7 cells	Inhibited ER Estrogen receptor expression	[62]
Renal carcinoma	In vitro	-	RCC4 and RCC10	Increased mitochondrial biogenesis	[69]
Hepatocellular carcinoma (HCC)	In vivo	Mice and Diethylnitrosamine model of Murine HCC	HCC cell lines	Promoted mitochondrial biogenesis and reduced ATP	[70]
Triple negative breast cancer (TNBC)	In vitro/in vivo	Nude mice	MDA-MB-231	Enhanced mitochondrial OXPHOS and elevated ROS generation	[71]
	In vitro	-	MDA-MB-231 and SUM-149 cells	Increased intracellular glutathione levels	[72]
	In vivo	MMTV-PyMT mice	Primary MECs	Regulated mitochondrial mass	[73,74]
Colorectal cancer	In vitro/in vivo	Oma1$^{-/-}$ mice	HCT116 cells	Increased mitochondrial ROS	[75]
Several human cancers	In vitro	-	A549, CCL39, HeLa, LS174, MCF7, PC3, ORL33, SKMel, and 786-O cells	Enlarged mitochondrial phenotype	[76]
Glioblastoma	In vitro/in vivo	Foxp3-$^{YFP\text{-}CRE}$ × HIF-1α $^{-fl/fl}$ mice	Murine glioma GL-261	Promoted fatty acids oxidation for mitochondrial metabolism	[64]
Cancer cachexia (CC)	In Vitro/in vivo	C26 mice model	Colon-26 (C26) adenocarcinoma	Affected the metabolic changes	[77]
Oral cancer	In vitro	-	Oral squamous cell carcinoma (OSCC)	Stimulated migration and invasion in the indicated cells	[78]

3.2.2. Hypoxia-Induced Mitochondrial ROS Production and Suppression

ROS are actively generated in the form of superoxide and hydroxyl free radicals as a by-product of OXPHOS and neutralized by antioxidant mechanism to ensure a proper cellular function [79]. Indeed, in breast cancer cell line (MDA-MB-231), the inhibition of HIF-1α expression using cardamonin repressed the mechanistic target of rapamycin (MTOR) pathway causing increased OXPHOS activity and enhanced ROS, which finally led to apoptosis [71]. Similarly, colorectal cancer cells under hypoxic conditions are characterized by activation of OMA1-OPA1 axis, which in turn increase the mtROS generation to stabilize HIF-1α, thereby promoting the glycolytic metabolism [75]. There is also evidence that mitigating ROS-mediated damage in hypoxia may promote the Warburg effect, an altered metabolism favoring enhanced uptake and use of glucose. Warburg effect has been also involved during the pathogenesis of cancer cachexia (CC), which is a complex pathological condition with metabolism dysregulation, affecting primarily the skeletal muscle [80]. The impaired mitochondrial homeostasis and metabolism has also been documented in various models of CC. Additionally, it has been suggested that HIF-1α may have an impact on the metabolic alteration in cancer cachexia (CC) [77]. Consistently, suppressing HIF-1α expression by Rhein and Emodin compounds mitigates cancer cell proliferation and CC in a dose- and time-dependent mode [81,82].

This indicates that the alleviation of ROS may promote the antioxidant system. Particularly, stimulation of AMP-activated protein kinase (AMPK) by mitochondrial ROS led to peroxisome proliferator-activated receptor gamma coactivator 1-alpha (PGC-1α)-dependent antioxidant response. Mechanically, AMPK-PGC-1α induced regulation of mitochondrial ROS mediates the HIF-1α stabilization [83]. Interestingly, opposite to earlier experimental observations, downregulation of superoxide dismutases (SOD2) expression has also been observed under hypoxia in a HIF-1 dependent manner in renal carcinoma cells [84]. Similarly, HIF-1 increased glutathione levels by upregulating glutamate cysteine ligase (GCLM) in breast cancer cells (MDA-MB-231 and SUM-149 cells) [72].

3.2.3. Hypoxia-Induced Mitochondrial Distribution and Dynamics

The mitochondrial distribution and dynamics are permitted by a correct balance between two opposite processes, that are the mitochondrial biogenesis and the mitochondrial fission. In addition to this, mitochondria removal mechanisms which permit the elimination of damaged and excessed mitochondria guarantees the preservation of a healthy mitochondrial pool. Among the diverse mechanisms regulating this mitochondrial turnover, the selective form of autophagy pathway, known as mitophagy [85], represents the primary line of intervention. Dysfunctions in such mitochondrial-related pathways may be cause and consequence of several disorders [86–89] and can be also related to a hypoxic condition. For example, it has been observed that cancer cells activate mitochondrial fusion to enlarge their mitochondrial population to evade from cell death stimuli. Consistently, inhibition of the mitochondrial fusion protein mitofusin-1 (MFN-1) reestablish a tubular network [76]. Other factors appear to influence the hypoxia-mediated fission of mitochondrial membranes. This is the case of the mitochondrial scaffolding protein A-kinase anchor protein 1 (AKAP121). During hypoxia, AKAP121 regulates mitochondrial dynamics through inhibition of phosphorylation of the main regulators of mitochondrial fusion events: the dynamin-related protein 1 (DRP1) and the fission 1 protein (FIS1). Interestingly, this work also accounted for the ubiquitin ligase SIAH2 a crucial role in regulation of hypoxia-mediated mitochondrial fission. Indeed, it is well documented that AKAP121 is a SIAH2 substrate and, of consequence, the availability of AKAP121 as well as the associated mitochondrial dynamics under hypoxia are highly dependent of SIAH2. Consistently, SIAH2 knockout (KO) cells displayed higher AKAP121 levels [90]. In addition, the mitochondrial biogenesis has important roles in cancer cells to adapt hypoxia. Microarray studies show that reconstitution of pVHL in renal carcinoma led to elevated mitochondrial mass, mitochondrial complexes activity and O_2 consumption rates. Mechanistic insights show MAX-interactor 1 (MXI1) expression reduces C-MYC-dependent expression of PGC-1α, which in turn inactivate mitochondrial biogenesis [69]. Similarly, ablation of high mobility group box 1 (HMGB1) protein in hepatocellular carcinoma subjected to hypoxia promoted mitochondrial biogenesis and reduced ATP. Mechanistic experiments suggest that the binding of HMGB1 to Toll-like receptor-9 in cytoplasm stimulate p38 which in turn led to phosphorylation of PGC-1α, which resulted in subsequent increase mitochondrial biogenesis [70]. There are several reasons given why tumor hypoxia maintain increased mitochondrial biogenesis. Thanks to this condition cancer cells result protected from cell damage induced by ROS, thereby providing a survival mechanism. Another benefit is to enhance their migratory and invasive properties to other zones with less hypoxia [91].

As reported above, central to keeping healthy mitochondrial population and guarantee and efficient turnover of the mitochondrial population is mitophagy [92]. The exact role of mitophagy in cancer remains unresolved, as it has divergent roles, either blocking or stimulating the cancer progression [93]. Interestingly, growing consensus documented an intimate relationship between mitophagy and hypoxia in regulate the tumor growth. Indeed, the mitophagy proteins BNIP3L/NIX, BNIP3, and FUNDC1 were found upregulated in response to hypoxia in cancer [94–96]. Furthermore, BNIP3 and NIX are target genes of HIF-1α and their upregulation stimulates mitophagy under hypoxia and inhibits cancer progression in MMTV-PyMT model under hypoxia through activated HIF [73,74]. In addition to this, the protein E2F3d, recently stated to localize to the OMM, revealed to augment the process of mitophagy during hypoxic exposure in cancer cells [97].

During hypoxia, HIF-1α regulates glucose, glutamine metabolisms, and lipid through various target genes, by which it enhances glycolysis by upregulating glucose transporters to induce the glycolysis flux and targeting PDK-1/3 to inhibit pyruvate changeover to acetyl CoA and stimulating LDH-A to turn the pyruvate to lactate. HIF-1α upregulates glutamine utilization and activates fatty acids for rapid cell growth and division. Moreover, when O_2 supply is limited, tumor cells are protected against excessive ROS and subsequently they resist to cell death.

4. Mitochondrial Dysfunction and Inflammation in CVDs and Cancer

4.1. Oxidative Stress and Mitochondria

Under physiologic conditions, other organelles in addition to mitochondria have the capacity to produce ROS such as peroxisome and endoplasmic reticulum (ER) [98]. Despite this, several lines of evidence ascribe to mtROS a predominant role in CVDs as well as in cancer (as reviewed in [99–101]). In the mitochondrion, ROS are produced by ETC in the IMM at the level of complex I and III during oxidative metabolism and cellular response to cytokines and bacterial invasion. Moreover, several enzymes here located or translocated following stimuli are considered additional sources of mtROS [102], and they include several flavoproteins (acyl-CoA dehydrogenase, glycerol α-phosphate dehydrogenase, α-ketoglutarate dehydrogenase) [103–105], the monoamine oxidases (MAOs) on OMM that produces ROS in the catabolism of neurotransmitters [106], aconitase [107], the ROS-generating NADPH oxidases (NOXs) [108], and p66shc in the intermembrane space (IMS) which oxidizes cytochrome c [109,110]. All these enzymes share the release of discrete amounts of superoxide (O_2^-) and hydrogen peroxide (H_2O_2) which have been linked to a consequent mitochondrial swelling and apoptosis [109]. Moreover, the O_2^- produced in these steps is able to react also with nitric oxide (NO) giving rise to reactive nitrogen species (RNS) production and consequent nitrosative stress, which involves mitochondria because affects their enzymatic activity, modifies mitochondrial respiration and increases mitochondria-mediated cell death [111].

Uncontrolled ROS and RNS production, generate oxidative stress with consequent mtDNA damage and oxidation of proteins and lipids of membranes. In the first case, ROS and RNS directly react with pyrimidine and purine bases, especially in the D-loop region [112]; here, this damage led to a significant decrease in mtDNA copies and of ETC function [113]. Otherwise, peroxidation of mitochondrial lipids and proteins alters the mitochondrial membrane potential (MMP), their energy production and triggers the opening of mPTP [105,109,110]. Subsequently, mitochondria result impaired, and several cell death pathways are activated.

mtROS scavenging systems exist and they represent the first line of defense against toxic ROS levels. Superoxide dismutase (SOD) is responsible for the conversion of O_{2-} in H_2O_2. SOD exists in three isoforms and, among them, SOD2 has a mitochondrial matrix localization, while SOD1 (Cu/Zn SOD) may be partially found in the intermembrane space. A second line of defense is covered by catalase, which splits in O_2 and water [114]. The action of this enzyme is flanked by the glutathione peroxidase/reductase (GSH-PX) and the peroxiredoxin/thioredoxin (PRX/Trx) systems which taking advantage from the reduced forms of GSH and PRX, they convert H_2O_2 in water. PRX3 and PRX5, as well as Trx2 are localized to mitochondria.

4.2. Inflammation and Mitochondria

It is currently recognized that ROS induces inflammation [115,116] and, in turn, inflammation further sustains mitochondrial dysfunction [117]. Mitochondria are organelles involved in the inflammatory process not only to produce ROS, but because in particular conditions (i.e., the opening of the mPTP or the permeabilization of the OMM) [118,119], they release into the cytosol several factors named as DAMPs (danger-associated molecular patterns). To date, many substances are recognized as mtDAMPs, and they include mtDNA [120], ATP [121], TFAM [122], NFP [123], succinate [124], and cardiolipin [125]. These molecules, once in the cytosol and sometimes in the extracellular milieu, are recognized by other molecules (adaptors or receptors) to trigger inflammation. There are different pools of receptors. Those recognizing mtDAMPs, which link mitochondria to inflammation, are the cytoplasmic nucleotide-binding domain leucine-rich repeat-containing receptors (NLRs). The most common and well-defined is the NLR and pyrin containing protein 3 (NLRP3). It exists as a multiprotein complex composed of the scaffold protein NLRP3. Normally localized in the cytosol, NLRP3 moves to the mitochondria and MAMs upon activation, and it regulates the innate immunity by recruiting the apoptosis-associated speck-like protein (ASC; an adaptor) and procaspase-1, which becomes matured in caspase-1 and is responsible for the activation of IL-1β and IL-18 [115]. The contribution of mitochondria in NLRP3-mediated inflammation is demonstrated by their recruitment in the cytoplasm of macrophages in a process mediated by microtubules [126] and the observation of NLRP3 relocalization to mitochondria following stressors. Indeed, under resting conditions, NLRP3 is normally localized at the ER, but after stimulation with mtDAMPs, NLRP3 clusters at mitochondria and at MAMs [124]. Thus, inflammation and mitochondria are directly related and together with ROS contribute to both CVDs and cancer.

4.3. Inflammation, Oxidative Stress, and Mitochondrial Dysfunction following Hypoxia in CVDs

Inflammation and oxidative-nitrosative stress are two main contributing causes in the onset and progression of CVDs [127]. In addition being crucial in the excitation-contraction coupling of the heart [128], mitochondria are at the crossroad between both routes, as these are often accompanied by an alteration of mitochondrial function. Inflammation has a key role in CVDs, as an example, atherosclerosis, plaque and vessel calcification, and post-ischemic pathologies. Atherosclerosis born from cholesterol, fat, and other substances deposition on the inner side of vessels. These deposits build up into the vessel in concomitance of endothelial dysfunction, the starting artery lesion, to lead to the plaque formation. Direct evidence of the regulation of lipid metabolism by hypoxia and HIF-a still lack in CVDs. However, it has been suggested that hypoxia-induced HIF-1α accumulation upregulates the expression of its target genes and affects the lipid metabolism of hypoxic macrophages in atherosclerosis, thereby providing evidence of a possible atherogenic role for hypoxia [129]. Lipids can be also modified by further oxidation from newly generated mtROS, and adhesion molecules (ICAM and VCAM) are activated to trigger the binding of inflammatory cells, such as monocytes. Often, cholesterol deposition represents the second hit in NLRP3 assembly for the production of interleukins and inflammation progress [130]. Inflammation is also responsible for the calcification of the plaque; indeed, activated macrophages promote the apoptosis of interstitial cells by releasing proinflammatory cytokines and promoting the release of vesicles reached in calcium (Ca^{2+}) and phosphates [131,132]. Similarly, an increase of aged mitochondria, reduced biogenesis, impaired mitophagy, and dysregulated cytosolic Ca^{2+} homeostasis are all associated with mitochondrial dysfunction [76,129,130].

On the other hand, mitochondrial oxidative stress has been found closely related to several risk factors for CVDs, and from studies in vitro and in vivo it has been found to be one of the main mediators of apoptosis in cardiomyocytes and endothelial dysfunctions [101,133]. The contribution of each of the above-mentioned ROS-producing enzymes in CVDs has been analyzed overtime in experimental settings including genetic manipulation of animal models. As an example, p66SHC is considered an important modulator of mitochondria-mediated cell death as it regulates MMP and ROS when localized to the mitochondria [134]. Accordingly, animals lacking p66SHC display reduced ROS and decreased cell death in a model of hypertension and HF [135]. Elevated levels of angiotensin II upregulate redox-sensitive pathways causing mtROS overproduction, cardiomyocyte damage, and finally hypertension and HF. The p66SHC$^{-/-}$ phenotype benefits in cardiac function are not only against the previously mentioned pathologies but also in a model of diabetic cardiomyopathy [136]. Likewise, the inhibition or deletion of isoform A of MAOs in murine models decreases mtROS production, preserves the heart against pressure overload, caspase 3 activation, and fetal gene reprogramming [137]. In addition, MAO-B targeting in a mouse model of induced pressure overload minimizes the worsening of the cardiac function and reduces death by preserving MMP and mitochondrial bioenergetics [138]. NOX isoforms, NOX4, constitute the main source of oxidative stress during HF [139]. In support of this, its levels are significantly elevated following chronic pressure overload. Consequently, oxidative stress, mitochondrial dysfunctions, and apoptosis are activated. Mice lacking NOX4 have lower left ventricular (LV) dysfunction with reduced oxidative stress and cardiomyocyte apoptosis [140].

The dysregulation of the antioxidant systems is the second aspect that demonstrates how mitochondrial oxidative stress is deleterious in heart disease. Studies from SOD2$^{-/-}$ mouse model reported either premature lethality [141] or deaths before 4 months due to HF. Indeed, the absence of this enzyme triggers excessive oxidative stress in mitochondria with the overproduction of 4-hydroxynonenal (4-HNE), produced from lipid peroxidation and a player in mitochondrial dysfunction, by acting via the severe impairment of the ETC [142].

The deriving phenotype consists of abnormalities in mitochondrial structure, decreased ejection fraction, dilated, and dysfunctional LV, all major phenotypes of dilated cardiomyopathy [142]. Deleterious effects are not limited to the complete deletion of SOD2. In a study in humans, the single Ala16Val SOD2 polymorphism has been investigated as associated with coronary heart disease (CHD) risk, in concomitance with decreased SOD2 activity and increased mtROS production [143]. In addition, endothelial dysfunction is increased when SOD2 is absent [144]. In the end, the inappropriate activity of GSH-PX and PRX/Trx systems enhance oxidative stress, which is the main cause of left ventricle (LV) contractile dysfunction and abnormalities in LV mass, under the action of angiotensin II [145]. If we consider these genotypes following I/R, they produce larger infarct size, alteration of cardiac contractility, cell death, and adverse events such as HF and dilated cardiomyopathies [146,147].

Overall, in understanding the importance of all these compensatory mechanisms in humans, an extensive meta-analysis on thousands of papers published and as many patients analyzed, reported an inverse and significant correlation between antioxidant systems activity (and circulating levels of enzymes involved in) and the most common risk factors of CHD [148].

Hypoxia is a bio-clinical condition in which either portion or whole tissues are subjected to a significant imbalance between O_2 consumption and blood perfusion [149]. It is often associated with nutrient deprivation. In normal conditions, the cardiovascular system is appointed to guarantee these functions in the human body. However, coronary artery disease (CAD) (i.e., atherosclerosis) can arise with aging and become responsible for a progressively decrease in the blood perfusion. As a consequence, CAD may result in ischemic heart disease (IHD), such as myocardial infarction (MI) and peripheral arterial diseases (PAD), in which hypoxia plays a predominant role. In addition, being that hypoxia is a condition that induces changes in the cardiovascular system, its chronic presence can also lead to hypertension and HF [150]. Thus, within certain limits and with different degrees, all CVDs mentioned in the previous paragraphs are linked to hypoxia. In normoxia, HIF-1α binds the pVHL for degradation by the proteosome; under hypoxic states, HIF-1α accumulates and translocates mainly to the nucleus where it dimerizes with beta subunit. Being a transcription factor, in the nucleus it triggers the expression of several genes involved in either adaptation or maladaptation to hypoxia [151]. In this context, whether cells and tissues find a means to adapt in the absence of O_2 levels, the reperfusion phase (if any) emphasizes the hypoxia-induced damages even more. Examples of adaptations induced by hypoxia are the HIF-1α and VEGF-mediated angiogenesis in post-MI hearts to ensure a compensated O_2 delivery in the infarcted tissue [57], and the downregulation of mitochondrial O_2 consumption through the activity of the PDK1, which is the enzyme that limits the use of pyruvate at mitochondrial level [48]. Here, the O_2 available for mitochondria increases and consequently cell death is reduced. Again, HIF-1α overexpression is important for mitochondrial metabolic adaptation to a persistent state of hypoxia in the heart, which distinguishes patients with cyanotic congenital heart disease (CCHD) [152]. Moreover, HIF-1α KO mice in which pressure overload is induced, immediately present severe traits of hypertrophy when compared to their wild-type littermates [60]. Note that it is no coincidence if hypoxic conditioning is considered as a non-pharmacological therapeutic intervention in the adaptation of the body against severe O_2 deprivation episodes [153]. However, chronic or sustained hypoxia states and all associated prosurvival pathways are better associated to cancer. In full-blown IHD, mechanical methods and pharmacological treatments are usually suddenly applied to solve the ischemic phase, also known as the cause of the necrotic core of the lesion. One of them is the reperfusion of the occluded vessels.

Mitochondria are the major consumers of O_2, and as consequence, they suffer from its absence. Whether in the normoxic condition the cell respire in the aerobic mode and produce big amount of ATP, under hypoxia ETC is inhibited in favor of glycolysis which is upregulated with consequent acidification of the environment, sodium and Ca^{2+} overload and breakdown of ATP production [154,155]. In the reperfusion phase, the reactivation of the ETC and the restoration of the MMP prompt further Ca^{2+} intake and a burst of mtROS production. These two events increase the susceptibility of mPTP to open [156]. Ca^{2+} enters the organelles by the mitochondrial calcium uniporter (MCU) complex, a highly selective channel in the IMM, ensuring Ca^{2+} uptake [157]. Indeed, cardiac-specific knockdown of MCU increases resistance against mPTP opening and reduces infarct size following ischemia [158,159]. On the other hand, most mtROS derive from an accurate and selective metabolic process which has been investigated by Chouchani and co-workers in 2014. By analyzing several ischemic and re-perfused tissues (i.e., brain, kidney, heart), they identified a mitochondrial metabolite that increased in concentration 19-fold during hypoxia: succinate. It accumulated proportionally to the time of the ischemic phase, and it was rapidly metabolized at reperfusion time, just when ROS increased. In this work, they supported the hypothesis that succinate comes from the reduction of fumarate by the reversal mode of succinate dehydrogenase (SDH) action. Indeed, fumarate also increases in ischemia through the activation of malate/aspartate shuttle and purine nucleotide cycle [160]. Succinate at reperfusion would produce mtROS by acting on complex III and inducing the reverse mode of complex I; in addition, it behaves like a DAMP to sustain inflammation. Thus, both Ca^{2+} and ROS constitute a second hit for the mPTP opening, following the ischemic priming phase. mPTP gives rise to and is accompanied by most mitochondrial dysfunctions that occur following hypoxia, including the mitochondrial permeability transition, the collapse of the MMP, and loss of cristae morphology and several proteins into the cytosol that trigger cell death and/or inflammation. In turn, mPTP opening stimulates mtROS production and all together facilitate the disassembly of super complexes of the ETC (i.e., complexes I+III+IV) further increasing ROS generation [161].

There is evidence of a strong upregulation of NLRP3-dependent inflammation following hypoxia both in cardiac fibroblasts and cardiomyocytes, which seems to have different but complementary roles. In cardiac fibroblasts, NLRP3 upregulation would sustain inflammation by IL−1β release and contribute to cardiac remodeling; in cardiomyocytes it induces pyroptosis by the solely activation of mature caspase 1 [162,163].

4.4. Hypoxia-Mediating Signaling Pathways and Cell Death in CVDs

Programmed cell death such as apoptosis, necrosis, and ferroptosis occur in cardiomyocytes and they are considered as central features in the pathogenesis of CVDs. In addition to the studies mentioned above on hypoxia-related oxidative stress and inflammation, several lines of evidence reported the involvement of hypoxia in mediating programmed cell death "apoptosis" in cardiomyocytes. Increased expression of HIF-1α upregulated the apoptotic effect in cardiac (H9C2) and renal ischemia (HK2) [164]. Furthermore, prolonged HIF-1α expression upregulates the activity of p53 tumor suppressor, and consequently, stimulates cardiomyocytes apoptosis following MI [58]. HIF-1α–mediated apoptosis may also have a central effect on HF. Indeed, as documented in some lines of evidence, a high level of HIF-1α expression was identified in the late phase of HF [28,165]. Notably, the induction of apoptosis in CVDs upon the HIF-1α axis is controversial, in which HIF-1α exerts a dual role in the heart during hypoxic exposure. In one hand, it acts as a cardioprotective through its stimulation to several genes and pathways to adjust to hypoxia. On the other, it stimulates cardiomyocyte's damage through the activation of various cell death pathways [166]. Hypoxia-induced HIF-1α accumulation blocks the apoptotic process by stimulating angiogenesis and declining fibrosis [167,168]. Along with that, another mechanism in which HIF-1α prevents cardiomyocytes loss is by giving rise to iNOS and cardiotrophin-1 (CT-1) expression [169]. HIF-1α exhibits a protective effect mostly during I/R injury [170] (Table 1). Similarly, during ischemic postconditioning, the increased level of HIF-1α expression improves MI and reduces cardiac damage [171,172]. It is not surprising that HIF-1α overexpression could alleviate apoptosis through the NFκB pathway under hypoxia in MI [173]. Consistently, other lines of evidence confirmed the cardioprotective effect of HIF-1 predomination against apoptosis via the impede of Bax [174]. Besides apoptosis, necroptosis and ferroptosis are nonapoptotic types of programmed cell death that occur in numerous disorders particularly in HF, I/R, and MI [174]. Hypoxia mediates various events including, oxidative stress, and mitochondrial deregulation thereby may stimulate cardiomyocyte necroptosis. Indeed, this process has been demonstrated to contribute to MI pathogenesis [175] and in the alteration of cardiac activity during chronic ischemia [176]. Necroptosis also has a prominent effect in cardiomyocytes loss during acute viral myocarditis, I/R injury, and atherosclerosis through its ligand death receptors including receptor-interacting protein 1, 3 (RIP1, 3) and mixed lineage kinase domain-like (MLKL) [177,178]. Recently, Karshovska and associates documented that a high level of HIF-1α regulation impairs the mitochondrial bioenergetic and elevates macrophage necroptosis in the atherosclerosis [179]. Another type of regulated cell death is ferroptosis, which consists of intracellular iron dependence. Its fundamental role has been effectively stated recently in various CVDs, including cardiomyopathy, HF, MI, and myocardial I/R [59,180–182]. The iron chelator deferoxamine (DFO) is an iron chelator able to mimic hypoxia and to upregulate HIF-1 that can impede the process of ferroptosis in cardiac cells and eventually alleviates cardiac injury [183,184]. DFO is also able to reduce MI and to improve cardiac activity in myocardial I/R models [185,186]. Strikingly, high levels of ferroptosis have been also observed during the reperfusion stage [187]. Several other lines of evidence document the role of programmed cell death in the modulation of inflammation, which may result in cardiomyocyte ferroptosis [188]. These current findings suggest that blocking ferroptosis may prevent cardiomyocytes death. However, the processes by which hypoxia modulates cardiomyocytes death remain to be thoroughly investigated. Further specific impact of hypoxia precisely HIF-1 on ferroptosis in the heart still to be well covered.

4.5. Inflammation, Oxidative Stress, and Mitochondrial Dysfunction following Hypoxia in Cancer Disease

Accumulating evidence demonstrates that inflammation is involved in all stages of tumorigenesis, including limitless replication, invasion and metastasis, apoptosis evasion, DNA damage, and angiogenesis. It remains unclear which molecular mechanism interconnects all these pathways. Recent scientific evidence suggest that the mitochondrial compartment may be the central platform for the regulation of the inflammatory response, which occurs during cancer development and growth. The ATP generated by mitochondria throughout OXPHOS is fundamental for the proliferation and differentiation of T cells, which are one of the main components of antitumor immunity [189]. However, different metabolic demands are required for each phenotypic stage of T cells. Indeed, upon their activation, T cells quickly shift to glycolysis, which guarantees a rapid energetic availability to support their growth and the production of biosynthetic factors. To boost the glycolytic pathway, the molecular axis composed of PI3K, AKT, and MTOR is the main executor. On one hand, PI3K-AKT-MTOR activates the avian myelocytomatosis virus oncogene cellular homolog (c-Myc) to increase the activity of key enzyme of glycolysis (such as the glucose transporters (GLUT)) [190] and of glutamine transporters (including glutaminase1) [190,191], and on the other, the PI3K-AKT-MTOR axis works to upregulate HIF-1α, which, at the same time, can inhibit the TCA cycle and upregulate the expression of glycolytic enzymes. In addition to this, HIF-1α-induced glycolysis is fundamental to control the subpopulation of T cells. Indeed, it regulates the differentiation of Th17 [192] and of CD8+ T cells [193]. By contrast, the transition of T cells from effector to memory states mainly relies on fatty acid oxidation (FAO). Recent studies demonstrated that during this transition phase, a small amount of T cells enforces FAO by reducing the MTOR pathway and activating the AMPK pathway [194]. In addition, AMPK and FAO support the development of CD8+ T cells. Indeed, by using the AMPK activator such as metformin, the amount of memory cells as well as the lipid oxidation were increased [195]. Consistent with this, in AMPK-null T cells, the generation memory CD8 T cells upon pathogen infection was defective [196]. Interestingly, the tumor necrosis factor (TNF) receptor-associated factor 6 (TRAF6) was also important for the FAO activation in memory cells. Mice lacking TRAF6 were unable to increase FAO and displayed defects in generation of memory cells [197]. T cell activation and proliferation may be also promoted by ROS. However, if the ROS levels become too great, some amount of T cells could undergo apoptosis, thus reducing their anticancer potential. Increased ROS production may also be provoked by proinflammatory soluble molecules (such as cytokines and chemokines) that are secreted following activation of innate immune cells and by macrophages and neutrophils, which can also produce RNS. In turn, ROS and RNS provoke serious damages at the mtDNA, causing dysfunction in the production and assembly of components of the mETC, thereby enhancing the ROS production in a dangerous loop reaction. Accordingly, loss of function of mETC is frequently associated with several cancer types, such as breast cancer [198], renal cell carcinoma [199], and thyroid carcinoma [200]. Furthermore, ROS and RNS may induce mutations of genes, and relative signaling pathways, involved in both tumor activation (oncogene) and suppression (tumor suppressor). Therefore, inflammation may lead to mutagenesis. Consistently, mutations in P53, GTPase Kras (KRAS), adenomatosis polyposis coli (APC), and wingless-related integration site (WNT) have been found in intestinal cancers characterized by chronic inflammation [201]. The inflammatory environment is also responsible to drive cell survival. Inflammatory mediators released by immune cells can converge on prosurvival pathways and increase cell proliferation and resistance to cell death. In this context, IL-1α and IL-1β promote IL-17A response, activate the NF-kB p65 subunit to cause colorectal cancer (CRC) initiation. Consistently, IL-17A was associated to poor diagnosis of CRC and inhibits the IL-1-dependent inflammation and prevented CRC development [202]. IL signaling is also responsible for promoting colitis-associated cancer. Here, IL-6 mediates cell survival and proliferation throughout the oncogenic transcription factor STAT3 [203]; alteration of the mitochondrial dynamics is an example. In fact,

DRP1-FIS1-mediated mitochondrial fission increases the mitochondrial damage and boosts the inflammasome recruitment [204]. In addition, the hypoxic condition activates NLRP3. At demonstration, hypoxia in prostate cancer cell lines increase NLRP3 levels throughout NF-kB [205].

Furthermore, the same HIF-1α can be regulated by NLRP3. In this case, it has been observed that the adaptor protein ASC associates and stabilizes HIF-1α to increase cell migration and metastasis in oral squamous cell carcinoma [78]. However, other investigations demonstrate that hypoxia may be a repressor of NLRP3. Indeed, it has been demonstrated that hypoxic condition attenuates inflammation by reducing NLRP3 expression and that the isoform β of HIF is necessary to control the expression of NLRP3 [206]. Interestingly, in this investigation, it has been demonstrated that the activation of NLRP3 required carnitine palmitoyltransferase 1A (CPT1A)-mediated enhancement of FAO [206]. Notably, CPT1A determines the entry of long-chain fatty acids into mitochondria, leading to FAO and excessive ROS production [207]. In addition, the function of NLRP3 in cancer is controversial: some evidence highlights a protective anti-tumorigenic role; others suggest pro-tumorigenic effects [208]. For example, in breast cancer, NLRP3 increases the tumor growth and metastasis by creating an inflammatory microenvironment [209,210]. In lung cancer, NLRP3 activates the prosurvival factors Akt, the extracellular signal-regulated kinase 1/(ERK1/2), and cAMP response element-binding protein (CREB) to increase cell migration, proliferation, and invasion of cancerous cells [211]. In melanoma, NLRP3 is constitutively activated [212], and its inhibition suppresses metastasis [213]. On the contrary, NLRP3 in colorectal cancer (CRC) inhibits tumor growth, senses tissue damage, and activates cell death mechanisms against the tumor cells [214,215]. In line with this, the absence of NLRP3 was correlated with the progression of hepatocellular carcinoma [216] (Figure 3). Besides, enhanced lipid peroxidation during hypoxia-mediated inflammation contributes to the stimulation of programmed cell death (apoptosis or necroptosis) in cancer cells, resulting in cellular injury and progression of multiple pathogenesis [217,218]. Notably, it has been disclosed that tumor cells have ample lipid contents [219]. As an example, in hepatocellular carcinoma (HCC), HIF-1α augments lipid stabilization by alleviating FAO [220]. Lipid metabolism is abundant in tumor cells and exhibits a critical effect in tumorigenesis, invasion, and metastasis. To date, studies report the significant involvement of hypoxia in modulating various aspects of lipid metabolism, which are crucial for the enhanced proliferation rate, and subsequently, cancer progression [221]. The role of HIF-1 in lipid metabolism attracted more attention recently. Indeed, inhibiting HIF-1 or its target genes involved in lipid stabilization lead to significant alleviation of proliferation as well as chemoresistance in multiple cancer disorders in response to hypoxic circumstances [218,222–225]. By contrast, the overexpression of HIF-1-dependent lipid metabolism target genes is associated with cancer malignancy [218,222].

4.6. Hypoxia-Mediated Signaling Pathways and Cell Death in Cancer

It is well known that most cancer cells are driven toward hypoxia-directed apoptosis. Hypoxia-induced HIF-1α is stated to be associated with cancer cell malignancy and chemoradiation therapy resistance [226,227]. It is well established that HIF-1α stimulates and protects cancer cells against apoptosis. For instance, high levels of HIF-1α stabilization in pancreatic cells under hypoxia elicit their ability to resist against the apoptotic event compared with the normoxic cells through the HIF-1α-enhanced PI3K/AKT signaling pathway [228]. Although the critical role of HIF-1 in response to hypoxia-stimulated apoptosis, its molecular process in question remains unclear. Necrosis death is also involved during hypoxia in cancer. Indeed, glioma cells undergo necrosis and display resistance against prolonged hypoxia. Interestingly, this event may block the apoptotic pathway [229]. Furthermore, emerging studies have stressed the crucial role of hypoxia in ferroptosis and, given the central role of the HIF-1α element upon hypoxia, its deficiency mitigates the ferroptosis sensitivity, as it modulates the transcriptional activity of genes responsible for iron metabolisms such as transferrin receptor 1 (TFR1) and Ferritin Heavy Chain

(FTH) [230]. High levels of HIF-1α protein increase the susceptibility to ferroptosis in renal carcinoma cells [231,232]. Notably, the tumor suppressor BRCA1-associated protein 1 (BAP1) is involved in apoptosis and ferroptosis in different cancer cells [233,234]. It has been demonstrated the role of BAP1 in attenuating cancer development by provoking the ferroptosis event via solute carrier family 7-member 11 (SLC7A11) inhibition [235]. Always regarding BAP1, it has been suggested that the depletion of this tumor suppressor causes enhanced HIF-1α expression through the modulation of NF-κB in uveal melanoma [236]. Taken together, given the importance of HIF-1α blockers to modulate ferroptosis in a context-dependent manner may open new directions for drug therapy [232]. Further research on the effect of hypoxia-induced HIF-1α stabilization on BAP1 would shed light on new molecular mechanisms of various events including cell death in cancer.

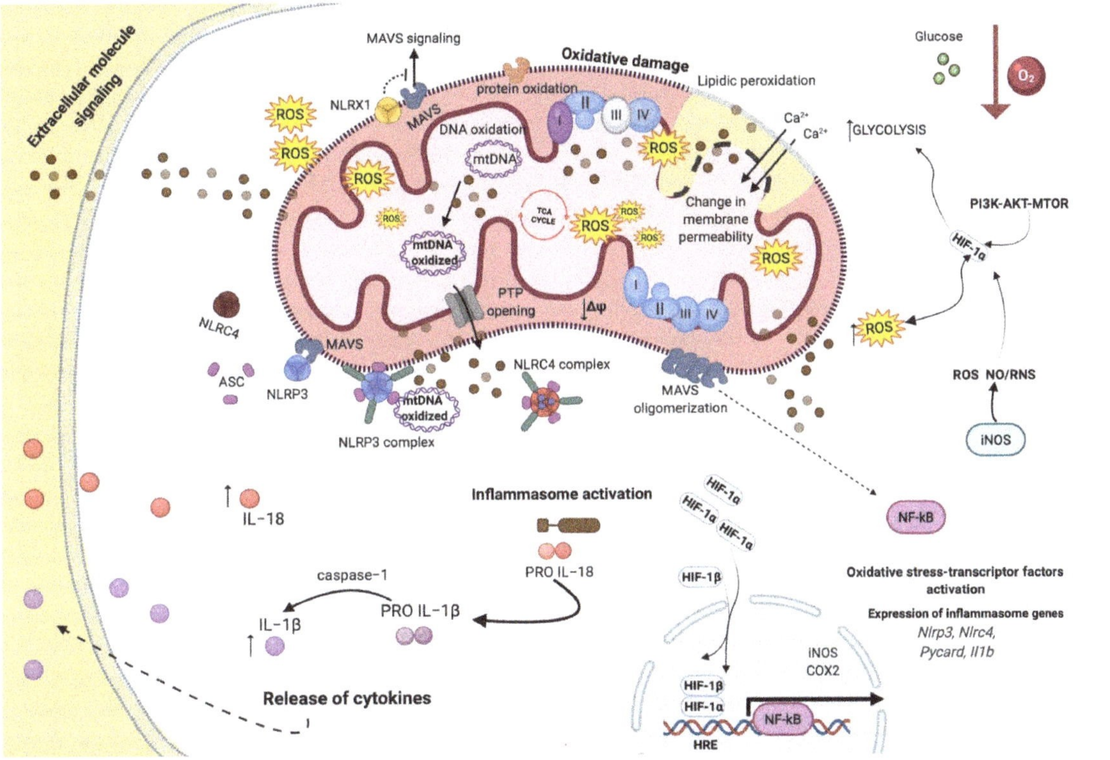

Figure 3. Inflammation and mitochondrial oxidative stress in response to hypoxic condition. During hypoxia, Hypoxic-inducible factor-1 α (HIF-1α) stimulates the transcription of target genes involved in inflammation and oxidative stress-transcription factors. Moreover, HIF-1α increased mitochondrial reactive oxygen species (mtROS) levels, activating nuclear factor kappa-light-chain-enhancer of activated B cells (NF-kB) transcription factor, stimulating the inflammasome genes expression, including NOD-, LRR- and pyrin domain-containing protein (NLRC)4, NLRP3, and interleukin 1β (IL1β) genes. Ultimately causing oxidative damage to the mitochondrial membrane, this event affects the membrane permeability, lipid peroxidation, and mtDNA, resulting in mitochondrial dysfunction. The FosfoInositide-3-Kinasi (PI3K)-protein kinase B (AKT)-mechanistic target of rapamycin (MTOR) pathway upregulates HIF-1α during hypoxia.

5. Novel Mechanisms and Therapeutic Targets in CVD and Cancer Disorders

HIF-1 plays critical roles in important aspects of cancer biology to allow and promote tumor cells to grow and survive in response to hypoxic conditions, including invasion, metastasis, angiogenesis, modulation of glucose and energy metabolism, and stem cell maintenance. Therefore, the poor prognosis of cancer patients results in the actual interest in studying HIF-1α as a therapeutic target in cancer disorder. Therefore, consistent with this concept, inhibiting HIF-1 or its related protein interactions has been demonstrated to block tumor proliferation. Different studies reported that HIF-1 inhibitors mitigate breast cancer metastasis in tumors sensitization to radiotherapy and mouse orthotopic transplant models [237,238]. Furthermore, several small molecules that repress HIF-1 directly or indirectly have been tested in clinical trials for different kinds of tumors [239].

By targeting HIF-1α, it is possible to intervene against malignant gliomas. Indeed, downregulation of HIF-1α by siRNA decreases both the level of matrix metalloproteinase (MMP)-2 as well as the functions of MMP-2 and MMP-9 decreasing the mobility of glioma cell via the impaired invasion-related molecules [240].

Small molecules blocking the expression and the functions of HIF-1 have been found effective to reduce the growth of solid tumors such as prostate and breast cancer. An example is the RNA antagonist EZN-2968, which represses HIF-1α mRNA expression [241] or PX-478, reduces the transcription of HIF-1α, blocks the translation through a VHL- and p53-independent pathway [242].

Furthermore, both PX-478 and EZN2968 elicit dose-dependent decreases in HIF-1α levels and VEGF expression, as well as the tumor size in DU145 xenograft models, and both small-molecule inhibitors were well tolerated in clinical activities [241,242]. Other molecules identified include geldanamycin, which reduces the bond of the heat shock protein 90 (HSP90) to HIF-1α to destabilize folding and increase proteasomal destruction [243]. Several HIF-1α prolylhydroxylase inhibitors that preclude VHL from binding to HIF-1α have also been developed and are now in late-stage clinical trials in disease in which HIF signaling is beneficial. For example, Roxadustat, which leads to increased endogenous erythropoietin generation, improved absorption of iron and anemia amelioration in chronic kidney disease (CKD) [244].

On the other hand, and in contrast to HIF-1α effect on cancer as a stimulator for the disease's progression, HIF-1α contributes to the cardiac protection in the majority of CVDs. This depends on the duration of hypoxia and the sustain of HIF-1α activation. Different studies documented that the HIF-1 transcription factor displays a dual role in CVDs, as a protective effect in acute exposure to hypoxia, which is rapidly accumulated and stimulates a cascade of downstream target genes transcription involved in mitochondrial metabolism regulation, angiogenesis, and cell functions, leading to cardioprotection against ischemic insults. Moreover, these mechanisms regulate cellular functions in the hypoxic field to the sparsely oxygen environment and sustain normal cellular homeostasis, which under hypoxic stress is crucial for the human body. The HIF-1 pathway has a pivotal role in repairing cardiac tissue through the angiogenesis activation in I/R [245]. Thereby, it conveys a novel target to develop innovative therapies for the treatment of ischemic diseases and reduction of reperfusion injury [6]. Consistently, recent pieces of evidence reported the contribution of HIF-1α in the therapeutic effect of certain natural compounds that attenuate myocardial I/R. One example may be found for Panax notoginseng saponins, which have protective roles against I/R via HIF-1α/BCL2/BNIP3 pathway, which in return upregulates mitochondrial autophagy [246].

Nevertheless, HIF-1 contributes to the pathogenesis of various disorders during chronic exposure to hypoxia. HIF-1 is considered a promoter of atherosclerosis development, and for this reason, it may not be regarded as the disease's therapeutic treatment. Angiogenesis induced by HIF-1 transcription is protective in the short term, but then it forms collateral vessels that may result in terrible consequences in atherosclerosis patients. Nevertheless, this hypothesis is based on current research data, which are mostly focusing on the HIF-1 signaling pathway, particularly the HIF-1α subunits. Further studies are needed on other members of the HIF family and the HIF-1α subunits. At the physiological level, there are numerous mechanisms whose modulation by these transcription factors may have great potential for treating CVDs such as atherosclerosis [247]. Pulmonary hypertension (PH) is another CVDs related to HIF-1 modulation. The proliferation of pulmonary arterial smooth muscle cells (PASMCs) and endothelial cells (Ecs) which are activated by HIF-1 and HIF-2 transcription, respectively, contributes toward the increase in pulmonary blood pressure [248]. Strategies that target the inhibition of HIF transcription in PH patients could be an interesting new perspective in the treatment of hypertension. Notably, it has been demonstrated that the HIF-2α translation inhibitor compound 76 inhibits HIF-2α, and it is capable of relieving pulmonary artery blood pressure in different models [249].

6. Conclusions and Future Perspectives

The mechanism of cellular response to O_2 deficiency is primarily regulated by the HIF-1α pathway. Biological studies of HIF-1α have improved the understanding of O_2 homeostasis and notably gained much attention recently in many research fields. In this current review, we highlighted the effect of hypoxia on mitochondrial (dys)function and inflammation in CVDs and cancer.

We have evidenced how HIF-1α signaling exhibits divergent effects in stimulating the disease progression or inducing protection after injury in different disease conditions. This dual role particularly happens when we consider HIF-1α in CVDs and cancer. Indeed, meanwhile, HIF-1α contributes to cardiac protection in the majority of CVDs, this hypoxic factor is highly associated to tumor progression, malignancy, and resistance to chemo-radiation therapy. Regardless of these divergent effects, significant improvements of the scientific research have permitted to propose the targeting of HIF-1α as a yielding strategy for the treatment of both CVDs and cancer.

Nevertheless, it is important to keep in mind key aspects when we are approaching treating cancer or CVDs with innovative therapies. Cancer progression contemplates different phases, which differ one from the other for several aspects. The term CVDs embraces at least 13 different conditions, which can affect both heart and blood vessels and that display its own clinical course. Furthermore, it is also fundamental to consider the heterogeneity caused by individual differences of the patients.

Only a deeper comprehension of the hypoxia-related mechanisms happening during the different phases of the tumor and in the single CVDs will really pave the way to endow and elucidate the great potential therapeutic targets and preventative approaches based on HIF-1α-modulation.

Author Contributions: E.B., S.P., G.M., A.E.K., M.P., M.D.S. and M.R.W. wrote the manuscript with input from P.P., C.G., E.B., and M.D.S. prepared display items (with https://biorender.com (accessed on 21 January 2022)) The figures are original and have not been published before. F.F., P.P., S.P. and C.G. reviewed the manuscript before submission. All authors have read and agreed to the published version of the manuscript.

Funding: P.P. is grateful to Camilla degli Scrovegni for continuous support. The Signal Transduction Laboratory is supported by the Italian Association for Cancer Research (AIRC; Grant IG-23670 to P.P. and Grant IG-19803 to C.G.), Associazione Ricerca Oncologica Sperimentale Estense (A-ROSE), Telethon (Grant GGP11139B to P.P.), the Ministry of Education, University and Research–Progetti di Rilevante Interesse Nazionale (Grant PRIN2017E5L5P3 to P.P. and Grant PRIN20177E9EPY to C.G.), the Italian Ministry of Health (Grant GR-2013-02356747 to C.G. and GR-2019-12369862 to G.M.), the European Research Council (Grant InflaPML 853057 to C.G.), and by local funds from the University of Ferrara to P.P. and C.G. M.R.W. was supported by the National Science Centre, Poland (Grant UMO-2018/29/B/NZ1/00589).

Institutional Review Board Statement: Not applicable.

Informed Consent Statement: Not applicable.

Data Availability Statement: Not applicable.

Conflicts of Interest: The authors declare no conflict of interest.

Abbreviations

2-HG	2-hydroxyglutarate
2-OG	2-oxoglutarate
4-HNE	4-hydroxynonenal
AKAP121	A-kinase anchor protein 1
APC	Adenomatosis Polyposis Coli
ARNT	Aryl hydrocarbon receptor nuclear translocator
ATP	Adenosine triphosphate
BAP1	BRCA1-associated protein 1
bHLH	Basic helix–loop–helix
BNIP3	BCL2 and adenovirus E1B 19-kD-interacting protein 3
C26	Colon cancer
Ca^{2+}	Calcium
CAD	Coronary artery disease
CBP/p300	CREB-binding protein
CC	Cancer cachexia
CCHD	Cyanotic congenital heart disease
CCL-39	Chinese hamster Lung fibroblasts
CH	Chronic hypoxia
CHD	Coronary heart disease
CKD	Chronic kidney disease
CPT1A	Carnitine palmitoyltransferase 1A
CVD	Cardiovascular disease
DFO	Deferoxamine
DRP1	Dynamin-related protein 1
Ecs	Endothelial cells
EPO	Erythropoietin
ER	Endoplasmic reticulum
ERK1/2	Extracellular signal-regulated kinase $\frac{1}{2}$
FAO	Fatty acid oxidation
FIH	Inhibiting HIF
Fis 1	Fission 1 mitochondrial protein
FTH	Ferritin Heavy Chain
FUNDC1	FUN14 Domain Containing 1
GCLM	Glutamate–cysteine ligase
H_2O_2	Hydrogen peroxide
HCC	Hepatocellular carcinoma
HCT 116	Human colorectal carcinoma
HF	Heart failure
HIF-1	Hypoxic-inducible factor-1
HIF-1α	Hypoxia-inducible factor 1-alpha
HMGB1	High mobility group box 1
HREs	Hypoxic-responsive elements
HSP70	Heat shock protein 70
HSP90	Heat shock protein 90
HUVEC	Human umbilical vein endothelial
IHD	Ischemic heart disease
IMM	Inner mitochondrial membrane
IMS	Mitochondrial intermembrane space
iNOS	Inducible nitric oxide synthase
ISCU	Iron-Sulfur Cluster Assembly Enzyme
JAK2/STAT3	Janus kinase/signal transducer and activator of transcription
KRAS	GTPase Kras
LDH A	Lactate dehydrogenase-A
LDH	Lactate dehydrogenase
LPS	Lipopolysaccharide
LS174	Human colonic adenocarcinoma cells

LV	Left ventricle
MAMs	Mitochondria-associated membranes
MAOs	Monoamine oxidases
MCF-7	Human breast cancer cells
MCU	Mitochondrial calcium uniporter
MDH	Malate dehydrogenase
MECs	mammary epithelial cells
mETC	Mitochondrial electron respiratory chain
MFN1	Mitofusin 1
MI	Myocardial infarction
miRs	MicroRNAs
MLKL	Mixed lineage kinase domain-like
MMP-2	Matrix metalloproteinase
MMP	Mitochondrial membrane potential
MMTV-PyMT	Mouse mammary tumor virus-polyoma middle tumor-antigen
mPTP	Mitochondrial permeability transition pore
MTOR	Rapamycin inhibitor rapamycin
mtROS	Mitochondrial ROS
MXI1	MAX-interactor 1
NLRs	Leucine-rich repeat-containing receptors
NOXs	NADPH oxidases
NRFs	Nuclear respiratory factors
O_2^-	Superoxide
O_2	Oxygen
ODDD	Oxygen-dependent degradation domain
OMM	Outer mitochondrial membrane
OSCC	Oral squamous cell carcinoma
OXPHOS	Oxidative phosphorylation
p70S6K	Phospho-p70 S6 Kinase
PAD	Peripheral arterial diseases
PASMCs	Pulmonary arterial smooth muscle cells
PCR	Polymerase chain reaction
PH	Pulmonary hypertension
PHDs	Propyl-hydroxylases
PGC-1α	peroxisome proliferator-activated receptor gamma coactivator 1-alpha
PI3K/AKT	Phosphoinositide-3-kinase/Akt
PKA	Protein kinase A
PKM2	M2 isoform of pyruvate kinase
PRX/Trx	peroxiredoxin/thioredoxin
pVHL	Von Hippel–Lindau
RIP1, 3	Receptor-interacting protein 1, 3
RNS	Reactive nitrogen species
ROS	Reactive oxygen species
SLC7A11	Solute carrier family 7-member 11
SOD	Superoxide dismutase
SUM-149	Triple-negative breast cancer cells
TCA	Tricarboxylic acid
TFR1	transferrin receptor 1
TGF-β	Transforming growth factor beta
TNBC	Triple negative breast cancer
TNF	Tumor necrosis factor
TRAF6	Receptor-associated factor 6
Tregs	regulatory T cells
VDAC1	Voltage-dependent anion-selective channel 1
VGF-A	vascular endothelial growth factor-A
WDR26	WD Repeat Domain 26
WNT	Wingless-related integration site
$\Delta\Psi_m$	Mitochondrial membrane potential

References

1. Masoudkabir, F.; Sarrafzadegan, N.; Gotay, C.; Ignaszewski, A.; Krahn, A.D.; Davis, M.K.; Franco, C.; Mani, A. Cardiovascular disease and cancer: Evidence for shared disease pathways and pharmacologic prevention. *Atherosclerosis* **2017**, *263*, 343–351. [CrossRef] [PubMed]
2. Gernaat, S.A.M.; Boer, J.M.A.; Bongard, D.V.D.; Maas, A.H.E.M.; Van Der Pol, C.C.; Bijlsma, R.M.; Grobbee, D.E.; Verkooijen, H.M.; Peeters, P.H. The risk of cardiovascular disease following breast cancer by Framingham risk score. *Breast Cancer Res. Treat.* **2018**, *170*, 119–127. [CrossRef] [PubMed]
3. Coviello, J.S. Cardiovascular and Cancer Risk: The Role of Cardio-oncology. *J. Adv. Pr. Oncol.* **2018**, *9*, 160–176.
4. De Boer, R.A.; Meijers, W.C.; Van Der Meer, P.; Van Veldhuisen, D.J. Cancer and heart disease: Associations and relations. *Eur. J. Hear. Fail.* **2019**, *21*, 1515–1525. [CrossRef] [PubMed]
5. Waypa, G.B.; Smith, K.A.; Schumacker, P.T. O_2 sensing, mitochondria and ROS signaling: The fog is lifting. *Mol. Asp. Med.* **2016**, *47–48*, 76–89. [CrossRef] [PubMed]
6. Liu, M.; Galli, G.; Wang, Y.; Fan, Q.; Wang, Z.; Wang, X.; Xiao, W. Novel Therapeutic Targets for Hypoxia-Related Cardiovascular Diseases: The Role of HIF-1. *Front. Physiol.* **2020**, *11*, 774. [CrossRef]
7. Bousquet, P.A.; Meltzer, S.; Sønstevold, L.; Esbensen, Y.; Dueland, S.; Flatmark, K.; Sitter, B.; Bathen, T.F.; Seierstad, T.; Redalen, K.R.; et al. Markers of Mitochondrial Metabolism in Tumor Hypoxia, Systemic Inflammation, and Adverse Outcome of Rectal Cancer. *Transl. Oncol.* **2019**, *12*, 76–83. [CrossRef]
8. Chen, P.-S.; Chiu, W.-T.; Hsu, P.-L.; Lin, S.-C.; Peng, I.-C.; Wang, C.-Y.; Tsai, S.-J. Pathophysiological implications of hypoxia in human diseases. *J. Biomed. Sci.* **2020**, *27*, 1–19. [CrossRef]
9. Kim, J.-W.; Tchernyshyov, I.; Semenza, G.L.; Dang, C.V. HIF-1-mediated expression of pyruvate dehydrogenase kinase: A metabolic switch required for cellular adaptation to hypoxia. *Cell Metab.* **2006**, *3*, 177–185. [CrossRef]
10. McClelland, G.B.; Brooks, G.A. Changes in MCT 1, MCT 4, and LDH expression are tissue specific in rats after long-term hypobaric hypoxia. *J. Appl. Physiol.* **2002**, *92*, 1573–1584. [CrossRef]
11. Tuder, R.M.; E Flook, B.; Voelkel, N.F. Increased gene expression for VEGF and the VEGF receptors KDR/Flk and Flt in lungs exposed to acute or to chronic hypoxia. Modulation of gene expression by nitric oxide. *J. Clin. Investig.* **1995**, *95*, 1798–1807. [CrossRef] [PubMed]
12. Semenza, G.L.; Wang, G.L. A nuclear factor induced by hypoxia via de novo protein synthesis binds to the human erythropoietin gene enhancer at a site required for transcriptional activation. *Mol. Cell. Biol.* **1992**, *12*, 5447–5454. [CrossRef]
13. Melillo, G.; Musso, T.; Sica, A.; Taylor, L.S.; Cox, G.W.; Varesio, L. A hypoxia-responsive element mediates a novel pathway of activation of the inducible nitric oxide synthase promoter. *J. Exp. Med.* **1995**, *182*, 1683–1693. [CrossRef]
14. Semenza, G.L. Hypoxia-Inducible Factor 1 (HIF-1) Pathway. *Sci. STKE* **2007**, *2007*, cm8. [CrossRef]
15. Nakayama, K.; Kataoka, N. Regulation of Gene Expression under Hypoxic Conditions. *Int. J. Mol. Sci.* **2019**, *20*, 3278. [CrossRef]
16. Wang, G.L.; Semenza, G.L. Purification and Characterization of Hypoxia-inducible Factor 1. *J. Biol. Chem.* **1995**, *270*, 1230–1237. [CrossRef] [PubMed]
17. Semenza, G.L. Hypoxia-Inducible Factors in Physiology and Medicine. *Cell* **2012**, *148*, 399–408. [CrossRef]
18. Semenza, G.L. Hypoxia-inducible factor 1: Regulator of mitochondrial metabolism and mediator of ischemic preconditioning. *Biochim. Biophys. Acta* **2011**, *1813*, 1263–1268. [CrossRef]
19. Lee, D.C.; Sohn, H.A.; Park, Z.-Y.; Oh, S.; Kang, Y.K.; Lee, K.-M.; Kang, M.; Jang, Y.J.; Yang, S.-J.; Hong, Y.K.; et al. A Lactate-Induced Response to Hypoxia. *Cell* **2015**, *161*, 595–609. [CrossRef]
20. Zheng, J.; Chen, P.; Zhong, J.; Cheng, Y.; Chen, H.; He, Y.; Chen, C. HIF-1α in myocardial ischemia-reperfusion injury (Review). *Mol. Med. Rep.* **2021**, *23*, 1–9. [CrossRef] [PubMed]
21. Ke, Q.; Costa, M. Hypoxia-Inducible Factor-1 (HIF-1). *Mol. Pharmacol.* **2006**, *70*, 1469–1480. [CrossRef] [PubMed]
22. Jiang, B.-H.; Rue, E.; Wang, G.L.; Roe, R.; Semenza, G.L. Dimerization, DNA Binding, and Transactivation Properties of Hypoxia-inducible Factor 1. *J. Biol. Chem.* **1996**, *271*, 17771–17778. [CrossRef] [PubMed]
23. Wang, G.L.; Jiang, B.-H.; Rue, E.A.; Semenza, G.L. Hypoxia-inducible factor 1 is a basic-helix-loop-helix-PAS heterodimer regulated by cellular O_2 tension. *Proc. Natl. Acad. Sci. USA* **1995**, *92*, 5510–5514. [CrossRef] [PubMed]
24. Bruick, R.K.; McKnight, S.L. A Conserved Family of Prolyl-4-Hydroxylases That Modify HIF. *Science* **2001**, *294*, 1337–1340. [CrossRef] [PubMed]
25. Jaakkola, P.; Mole, D.R.; Tian, Y.-M.; Wilson, M.I.; Gielbert, J.; Gaskell, S.J.; von Kriegsheim, A.; Hebestreit, H.F.; Mukherji, M.; Schofield, C.J.; et al. Targeting of HIF-alpha to the von Hippel-Lindau Ubiquitylation Complex by O_2-Regulated Prolyl Hydroxylation. *Science* **2001**, *292*, 468–472. [CrossRef]
26. Luo, W.; Hu, H.; Chang, R.; Zhong, J.; Knabel, M.; O'Meally, R.; Cole, R.N.; Pandey, A.; Semenza, G.L. Pyruvate Kinase M2 Is a PHD3-Stimulated Coactivator for Hypoxia-Inducible Factor 1. *Cell* **2011**, *145*, 732–744. [CrossRef]
27. Tian, X.; Zhou, N.; Yuan, J.; Lu, L.; Zhang, Q.; Wei, M.; Zou, Y.; Yuan, L. Heat shock transcription factor 1 regulates exercise-induced myocardial angiogenesis after pressure overload via HIF-1α/VEGF pathway. *J. Cell. Mol. Med.* **2020**, *24*, 2178–2188. [CrossRef]
28. Hölscher, M.; Schäfer, K.; Krull, S.; Farhat, K.; Hesse, A.; Silter, M.; Lin, Y.; Pichler, B.J.; Thistlethwaite, P.; El-Armouche, A.; et al. Unfavourable consequences of chronic cardiac HIF-1α stabilization. *Cardiovasc. Res.* **2012**, *94*, 77–86. [CrossRef]

29. Zhang, Y.; Xu, Y.; Zhou, K.; Kao, G.; Yan, M.; Xiao, J. Hypoxia-inducible transcription factor-1α inhibition by topotecan protects against lipopolysaccharide-induced inflammation and apoptosis of cardiomyocytes. *BioMed. Eng. OnLine* **2021**, *20*, 88. [CrossRef] [PubMed]
30. Xie, L.; Pi, X.; Wang, Z.; He, J.; Willis, M.S.; Patterson, C. Depletion of PHD3 protects heart from ischemia/reperfusion injury by inhibiting cardiomyocyte apoptosis. *J. Mol. Cell. Cardiol.* **2015**, *80*, 156–165. [CrossRef]
31. Neckar, J.; Hsu, A.; Khan, A.H.; Gross, G.J.; Nithipatikom, K.; Cyprova, M.; Benak, D.; Hlavackova, M.; Sotáková-Kašparová, D.; Falck, J.R.; et al. Infarct size-limiting effect of epoxyeicosatrienoic acid analog EET-B is mediated by hypoxia-inducible factor-1α via downregulation of prolyl hydroxylase 3. *Am. J. Physiol. Circ. Physiol.* **2018**, *315*, H1148–H1158. [CrossRef]
32. Date, T.; Mochizuki, S.; Belanger, A.J.; Yamakawa, M.; Luo, Z.; Vincent, K.A.; Cheng, S.H.; Gregory, R.J.; Jiang, C. Expression of constitutively stable hybrid hypoxia-inducible factor-1α protects cultured rat cardiomyocytes against simulated ischemia-reperfusion injury. *Am. J. Physiol. Physiol.* **2005**, *288*, C314–C320. [CrossRef]
33. Morand, J.; Arnaud, C.; Pépin, J.L.; Godin-Ribuot, D. Chronic intermittent hypoxia promotes myocardial ischemia-related ventricular arrhythmias and sudden cardiac death. *Sci. Rep.* **2018**, *8*, 1–8. [CrossRef]
34. Meerson, F.Z.; Beloshitskiĭ, P.V.; Vorontsova, E.I.; Ustinova, E.E.; Rozhitskaia, I.I. Effect of adaptation to continuous and intermittent hypoxia on heart resistance to ischemic and reperfusion arrhythmias. *Patol. Fiziol. Eksp Ter.* **1989**, 48–50. (In Russian)
35. Belaidi, E.; Béguin, P.C.; Lévy, P.; Ribuot, C.; Godin-Ribuot, D. Prevention of HIF-1 activation and iNOS gene targeting by low-dose cadmium results in loss of myocardial hypoxic preconditioning in the rat. *Am. J. Physiol. Circ. Physiol.* **2008**, *294*, H901–H908. [CrossRef]
36. Xi, L.; Tekin, D.; Gursoy, E.; Salloum, F.; Levasseur, J.E.; Kukreja, R.C. Evidence that NOS2 acts as a trigger and mediator of late preconditioning induced by acute systemic hypoxia. *Am. J. Physiol. Circ. Physiol.* **2002**, *283*, H5–H12. [CrossRef]
37. Ding, H.-L.; Zhu, H.-F.; Dong, J.-W.; Zhu, W.-Z.; Yang, W.-W.; Yang, H.-T.; Zhou, Z.-N. Inducible nitric oxide synthase contributes to intermittent hypoxia against ischemia/reperfusion injury1. *Acta Pharmacol. Sin.* **2005**, *26*, 315–322. [CrossRef] [PubMed]
38. Minamishima, Y.A.; Moslehi, J.; Bardeesy, N.; Cullen, D.; Bronson, R.T.; Kaelin, W.G. Somatic inactivation of the PHD2 prolyl hydroxylase causes polycythemia and congestive heart failure. *Blood* **2008**, *111*, 3236–3244. [CrossRef] [PubMed]
39. Zhang, Y.; Liu, D.; Hu, H.; Zhang, P.; Xie, R.; Cui, W. HIF-1α/BNIP3 signaling pathway-induced-autophagy plays protective role during myocardial ischemia-reperfusion injury. *Biomed. Pharmacother.* **2019**, *120*, 109464. [CrossRef] [PubMed]
40. Chen, Y.-F.; Pandey, S.; Day, C.H.; Chen, Y.-F.; Jiang, A.-Z.; Ho, T.-J.; Chen, R.-J.; Padma, V.V.; Kuo, W.-W.; Huang, C.-Y. Synergistic effect of HIF-1α and FoxO3a trigger cardiomyocyte apoptosis under hyperglycemic ischemia condition. *J. Cell. Physiol.* **2018**, *233*, 3660–3671. [CrossRef]
41. Dorn, G.W.; Kirshenbaum, L.A. Cardiac reanimation: Targeting cardiomyocyte death by BNIP3 and NIX/BNIP3L. *Oncogene* **2008**, *27*, S158–S167. [CrossRef]
42. Xin, T.; Lv, W.; Liu, D.; Jing, Y.; Hu, F. Opa1 Reduces Hypoxia-Induced Cardiomyocyte Death by Improving Mitochondrial Quality Control. *Front. Cell Dev. Biol.* **2020**, *8*, 853. [CrossRef]
43. Zhou, H.; Zhu, P.; Wang, J.; Zhu, H.; Ren, J.; Chen, Y. Pathogenesis of cardiac ischemia reperfusion injury is associated with CK2α-disturbed mitochondrial homeostasis via suppression of FUNDC1-related mitophagy. *Cell Death Differ.* **2018**, *25*, 1080–1093. [CrossRef]
44. Zhou, H.; Zhu, P.; Guo, J.; Hu, N.; Wang, S.; Li, D.; Hu, S.; Ren, J.; Cao, F.; Chen, Y. Ripk3 induces mitochondrial apoptosis via inhibition of FUNDC1 mitophagy in cardiac IR injury. *Redox Biol.* **2017**, *13*, 498–507. [CrossRef] [PubMed]
45. Feng, Y.; Zhao, J.; Hou, H.; Zhang, H.; Jiao, Y.; Wang, J.; Wang, Y.; Sun, Y. WDR26 promotes mitophagy of cardiomyocytes induced by hypoxia through Parkin translocation. *Acta Biochim. Biophys. Sin.* **2016**, *48*, 1075–1084. [CrossRef]
46. Zhao, J.-F.; Rodger, C.E.; Allen, G.F.G.; Weidlich, S.; Ganley, I.G. HIF1α-dependent mitophagy facilitates cardiomyoblast differentiation. *Cell Stress* **2020**, *4*, 99–113. [CrossRef] [PubMed]
47. Chowdhury, A.; Aich, A.; Jain, G.; Wozny, K.; Lüchtenborg, C.; Hartmann, M.; Bernhard, O.; Balleiniger, M.; Alfar, E.A.; Zieseniss, A.; et al. Defective Mitochondrial Cardiolipin Remodeling Dampens HIF-1α Expression in Hypoxia. *Cell Rep.* **2018**, *25*, 561–570.e6. [CrossRef] [PubMed]
48. Papandreou, I.; Cairns, R.A.; Fontana, L.; Lim, A.L.; Denko, N.C. HIF-1 mediates adaptation to hypoxia by actively downregulating mitochondrial oxygen consumption. *Cell Metab.* **2006**, *3*, 187–197. [CrossRef] [PubMed]
49. Gomes, A.P.; Price, N.L.; Ling, A.J.Y.; Moslehi, J.J.; Montgomery, M.K.; Rajman, L.; White, J.P.; Teodoro, J.S.; Wrann, C.D.; Hubbard, B.P.; et al. Declining NAD+ Induces a Pseudohypoxic State Disrupting Nuclear-Mitochondrial Communication during Aging. *Cell* **2013**, *155*, 1624–1638. [CrossRef]
50. Fuhrmann, D.; Brüne, B. Mitochondrial composition and function under the control of hypoxia. *Redox Biol.* **2017**, *12*, 208–215. [CrossRef] [PubMed]
51. Koutnikova, H.; Campuzano, V.; Foury, F.; Dollé, P.; Cazzalini, O.; Koenig, M. Studies of human, mouse and yeast homologues indicate a mitochondrial function for frataxin. *Nat. Genet.* **1997**, *16*, 345–351. [CrossRef]
52. Nanayakkara, G.; Alasmari, A.; Mouli, S.; Eldoumani, H.; Quindry, J.C.; McGinnis, G.; Fu, X.; Berlin, A.; Peters, B.; Zhong, J.; et al. Cardioprotective HIF-1α-frataxin signaling against ischemia-reperfusion injury. *Am. J. Physiol. Circ. Physiol.* **2015**, *309*, H867–H879. [CrossRef] [PubMed]

53. Rane, S.; He, M.; Sayed, D.; Vashistha, H.; Malhotra, A.; Sadoshima, J.; Vatner, D.E.; Vatner, S.F.; Abdellatif, M. Downregulation of MiR-199a Derepresses Hypoxia-Inducible Factor-1α and Sirtuin 1 and Recapitulates Hypoxia Preconditioning in Cardiac Myocytes. *Circ. Res.* **2009**, *104*, 879–886. [CrossRef] [PubMed]
54. Briston, T.; Yang, J.; Ashcroft, M. HIF-1α localization with mitochondria. *Cell Cycle* **2011**, *10*, 4170–4171. [CrossRef] [PubMed]
55. Mylonis, I.; Kourti, M.; Samiotaki, M.; Panayotou, G.; Simos, G. Mortalin-mediated and ERK-controlled targeting of HIF-1α to mitochondria confers resistance to apoptosis under hypoxia. *J. Cell Sci.* **2017**, *130*, 466–479. [CrossRef] [PubMed]
56. Li, H.-S.; Zhou, Y.-N.; Li, L.; Li, S.-F.; Long, D.; Chen, X.-L.; Zhang, J.-B.; Feng, L.; Li, Y.-P. HIF-1α protects against oxidative stress by directly targeting mitochondria. *Redox Biol.* **2019**, *25*, 101109. [CrossRef]
57. Du, Y.; Ge, Y.; Xu, Z.-H.; Aa, N.; Gu, X.; Meng, H.; Lin, Z.; Zhu, N.; Shi, J.; Zhuang, R.; et al. Hypoxia-Inducible Factor 1 alpha (HIF-1α)/Vascular Endothelial Growth Factor (VEGF) Pathway Participates in Angiogenesis of Myocardial Infarction in Muscone-Treated Mice: Preliminary Study. *Med. Sci. Monit.* **2018**, *24*, 8870–8877. [CrossRef]
58. Ikeda, M.; Ide, T.; Tadokoro, T.; Miyamoto, H.D.; Ikeda, S.; Okabe, K.; Ishikita, A.; Sato, M.; Abe, K.; Furusawa, S.; et al. Excessive Hypoxia-Inducible Factor-1α Expression Induces Cardiac Rupture via p53-Dependent Apoptosis After Myocardial Infarction. *J. Am. Hear. Assoc.* **2021**, *10*, 020895. [CrossRef]
59. Liu, Y.; Zeng, L.; Yang, Y.; Chen, C.; Wang, D.; Wang, H. Acyl-CoA thioesterase 1 prevents cardiomyocytes from Doxorubicin-induced ferroptosis via shaping the lipid composition. *Cell Death Dis.* **2020**, *11*, 756. [CrossRef]
60. Wei, H.; Bedja, D.; Koitabashi, N.; Xing, D.; Chen, J.; Fox-Talbot, K.; Rouf, R.; Chen, S.; Steenbergen, C.; Harmon, J.W.; et al. Endothelial expression of hypoxia-inducible factor 1 protects the murine heart and aorta from pressure overload by suppression of TGF- signaling. *Proc. Natl. Acad. Sci.* **2012**, *109*, E841–E850. [CrossRef]
61. Stamati, K.; Mudera, V.; Cheema, U. Evolution of oxygen utilization in multicellular organisms and implications for cell signalling in tissue engineering. *J. Tissue Eng.* **2011**, *2*, 2041731411432365. [CrossRef]
62. Wolff, M.; Kosyna, F.K.; Dunst, J.; Jelkmann, W.; Depping, R. Impact of hypoxia inducible factors on estrogen receptor expression in breast cancer cells. *Arch. Biochem. Biophys.* **2017**, *613*, 23–30. [CrossRef]
63. Nadtochiy, S.M.; Schafer, X.; Fu, D.; Nehrke, K.; Munger, J.; Brookes, P.S. Acidic pH Is a Metabolic Switch for 2-Hydroxyglutarate Generation and Signaling. *J. Biol. Chem.* **2016**, *291*, 20188–20197. [CrossRef]
64. Miska, J.; Lee-Chang, C.; Rashidi, A.; Muroski, M.E.; Chang, A.L.; Lopez-Rosas, A.; Zhang, P.; Panek, W.K.; Cordero, A.; Han, Y.; et al. HIF-1α Is a Metabolic Switch between Glycolytic-Driven Migration and Oxidative Phosphorylation-Driven Immunosuppression of Tregs in Glioblastoma. *Cell Rep.* **2019**, *27*, 226–237.e4. [CrossRef] [PubMed]
65. Fukuda, R.; Zhang, H.; Kim, J.-W.; Shimoda, L.; Dang, C.V.; Semenza, G.L. HIF-1 Regulates Cytochrome Oxidase Subunits to Optimize Efficiency of Respiration in Hypoxic Cells. *Cell* **2007**, *129*, 111–122. [CrossRef] [PubMed]
66. Rodríguez-Enríquez, S.; Carreño-Fuentes, L.; Gallardo-Pérez, J.C.; Saavedra, E.; Quezada, H.; Vega, A.; Marín-Hernández, A.; Olín-Sandoval, V.; Torres-Márquez, M.E.; Moreno-Sánchez, R. Oxidative phosphorylation is impaired by prolonged hypoxia in breast and possibly in cervix carcinoma. *Int. J. Biochem. Cell Biol.* **2010**, *42*, 1744–1751. [CrossRef] [PubMed]
67. Chan, S.Y.; Zhang, Y.-Y.; Hemann, C.; Mahoney, C.E.; Zweier, J.L.; Loscalzo, J. MicroRNA-210 Controls Mitochondrial Metabolism during Hypoxia by Repressing the Iron-Sulfur Cluster Assembly Proteins ISCU1/2. *Cell Metab.* **2009**, *10*, 273–284. [CrossRef]
68. Guarino, F.; Zinghirino, F.; Mela, L.; Pappalardo, X.G.; Ichas, F.; De Pinto, V.; Messina, A. NRF-1 and HIF-1α contribute to modulation of human VDAC1 gene promoter during starvation and hypoxia in HeLa cells. *Biochim. Biophys. Acta* **2020**, *1861*, 148289. [CrossRef] [PubMed]
69. Zhang, H.; Gao, P.; Fukuda, R.; Kumar, G.; Krishnamachary, B.; Zeller, K.I.; Dang, C.; Semenza, G.L. HIF-1 Inhibits Mitochondrial Biogenesis and Cellular Respiration in VHL-Deficient Renal Cell Carcinoma by Repression of C-MYC Activity. *Cancer Cell* **2007**, *11*, 407–420. [CrossRef] [PubMed]
70. Tohme, S.; Yazdani, H.O.; Liu, Y.; Loughran, P.; Van Der Windt, D.J.; Huang, H.; Simmons, R.L.; Shiva, S.; Tai, S.; Tsung, A. Hypoxia mediates mitochondrial biogenesis in hepatocellular carcinoma to promote tumor growth through HMGB1 and TLR9 interaction. *Hepatol.* **2017**, *66*, 182–197. [CrossRef]
71. Jin, J.; Qiu, S.; Wang, P.; Liang, X.; Huang, F.; Wu, H.; Zhang, B.; Zhang, W.; Tian, X.; Xu, R.; et al. Cardamonin inhibits breast cancer growth by repressing HIF-1α-dependent metabolic reprogramming. *J. Exp. Clin. Cancer Res.* **2019**, *38*, 1–16. [CrossRef] [PubMed]
72. Lu, H.; Samanta, D.; Xiang, L.; Zhang, H.; Hu, H.; Chen, I.; Bullen, J.W.; Semenza, G.L. Chemotherapy triggers HIF-1–dependent glutathione synthesis and copper chelation that induces the breast cancer stem cell phenotype. *Proc. Natl. Acad. Sci.* **2015**, *112*, E4600–E4609. [CrossRef] [PubMed]
73. Chourasia, A.H.; Tracy, K.; Frankenberger, C.; Boland, M.L.; Sharifi, M.; E Drake, L.; Sachleben, J.R.; Asara, J.M.; Locasale, J.; Karczmar, G.S.; et al. Mitophagy defects arising from BNip3 loss promote mammary tumor progression to metastasis. *EMBO Rep.* **2015**, *16*, 1145–1163. [CrossRef] [PubMed]
74. Castelli, S.; Ciccarone, F.; Tavian, D.; Ciriolo, M.R. ROS-dependent HIF1α activation under forced lipid catabolism entails glycolysis and mitophagy as mediators of higher proliferation rate in cervical cancer cells. *J. Exp. Clin. Cancer Res.* **2021**, *40*, 1–18. [CrossRef] [PubMed]
75. Wu, Z.; Zuo, M.; Zeng, L.; Cui, K.; Liu, B.; Yan, C.; Chen, L.; Dong, J.; Shangguan, F.; Hu, W.; et al. OMA1 reprograms metabolism under hypoxia to promote colorectal cancer development. *EMBO Rep.* **2021**, *22*, e50827. [CrossRef]

76. Chiche, J.; Rouleau, M.; Gounon, P.; Brahimi-Horn, M.C.; Pouyssã©Gur, J.; Mazure, N.M. Hypoxic enlarged mitochondria protect cancer cells from apoptotic stimuli. *J. Cell. Physiol.* **2009**, *222*, 648–657. [CrossRef]
77. Devine, R.; Bicer, S.; Reiser, P.J.; Wold, L.E. Increased hypoxia-inducible factor-1α in striated muscle of tumor-bearing mice. *Am. J. Physiol. Circ. Physiol.* **2017**, *312*, H1154–H1162. [CrossRef] [PubMed]
78. Wu, C.-S.; Chang, I.Y.-F.; Hung, J.-L.; Liao, W.-C.; Lai, Y.-R.; Chang, K.-P.; Li, H.-P.; Chang, Y.-S. ASC modulates HIF-1α stability and induces cell mobility in OSCC. *Cell Death Dis.* **2020**, *11*, 1–11. [CrossRef] [PubMed]
79. Coimbra-Costa, D.; Alva, N.; Duran, M.; Carbonell, T.; Rama, R. Oxidative stress and apoptosis after acute respiratory hypoxia and reoxygenation in rat brain. *Redox Biol.* **2017**, *12*, 216–225. [CrossRef] [PubMed]
80. Liberti, M.V.; Locasale, J.W. The Warburg Effect: How Does it Benefit Cancer Cells? *Trends Biochem. Sci.* **2016**, *41*, 211–218. [CrossRef]
81. Akkol, E.; Tatlı, I.; Karatoprak, G.; Ağar, O.; Yücel, Ç.; Sobarzo-Sánchez, E.; Capasso, R. Is Emodin with Anticancer Effects Completely Innocent? Two Sides of the Coin. *Cancers* **2021**, *13*, 2733. [CrossRef]
82. Hu, L.; Cui, R.; Liu, H.; Wang, F. Emodin and rhein decrease levels of hypoxia-inducible factor-1α in human pancreatic cancer cells and attenuate cancer cachexia in athymic mice carrying these cells. *Oncotarget* **2017**, *8*, 88008–88020. [CrossRef] [PubMed]
83. Rabinovitch, R.C.; Samborska, B.; Faubert, B.; Ma, E.H.; Gravel, S.-P.; Andrzejewski, S.; Raissi, T.C.; Pause, A.; St.-Pierre, J.; Jones, R.G. AMPK Maintains Cellular Metabolic Homeostasis through Regulation of Mitochondrial Reactive Oxygen Species. *Cell Rep.* **2017**, *21*, 1–9. [CrossRef] [PubMed]
84. Gao, Y.-H.; Li, C.-X.; Shen, S.-M.; Li, H.; Chen, G.-Q.; Wei, Q.; Wang, L.-S. Hypoxia-inducible factor 1α mediates the down-regulation of superoxide dismutase 2 in von Hippel–Lindau deficient renal clear cell carcinoma. *Biochem. Biophys. Res. Commun.* **2013**, *435*, 46–51. [CrossRef] [PubMed]
85. Patergnani, S.; Missiroli, S.; Morciano, G.; Perrone, M.; Mantovani, C.M.; Anania, G.; Fiorica, F.; Pinton, P.; Giorgi, C. Understanding the Role of Autophagy in Cancer Formation and Progression Is a Real Opportunity to Treat and Cure Human Cancers. *Cancers* **2021**, *13*, 5622. [CrossRef]
86. Patergnani, S.; Bonora, M.; Ingusci, S.; Previati, M.; Marchi, S.; Zucchini, S.; Perrone, M.; Wieckowski, M.R.; Castellazzi, M.; Pugliatti, M.; et al. Antipsychotic drugs counteract autophagy and mitophagy in multiple sclerosis. *Proc. Natl. Acad. Sci.* **2021**, *118*, 2020078118. [CrossRef] [PubMed]
87. Song, M.; Franco, A.; Fleischer, J.A.; Zhang, L.; Dorn, G.W. Abrogating Mitochondrial Dynamics in Mouse Hearts Accelerates Mitochondrial Senescence. *Cell Metab.* **2017**, *26*, 872–883.e5. [CrossRef]
88. Torrano, V.; Valcarcel-Jimenez, L.; Cortazar, A.R.; Liu, X.; Urosevic, J.; Castillo-Martin, M.; Fernández-Ruiz, S.; Morciano, G.; Caro-Maldonado, A.; Guiu, M.; et al. The metabolic co-regulator PGC1α suppresses prostate cancer metastasis. *Nat. Cell Biol.* **2016**, *18*, 645–656. [CrossRef] [PubMed]
89. Sebastian, D.; Hernandez-Alvarez, M.I.; Segales, J.; Sorianello, E.; Muñoz, J.P.; Sala, D.; Waget, A.; Liesa, M.; Paz, J.C.; Gopalacharyulu, P.; et al. Mitofusin 2 (Mfn2) links mitochondrial and endoplasmic reticulum function with insulin signaling and is essential for normal glucose homeostasis. *Proc. Natl. Acad. Sci.* **2012**, *109*, 5523–5528. [CrossRef]
90. Kim, H.; Scimia, M.C.; Wilkinson, D.; Trelles, R.D.; Wood, M.R.; Bowtell, D.; Dillin, A.; Mercola, M.; Ronai, Z.A. Fine-Tuning of Drp1/Fis1 Availability by AKAP121/Siah2 Regulates Mitochondrial Adaptation to Hypoxia. *Mol. Cell* **2011**, *44*, 532–544. [CrossRef] [PubMed]
91. LeBleu, V.S.; O'Connell, J.T.; Gonzalez Herrera, K.N.G.; Wikman, H.; Pantel, K.; Haigis, M.C.; De Carvalho, F.M.; Damascena, A.; Domingos Chinen, L.T.; Rocha, R.M.; et al. PGC-1α mediates mitochondrial biogenesis and oxidative phosphorylation in cancer cells to promote metastasis. *Nat. Cell Biol.* **2014**, *16*, 992–1003. [CrossRef] [PubMed]
92. Palikaras, K.; Lionaki, E.; Tavernarakis, N. Balancing mitochondrial biogenesis and mitophagy to maintain energy metabolism homeostasis. *Cell Death Differ.* **2015**, *22*, 1399–1401. [CrossRef] [PubMed]
93. Youle, R.J.; Narendra, D.P. Mechanisms of mitophagy. *Nat. Rev. Mol. Cell Biol.* **2011**, *12*, 9–14. [CrossRef] [PubMed]
94. Bellot, G.; Garcia-Medina, R.; Gounon, P.; Chiche, J.; Roux, D.; Pouysségur, J.; Mazure, N.M. Hypoxia-Induced Autophagy Is Mediated through Hypoxia-Inducible Factor Induction of BNIP3 and BNIP3L via Their BH3 Domains. *Mol. Cell. Biol.* **2009**, *29*, 2570–2581. [CrossRef] [PubMed]
95. Sowter, H.M.; Ratcliffe, P.J.; Watson, P.; Greenberg, A.H.; Harris, A.L. HIF-1-dependent regulation of hypoxic induction of the cell death factors BNIP3 and NIX in human tumors. *Cancer Res.* **2001**, *61*, 6669–6673.
96. Liu, L.; Feng, D.; Chen, G.; Chen, M.; Zheng, Q.; Song, P.; Ma, Q.; Zhu, C.; Wang, R.; Qi, W.; et al. Mitochondrial outer-membrane protein FUNDC1 mediates hypoxia-induced mitophagy in mammalian cells. *Nat. Cell Biol.* **2014**, *14*, 177–185. [CrossRef]
97. Araki, K.; Kawauchi, K.; Sugimoto, W.; Tsuda, D.; Oda, H.; Yoshida, R.; Ohtani, K. Mitochondrial protein E2F3d, a distinctive E2F3 product, mediates hypoxia-induced mitophagy in cancer cells. *Commun. Biol.* **2019**, *2*, 3. [CrossRef]
98. Di Meo, S.; Reed, T.T.; Venditti, P.; Victor, V.M. Role of ROS and RNS Sources in Physiological and Pathological Conditions. *Oxid. Med. Cell. Longev.* **2016**, *2016*, 1245049. [CrossRef]
99. Dikalov, S.I.; Nazarewicz, R.R. Angiotensin II-Induced Production of Mitochondrial Reactive Oxygen Species: Potential Mechanisms and Relevance for Cardiovascular Disease. *Antioxid. Redox Signal.* **2013**, *19*, 1085–1094. [CrossRef]
100. Münzel, T.; Camici, G.G.; Maack, C.; Bonetti, N.R.; Fuster, V.; Kovacic, J.C. Impact of Oxidative Stress on the Heart and Vasculature. *J. Am. Coll. Cardiol.* **2017**, *70*, 212–229. [CrossRef]

101. Lenaz, G. The Mitochondrial Production of Reactive Oxygen Species: Mechanisms and Implications in Human Pathology. *IUBMB Life* **2001**, *52*, 159–164. [CrossRef] [PubMed]
102. Lambertucci, R.H.; Hirabara, S.M.; Silveira, L.D.R.; Levada-Pires, A.C.; Curi, R.; Pithon-Curi, T.C. Palmitate increases superoxide production through mitochondrial electron transport chain and NADPH oxidase activity in skeletal muscle cells. *J. Cell. Physiol.* **2008**, *216*, 796–804. [CrossRef]
103. Di Lisa, F.; Kaludercic, N.; Carpi, A.; Menabò, R.; Giorgio, M. Mitochondrial pathways for ROS formation and myocardial injury: The relevance of p66Shc and monoamine oxidase. *Basic Res. Cardiol.* **2009**, *104*, 131–139. [CrossRef]
104. Starkov, A.; Fiskum, G.; Chinopoulos, C.; Lorenzo, B.J.; Browne, S.E.; Patel, M.S.; Beal, M.F. Mitochondrial -Ketoglutarate Dehydrogenase Complex Generates Reactive Oxygen Species. *J. Neurosci.* **2004**, *24*, 7779–7788. [CrossRef]
105. St-Pierre, J.; Buckingham, J.A.; Roebuck, S.J.; Brand, M. Topology of Superoxide Production from Different Sites in the Mitochondrial Electron Transport Chain. *J. Biol. Chem.* **2002**, *277*, 44784–44790. [CrossRef]
106. Vásquez-Vivar, J.; Kalyanaraman, B.; Kennedy, M.C. Mitochondrial Aconitase Is a Source of Hydroxyl Radical. *J. Biol. Chem.* **2000**, *275*, 14064–14069. [CrossRef] [PubMed]
107. Weyemi, U.; Dupuy, C. The emerging role of ROS-generating NADPH oxidase NOX4 in DNA-damage responses. *Mutat. Res. Mutat. Res.* **2012**, *751*, 77–81. [CrossRef]
108. Giorgio, M.; Migliaccio, E.; Orsini, F.; Paolucci, D.; Moroni, M.; Contursi, C.; Pelliccia, G.; Luzi, L.; Minucci, S.; Marcaccio, M.; et al. Electron Transfer between Cytochrome c and p66Shc Generates Reactive Oxygen Species that Trigger Mitochondrial Apoptosis. *Cell* **2005**, *122*, 221–233. [CrossRef]
109. Giorgi, C.; Marchi, S.; Simoes, I.C.; Ren, Z.; Morciano, G.; Perrone, M.; Patalas-Krawczyk, P.; Borchard, S.; Jędrak, P.; Pierzynowska, K.; et al. Mitochondria and Reactive Oxygen Species in Aging and Age-Related Diseases. *Int. Rev. Cell Mol. Biol.* **2018**, *340*, 209–344. [CrossRef] [PubMed]
110. Arnaiz, S.L.; Coronel, M.F.; Boveris, A. Nitric Oxide, Superoxide, and Hydrogen Peroxide Production in Brain Mitochondria after Haloperidol Treatment. *Nitric Oxide* **1999**, *3*, 235–243. [CrossRef]
111. Tewari, S.; Santos, J.M.; Kowluru, R.A. Damaged Mitochondrial DNA Replication System and the Development of Diabetic Retinopathy. *Antioxid. Redox Signal.* **2012**, *17*, 492–504. [CrossRef] [PubMed]
112. Tewari, S.; Zhong, Q.; Santos, J.M.; A Kowluru, R. Mitochondria DNA Replication and DNA Methylation in the Metabolic Memory Associated with Continued Progression of Diabetic Retinopathy. *Investig. Ophthalmol. Vis. Sci.* **2012**, *53*, 4881–4888. [CrossRef]
113. Morciano, G.; Naumova, N.; Koprowski, P.; Valente, S.; Sardão, V.A.; Potes, Y.; Rimessi, A.; Wieckowski, M.R.; Oliveira, P.J. The mitochondrial permeability transition pore: An evolving concept critical for cell life and death. *Biol. Rev.* **2021**, *96*, 2489–2521. [CrossRef]
114. Radi, R.; Turrens, J.; Chang, L.; Bush, K.; Crapo, J.; Freeman, B. Detection of catalase in rat heart mitochondria. *J. Biol. Chem.* **1991**, *266*, 22028–22034. [CrossRef]
115. Zhou, R.; Yazdi, A.S.; Menu, P.; Tschopp, J. A role for mitochondria in NLRP3 inflammasome activation. *Nature* **2011**, *469*, 221–225. [CrossRef]
116. Zorov, D.B.; Juhaszova, M.; Sollott, S.J. Mitochondrial Reactive Oxygen Species (ROS) and ROS-Induced ROS Release. *Physiol. Rev.* **2014**, *94*, 909–950. [CrossRef]
117. Patergnani, S.; Bouhamida, E.; Leo, S.; Pinton, P.; Rimessi, A. Mitochondrial Oxidative Stress and "Mito-Inflammation": Actors in the Diseases. *Biomedicines* **2021**, *9*, 216. [CrossRef]
118. Bonora, M.; Wieckowski, M.; Sinclair, D.; Kroemer, G.; Pinton, P.; Galluzzi, L. Targeting mitochondria for cardiovascular disorders: Therapeutic potential and obstacles. *Nat. Rev. Cardiol.* **2019**, *16*, 33–55. [CrossRef]
119. Bonora, M.; Pinton, P. A New Current for the Mitochondrial Permeability Transition. *Trends Biochem. Sci.* **2019**, *44*, 559–561. [CrossRef] [PubMed]
120. Zhong, Z.; Liang, S.; Sanchez-Lopez, E.; He, F.; Shalapour, S.; Lin, X.-J.; Wong, J.; Ding, S.; Seki, E.; Schnabl, B.; et al. New mitochondrial DNA synthesis enables NLRP3 inflammasome activation. *Nature* **2018**, *560*, 198–203. [CrossRef] [PubMed]
121. Idzko, M.; Hammad, H.; Van Nimwegen, M.; Kool, M.; Willart, M.A.M.; Muskens, F.; Hoogsteden, H.C.; Luttmann, W.; Ferrari, D.; Di Virgilio, F.; et al. Extracellular ATP triggers and maintains asthmatic airway inflammation by activating dendritic cells. *Nat. Med.* **2007**, *13*, 913–919. [CrossRef] [PubMed]
122. Wang, P.; Chaung, W.W.; Wu, R.; Ji, Y.; Dong, W. Mitochondrial transcription factor A is a proinflammatory mediator in hemorrhagic shock. *Int. J. Mol. Med.* **2012**, *30*, 199–203. [CrossRef] [PubMed]
123. Zhang, Q.; Raoof, M.; Chen, Y.; Sumi, Y.; Sursal, T.; Junger, W.; Brohi, K.; Itagaki, K.; Hauser, C.J. Circulating mitochondrial DAMPs cause inflammatory responses to injury. *Nature* **2010**, *464*, 104–107. [CrossRef] [PubMed]
124. Nakahira, K.; Hisata, S.; Choi, A.M. The Roles of Mitochondrial Damage-Associated Molecular Patterns in Diseases. *Antioxid. Redox Signal.* **2015**, *23*, 1329–1350. [CrossRef] [PubMed]
125. Tuominen, A.; Miller, Y.I.; Hansen, L.F.; Kesäniemi, Y.A.; Witztum, J.L.; Hörkkö, S. A Natural Antibody to Oxidized Cardiolipin Binds to Oxidized Low-Density Lipoprotein, Apoptotic Cells, and Atherosclerotic Lesions. *Arter. Thromb. Vasc. Biol.* **2006**, *26*, 2096–2102. [CrossRef] [PubMed]

126. Misawa, T.; Takahama, M.; Kozaki, T.; Lee, H.; Zou, J.; Saitoh, T.; Akira, S. Microtubule-driven spatial arrangement of mitochondria promotes activation of the NLRP3 inflammasome. *Nat. Immunol.* **2013**, *14*, 454–460. [CrossRef]
127. Senoner, T.; Dichtl, W. Oxidative Stress in Cardiovascular Diseases: Still a Therapeutic Target? *Nutrients* **2019**, *11*, 2090. [CrossRef]
128. Morciano, G.; Vitto, V.; Bouhamida, E.; Giorgi, C.; Pinton, P. Mitochondrial Bioenergetics and Dynamism in the Failing Heart. *Life* **2021**, *11*, 436. [CrossRef]
129. Parathath, S.; Mick, S.L.; Feig, J.E.; Joaquin, V.; Grauer, L.; Habiel, D.M.; Gassmann, M.; Gardner, L.; Fisher, E.A. Hypoxia Is Present in Murine Atherosclerotic Plaques and Has Multiple Adverse Effects on Macrophage Lipid Metabolism. *Circ. Res.* **2011**, *109*, 1141–1152. [CrossRef]
130. Jin, Y.; Fu, J. Novel Insights Into the NLRP3 Inflammasome in Atherosclerosis. *J. Am. Hear. Assoc.* **2019**, *8*, e012219. [CrossRef]
131. Proudfoot, D.; Skepper, J.N.; Hegyi, L.; Bennett, M.R.; Shanahan, C.; Weissberg, P.L. Apoptosis Regulates Human Vascular Calcification In Vitro. *Circ. Res.* **2000**, *87*, 1055–1062. [CrossRef] [PubMed]
132. Hutcheson, J.D.; Goettsch, C.; Bertazzo, S.; Maldonado, N.; Ruiz, J.L.; Bin Goh, W.W.; Yabusaki, K.; Faits, T.; Bouten, C.; Franck, G.; et al. Genesis and growth of extracellular-vesicle-derived microcalcification in atherosclerotic plaques. *Nat. Mater.* **2016**, *15*, 335–343. [CrossRef] [PubMed]
133. Harrison, C.M.; Pompilius, M.; Pinkerton, K.E.; Ballinger, S.W. Mitochondrial Oxidative Stress Significantly Influences Atherogenic Risk and Cytokine-Induced Oxidant Production. *Environ. Heal. Perspect.* **2011**, *119*, 676–681. [CrossRef]
134. Orsini, F.; Migliaccio, E.; Moroni, M.; Contursi, C.; Raker, V.A.; Piccini, D.; Martin-Padura, I.; Pelliccia, G.; Trinei, M.; Bono, M.; et al. The Life Span Determinant p66Shc Localizes to Mitochondria Where It Associates with Mitochondrial Heat Shock Protein 70 and Regulates Trans-membrane Potential. *J. Biol. Chem.* **2004**, *279*, 25689–25695. [CrossRef]
135. Graiani, G.; Lagrasta, C.; Migliaccio, E.; Spillmann, F.; Meloni, M.; Madeddu, P.; Quaini, F.; Padura, I.M.; Lanfrancone, L.; Pelicci, P.; et al. Genetic Deletion of the p66 Shc Adaptor Protein Protects from Angiotensin II–Induced Myocardial Damage. *Hypertension* **2005**, *46*, 433–440. [CrossRef] [PubMed]
136. Rota, M.; LeCapitaine, N.; Hosoda, T.; Boni, A.; De Angelis, A.; Padin-Iruegas, M.E.; Esposito, G.; Vitale, S.; Urbanek, K.; Casarsa, C.; et al. Diabetes Promotes Cardiac Stem Cell Aging and Heart Failure, Which Are Prevented by Deletion of the p66 shc Gene. *Circ. Res.* **2006**, *99*, 42–52. [CrossRef]
137. Kaludercic, N.; Takimoto, E.; Nagayama, T.; Feng, N.; Lai, E.W.; Bedja, D.; Chen, K.; Gabrielson, K.L.; Blakely, R.D.; Shih, J.C.; et al. Monoamine Oxidase A–Mediated Enhanced Catabolism of Norepinephrine Contributes to Adverse Remodeling and Pump Failure in Hearts with Pressure Overload. *Circ. Res.* **2010**, *106*, 193–202. [CrossRef]
138. Kaludercic, N.; Carpi, A.; Nagayama, T.; Sivakumaran, V.; Zhu, G.; Lai, E.W.; Bedja, D.; De Mario, A.; Chen, K.; Gabrielson, K.L.; et al. Monoamine Oxidase B Prompts Mitochondrial and Cardiac Dysfunction in Pressure Overloaded Hearts. *Antioxid. Redox Signal.* **2014**, *20*, 267–280. [CrossRef]
139. Kuroda, J.; Ago, T.; Matsushima, S.; Zhai, P.; Schneider, M.; Sadoshima, J. NADPH oxidase 4 (Nox4) is a major source of oxidative stress in the failing heart. *Proc. Natl. Acad. Sci.* **2010**, *107*, 15565–15570. [CrossRef]
140. Ikegami, T.; Suzuki, Y.-I.; Shimizu, T.; Isono, K.; Koseki, H.; Shirasawa, T. Model mice for tissue-specific deletion of the manganese superoxide dismutase (MnSOD) gene. *Biochem. Biophys. Res. Commun.* **2002**, *296*, 729–736. [CrossRef]
141. Sharma, S.; Bhattarai, S.; Ara, H.; Sun, G.; Clair, D.K.S.; Bhuiyan, S.; Kevil, C.; Watts, M.N.; Dominic, P.; Shimizu, T.; et al. SOD2 deficiency in cardiomyocytes defines defective mitochondrial bioenergetics as a cause of lethal dilated cardiomyopathy. *Redox Biol.* **2020**, *37*, 101740. [CrossRef] [PubMed]
142. Souiden, Y.; Mallouli, H.; Meskhi, S.; Chaabouni, Y.; Rebai, A.; Chéour, F.; Mahdouani, K. MnSOD and GPx1 polymorphism relationship with coronary heart disease risk and severity. *Biol. Res.* **2016**, *49*, 22. [CrossRef]
143. Faraci, F.M.; Didion, S.P. Vascular Protection. *Arter. Thromb. Vasc. Biol.* **2004**, *24*, 1367–1373. [CrossRef] [PubMed]
144. Ardanaz, N.; Yang, X.-P.; Cifuentes, M.E.; Haurani, M.J.; Jackson, K.W.; Liao, T.-D.; Carretero, O.A.; Pagano, P.J. Lack of Glutathione Peroxidase 1 Accelerates Cardiac-Specific Hypertrophy and Dysfunction in Angiotensin II Hypertension. *Hypertension* **2010**, *55*, 116–123. [CrossRef] [PubMed]
145. Chen, Z.; Chua, C.C.; Gao, J.; Chua, K.-W.; Ho, Y.-S.; Hamdy, R.C.; Chua, B.H.L. Prevention of ischemia/reperfusion-induced cardiac apoptosis and injury by melatonin is independent of glutathione peroxdiase 1. *J. Pineal Res.* **2009**, *46*, 235–241. [CrossRef]
146. Matsushima, S.; Ide, T.; Yamato, M.; Matsusaka, H.; Hattori, F.; Ikeuchi, M.; Kubota, T.; Sunagawa, K.; Hasegawa, Y.; Kurihara, T.; et al. Overexpression of Mitochondrial Peroxiredoxin-3 Prevents Left Ventricular Remodeling and Failure After Myocardial Infarction in Mice. *Circulation* **2006**, *113*, 1779–1786. [CrossRef] [PubMed]
147. Huang, Q.; Zhou, H.J.; Zhang, H.; Huang, Y.; Hinojosa-Kirschenbaum, F.; Fan, P.; Yao, L.; Belardinelli, L.; Tellides, G.; Giordano, F.J.; et al. Thioredoxin-2 Inhibits Mitochondrial Reactive Oxygen Species Generation and Apoptosis Stress Kinase-1 Activity to Maintain Cardiac Function. *Circulation* **2015**, *131*, 1082–1097. [CrossRef]
148. Flores-Mateo, G.; Carrillo-Santisteve, P.; Elosua, R.; Guallar, E.; Marrugat, J.; Bleys, J.; Covas, M.-I. Antioxidant Enzyme Activity and Coronary Heart Disease: Meta-analyses of Observational Studies. *Am. J. Epidemiol.* **2009**, *170*, 135–147. [CrossRef]
149. Tonet, E.; Bernucci, D.; Morciano, G.; Campo, G. Pharmacological protection of reperfusion injury in ST-segment elevation myocardial infarction. Gone with the wind? *Postepy Kardiol Interwencyjnej* **2018**, *14*, 5–8. [CrossRef]
150. Calbet, J.A. Chronic hypoxia increases blood pressure and noradrenaline spillover in healthy humans. *J. Physiol.* **2003**, *551*, 379–386. [CrossRef]

151. Lee, S.H.; Wolf, P.L.; Escudero, R.; Deutsch, R.; Jamieson, S.W.; Thistlethwaite, P.A. Early Expression of Angiogenesis Factors in Acute Myocardial Ischemia and Infarction. *New Engl. J. Med.* **2000**, *342*, 626–633. [CrossRef] [PubMed]
152. Liu, Y.; Luo, Q.; Su, Z.; Xing, J.; Wu, J.; Xiang, L.; Huang, Y.; Pan, H.; Wu, X.; Zhang, X.; et al. Suppression of Myocardial Hypoxia-Inducible Factor-1α Compromises Metabolic Adaptation and Impairs Cardiac Function in Patients with Cyanotic Congenital Heart Disease During Puberty. *Circulation* **2021**, *143*, 2254–2272. [CrossRef]
153. Murry, C.E.; Jennings, R.B.; Reimer, K.A. Preconditioning with ischemia: A delay of lethal cell injury in ischemic myocardium. *Circulation* **1986**, *74*, 1124–1136. [CrossRef]
154. Morciano, G.; Giorgi, C.; Bonora, M.; Punzetti, S.; Pavasini, R.; Wieckowski, M.; Campo, G.; Pinton, P. Molecular identity of the mitochondrial permeability transition pore and its role in ischemia-reperfusion injury. *J. Mol. Cell. Cardiol.* **2015**, *78*, 142–153. [CrossRef]
155. Giorgi, C.; Marchi, S.; Pinton, P. The machineries, regulation and cellular functions of mitochondrial calcium. *Nat. Rev. Mol. Cell Biol.* **2018**, *19*, 713–730. [CrossRef]
156. Bonora, M.; Giorgi, C.; Pinton, P. Molecular mechanisms and consequences of mitochondrial permeability transition. *Nat. Rev. Mol. Cell Biol.* **2021**, 1–20. [CrossRef] [PubMed]
157. Modesti, L.; Danese, A.; Vitto, V.A.M.; Ramaccini, D.; Aguiari, G.; Gafà, R.; Lanza, G.; Giorgi, C.; Pinton, P. Mitochondrial Ca^{2+} Signaling in Health, Disease and Therapy. *Cells* **2021**, *10*, 1317. [CrossRef] [PubMed]
158. Luongo, T.; Lambert, J.; Yuan, A.; Zhang, X.; Gross, P.; Song, J.; Shanmughapriya, S.; Gao, E.; Jain, M.; Houser, S.R.; et al. The Mitochondrial Calcium Uniporter Matches Energetic Supply with Cardiac Workload during Stress and Modulates Permeability Transition. *Cell Rep.* **2015**, *12*, 23–34. [CrossRef]
159. Kwong, J.Q.; Lu, X.; Correll, R.N.; Schwanekamp, J.; Vagnozzi, R.J.; Sargent, M.A.; York, A.J.; Zhang, J.; Bers, D.; Molkentin, J.D. The Mitochondrial Calcium Uniporter Selectively Matches Metabolic Output to Acute Contractile Stress in the Heart. *Cell Rep.* **2015**, *12*, 15–22. [CrossRef]
160. Chouchani, E.T.; Pell, V.R.; Gaude, E.; Aksentijevic, D.; Sundier, S.Y.; Robb, E.L.; Logan, A.; Nadtochiy, S.M.; Ord, E.N.J.; Smith, A.C.; et al. Ischaemic accumulation of succinate controls reperfusion injury through mitochondrial ROS. *Nature* **2014**, *515*, 431–435. [CrossRef]
161. Jang, S.; Lewis, T.S.; Powers, C.; Khuchua, Z.; Baines, C.P.; Wipf, P.; Javadov, S. Elucidating Mitochondrial Electron Transport Chain Supercomplexes in the Heart During Ischemia–Reperfusion. *Antioxid. Redox Signal.* **2017**, *27*, 57–69. [CrossRef] [PubMed]
162. Mezzaroma, E.; Toldo, S.; Farkas, D.; Seropian, I.M.; Van Tassell, B.W.; Salloum, F.; Kannan, H.R.; Menna, A.C.; Voelkel, N.F.; Abbate, A. The inflammasome promotes adverse cardiac remodeling following acute myocardial infarction in the mouse. *Proc. Natl. Acad. Sci.* **2011**, *108*, 19725–19730. [CrossRef] [PubMed]
163. Sandanger, Ø.; Ranheim, T.; Vinge, L.E.; Bliksøen, M.; Alfsnes, K.; Finsen, A.V.; Dahl, C.P.; Askevold, E.T.; Florholmen, G.; Christensen, G.; et al. The NLRP3 inflammasome is up-regulated in cardiac fibroblasts and mediates myocardial ischaemia–reperfusion injury. *Cardiovasc. Res.* **2013**, *99*, 164–174. [CrossRef] [PubMed]
164. Malhotra, R.; Tyson, D.W.; Rosevear, H.M.; Brosius, F.C. Hypoxia-inducible factor-1alpha is a critical mediator of hypoxia induced apoptosis in cardiac H9c2 and kidney epithelial HK-2 cells. *BMC Cardiovasc. Disord.* **2008**, *8*, 9. [CrossRef]
165. Moslehi, J.; Minamishima, Y.A.; Shi, J.; Neuberg, D.; Charytan, D.; Padera, R.F.; Signoretti, S.; Liao, R.; Kaelin, W.G. Loss of Hypoxia-Inducible Factor Prolyl Hydroxylase Activity in Cardiomyocytes Phenocopies Ischemic Cardiomyopathy. *Circulation* **2010**, *122*, 1004–1016. [CrossRef]
166. Loor, G.; Schumacker, P.T. Role of hypoxia-inducible factor in cell survival during myocardial ischemia–reperfusion. *Cell Death Differ.* **2008**, *15*, 686–690. [CrossRef] [PubMed]
167. Li, B.; Li, M.; Li, X.; Li, H.; Lai, Y.; Huang, S.; He, X.; Si, X.; Zheng, H.; Liao, W.; et al. Sirt1-inducible deacetylation of p21 promotes cardiomyocyte proliferation. *Aging* **2019**, *11*, 12546–12567. [CrossRef] [PubMed]
168. Luo, G.; Jian, Z.; Zhu, Y.; Chen, B.; Ma, R.; Tang, F.; Xiao, Y. Sirt1 promotes autophagy and inhibits apoptosis to protect cardiomyocytes from hypoxic stress. *Int. J. Mol. Med.* **2019**, *43*, 2033–2043. [CrossRef]
169. Robador, P.A.; José, G.S.; Rodríguez, C.; Guadall, A.; Moreno, M.U.; Beaumont, J.; Fortuño, A.; Díez, J.; Martínez-González, J.; Zalba, G. HIF-1-mediated up-regulation of cardiotrophin-1 is involved in the survival response of cardiomyocytes to hypoxia. *Cardiovasc. Res.* **2011**, *92*, 247–255. [CrossRef]
170. Tennant, D.; Howell, N.J. The role of HIFs in ischemia-reperfusion injury. *Hypoxia* **2014**, *2*, 107–115. [CrossRef]
171. Wan, D.-Y.; Zhang, Z.; Yang, H.-H. Cardioprotective effect of miR-214 in myocardial ischemic postconditioning by down-regulation of hypoxia inducible factor 1, alpha subunit inhibitor. *Cell. Mol. Biol.* **2015**, *61*, 1–6.
172. Stanley, M.; Crowcroft, N.; Quigley, J.; Parkinson, E. Responses of human cervical keratinocytes in vitro to tumour promoters and diethylstilboestrol. *Carcinogenesis* **1985**, *6*, 1011–1015. [CrossRef] [PubMed]
173. Chaudhuri, R.D.; Banik, A.; Mandal, B.; Sarkar, S. Cardiac-specific overexpression of HIF-1α during acute myocardial infarction ameliorates cardiomyocyte apoptosis via differential regulation of hypoxia-inducible pro-apoptotic and anti-oxidative genes. *Biochem. Biophys. Res. Commun.* **2021**, *537*, 100–108. [CrossRef]
174. Yang, B.; He, K.; Zheng, F.; Wan, L.; Yu, X.; Wang, X.; Zhao, D.; Bai, Y.; Chu, W.; Sun, Y.; et al. Over-expression of hypoxia-inducible factor-1 alpha in vitro protects the cardiac fibroblasts from hypoxia-induced apoptosis. *J. Cardiovasc. Med.* **2014**, *15*, 579–586. [CrossRef]

175. Zhu, H.; Sun, A. Programmed necrosis in heart disease: Molecular mechanisms and clinical implications. *J. Mol. Cell. Cardiol.* **2018**, *116*, 125–134. [CrossRef]
176. Piamsiri, C.; Maneechote, C.; Siri-Angkul, N.; Chattipakorn, S.C.; Chattipakorn, N. Targeting necroptosis as therapeutic potential in chronic myocardial infarction. *J. Biomed. Sci.* **2021**, *28*, 1–13. [CrossRef]
177. Karunakaran, D.; Geoffrion, M.; Wei, L.; Gan, W.; Richards, L.; Shangari, P.; DeKemp, E.M.; Beanlands, R.A.; Perisic, L.; Maegdefessel, L.; et al. Targeting macrophage necroptosis for therapeutic and diagnostic interventions in atherosclerosis. *Sci. Adv.* **2016**, *2*, e1600224. [CrossRef]
178. Zhang, T.; Zhang, Y.; Cui, M.; Jin, L.; Wang, Y.; Lv, F.; Liu, Y.; Zheng, W.; Shang, H.; Zhang, J.; et al. CaMKII is a RIP3 substrate mediating ischemia- and oxidative stress–induced myocardial necroptosis. *Nat. Med.* **2016**, *22*, 175–182. [CrossRef] [PubMed]
179. Karshovska, E.; Wei, Y.; Subramanian, P.; Mohibullah, R.; Geißler, C.; Baatsch, I.; Popal, A.; Campos, J.C.; Exner, N.; Schober, A. HIF-1α (Hypoxia-Inducible Factor-1α) Promotes Macrophage Necroptosis by Regulating miR-210 and miR-383. *Arter. Thromb. Vasc. Biol.* **2020**, *40*, 583–596. [CrossRef] [PubMed]
180. Liu, B.; Zhao, C.; Li, H.; Chen, X.; Ding, Y.; Xu, S. Puerarin protects against heart failure induced by pressure overload through mitigation of ferroptosis. *Biochem. Biophys. Res. Commun.* **2018**, *497*, 233–240. [CrossRef]
181. Chen, X.; Xu, S.; Zhao, C.; Liu, B. Role of TLR4/NADPH oxidase 4 pathway in promoting cell death through autophagy and ferroptosis during heart failure. *Biochem. Biophys. Res. Commun.* **2019**, *516*, 37–43. [CrossRef]
182. Yoshimura, C.; Nagasaka, A.; Kurose, H.; Nakaya, M. Efferocytosis during myocardial infarction. *J. Biochem.* **2020**, *168*, 1–6. [CrossRef]
183. Davis, C.K.; Jain, S.A.; Bae, O.-N.; Majid, A.; Rajanikant, G.K. Hypoxia Mimetic Agents for Ischemic Stroke. *Front. Cell Dev. Biol.* **2019**, *6*, 175. [CrossRef]
184. Yu, Y.; Yan, Y.; Niu, F.; Wang, Y.; Chen, X.; Su, G.; Liu, Y.; Zhao, X.; Qian, L.; Liu, P.; et al. Ferroptosis: A cell death connecting oxidative stress, inflammation and cardiovascular diseases. *Cell Death Discov.* **2021**, *7*, 1–10. [CrossRef] [PubMed]
185. Li, W.; Li, W.; Leng, Y.; Xiong, Y.; Xia, Z. Ferroptosis Is Involved in Diabetes Myocardial Ischemia/Reperfusion Injury Through Endoplasmic Reticulum Stress. *DNA Cell Biol.* **2020**, *39*, 210–225. [CrossRef]
186. Gao, M.; Monian, P.; Quadri, N.; Ramasamy, R.; Jiang, X. Glutaminolysis and Transferrin Regulate Ferroptosis. *Mol. Cell* **2015**, *59*, 298–308. [CrossRef] [PubMed]
187. Tang, L.-J.; Luo, X.-J.; Tu, H.; Chen, H.; Xiong, X.-M.; Li, N.-S.; Peng, J. Ferroptosis occurs in phase of reperfusion but not ischemia in rat heart following ischemia or ischemia/reperfusion. *Naunyn-Schmiedebergs Arch. fur Exp. Pathol. Pharmakol.* **2021**, *394*, 401–410. [CrossRef] [PubMed]
188. Tay, W.T.; Fang, Y.-H.; Beh, S.T.; Liu, Y.-W.; Hsu, L.-W.; Yen, C.-J.; Liu, P.-Y. Programmed Cell Death-1: Programmed Cell Death-Ligand 1 Interaction Protects Human Cardiomyocytes Against T-Cell Mediated Inflammation and Apoptosis Response In Vitro. *Int. J. Mol. Sci.* **2020**, *21*, 2399. [CrossRef]
189. Van der Windt, G.J.W.; O'Sullivan, D.; Everts, B.; Huang, S.C.-C.; Buck, M.; Curtis, J.D.; Chang, C.-H.; Smith, A.M.; Ai, T.; Faubert, B.; et al. CD8 memory T cells have a bioenergetic advantage that underlies their rapid recall ability. *Proc. Natl. Acad. Sci. USA* **2013**, *110*, 14336–14341. [CrossRef]
190. Wang, R.; Dillon, C.P.; Shi, L.Z.; Milasta, S.; Carter, R.; Finkelstein, D.; McCormick, L.L.; Fitzgerald, P.; Chi, H.; Munger, J.; et al. The Transcription Factor Myc Controls Metabolic Reprogramming upon T Lymphocyte Activation. *Immunity* **2011**, *35*, 871–882. [CrossRef]
191. Gao, P.; Tchernyshyov, I.; Chang, T.-C.; Lee, Y.-S.; Kita, K.; Ochi, T.; Zeller, K.I.; De Marzo, A.M.; Van Eyk, J.E.; Mendell, J.T.; et al. c-Myc suppression of miR-23a/b enhances mitochondrial glutaminase expression and glutamine metabolism. *Nature* **2009**, *458*, 762–765. [CrossRef] [PubMed]
192. Shi, L.Z.; Wang, R.; Huang, G.; Vogel, P.; Neale, G.; Green, D.R.; Chi, H. HIF1α–dependent glycolytic pathway orchestrates a metabolic checkpoint for the differentiation of TH17 and Treg cells. *J. Exp. Med.* **2011**, *208*, 1367–1376. [CrossRef] [PubMed]
193. Finlay, D.K.; Rosenzweig, E.; Sinclair, L.V.; Carmen, F.C.; Hukelmann, J.L.; Rolf, J.; Panteleyev, A.A.; Okkenhaug, K.; Cantrell, D.A. PDK1 regulation of mTOR and hypoxia-inducible factor 1 integrate metabolism and migration of CD8+ T cells. *J. Exp. Med.* **2012**, *209*, 2441–2453. [CrossRef] [PubMed]
194. Gualdoni, G.A.; Mayer, K.A.; Göschl, L.; Boucheron, N.; Ellmeier, W.; Zlabinger, G.J. The AMP analog AICAR modulates the T reg /T h 17 axis through enhancement of fatty acid oxidation. *FASEB J.* **2016**, *30*, 3800–3809. [CrossRef]
195. Sun, Y.; Tian, T.; Gao, J.; Liu, X.; Hou, H.; Cao, R.; Li, B.; Quan, M.; Guo, L. Metformin ameliorates the development of experimental autoimmune encephalomyelitis by regulating T helper 17 and regulatory T cells in mice. *J. Neuroimmunol.* **2016**, *292*, 58–67. [CrossRef]
196. Rolf, J.; Zarrouk, M.; Finlay, D.K.; Foretz, M.; Viollet, B.; Cantrell, D.A. AMPKα1: A glucose sensor that controls CD8 T-cell memory. *Eur. J. Immunol.* **2013**, *43*, 889–896. [CrossRef]
197. Pearce, E.L.; Walsh, M.C.; Cejas, P.J.; Harms, G.M.; Shen, H.; Wang, L.-S.; Jones, R.G.; Choi, Y. Enhancing CD8 T-cell memory by modulating fatty acid metabolism. *Nature* **2009**, *460*, 103–107. [CrossRef]
198. Putignani, L.; Raffa, S.; Pescosolido, R.; Aimati, L.; Signore, F.; Torrisi, M.R.; Grammatico, P. Alteration of expression levels of the oxidative phosphorylation system (OXPHOS) in breast cancer cell mitochondria. *Breast Cancer Res. Treat.* **2007**, *110*, 439–452. [CrossRef]

199. Simonnet, H.; Alazard, N.; Pfeiffer, K.; Gallou, C.; Beroud, C.; Demont, J.; Bouvier, R.; Schägger, H.; Godinot, C. Low mitochondrial respiratory chain content correlates with tumor aggressiveness in renal cell carcinoma. *Carcinog.* **2002**, *23*, 759–768. [CrossRef]
200. Bonora, E.; Porcelli, A.M.; Gasparre, G.; Biondi, A.; Ghelli, A.; Carelli, V.; Baracca, A.; Tallini, G.; Martinuzzi, A.; Lenaz, G.; et al. Defective Oxidative Phosphorylation in Thyroid Oncocytic Carcinoma Is Associated with Pathogenic Mitochondrial DNA Mutations Affecting Complexes I and III. *Cancer Res.* **2006**, *66*, 6087–6096. [CrossRef]
201. Robles, A.I.; Traverso, G.; Zhang, M.; Roberts, N.J.; Khan, M.A.; Joseph, C.; Lauwers, G.Y.; Selaru, F.M.; Popoli, M.; Pittman, M.E.; et al. Whole-Exome Sequencing Analyses of Inflammatory Bowel Disease−Associated Colorectal Cancers. *Gastroenterology* **2016**, *150*, 931–943. [CrossRef] [PubMed]
202. Dmitrieva-Posocco, O.; Dzutsev, A.; Posocco, D.F.; Hou, V.; Yuan, W.; Thovarai, V.; Mufazalov, I.; Gunzer, M.; Shilovskiy, I.; Khaitov, M.; et al. Cell-Type-Specific Responses to Interleukin-1 Control Microbial Invasion and Tumor-Elicited Inflammation in Colorectal Cancer. *Immunity* **2019**, *50*, 166–180. [CrossRef] [PubMed]
203. Grivennikov, S.; Karin, E.; Terzic, J.; Mucida, D.; Yu, G.-Y.; Vallabhapurapu, S.; Scheller, J.; Rose-John, S.; Cheroutre, H.; Eckmann, L.; et al. IL-6 and Stat3 Are Required for Survival of Intestinal Epithelial Cells and Development of Colitis-Associated Cancer. *Cancer Cell* **2009**, *15*, 103–113. [CrossRef] [PubMed]
204. Joshi, A.U.; Minhas, P.S.; Liddelow, S.A.; Haileselassie, B.; Andreasson, K.I.; Dorn, G.W., II; Mochly-Rosen, D. Fragmented mitochondria released from microglia trigger A1 astrocytic response and propagate inflammatory neurodegeneration. *Nat. Neurosci.* **2019**, *22*, 1635–1648. [CrossRef] [PubMed]
205. Panchanathan, R.; Liu, H.; Choubey, D. Hypoxia primes human normal prostate epithelial cells and cancer cell lines for the NLRP3 and AIM2 inflammasome activation. *Oncotarget* **2016**, *7*, 28183–28194. [CrossRef] [PubMed]
206. Li, X.; Zhang, X.; Xia, J.; Zhang, L.; Chen, B.; Lian, G.; Yun, C.; Yang, J.; Yan, Y.; Wang, P.; et al. Macrophage HIF-2α suppresses NLRP3 inflammasome activation and alleviates insulin resistance. *Cell Rep.* **2021**, *36*, 109607. [CrossRef]
207. Moon, J.-S.; Nakahira, K.; Chung, K.-P.; DeNicola, G.; Koo, M.J.; Pabón, M.A.; Rooney, K.T.; Yoon, J.-H.; Ryter, S.W.; Stout-Delgado, H.; et al. NOX4-dependent fatty acid oxidation promotes NLRP3 inflammasome activation in macrophages. *Nat. Med.* **2016**, *22*, 1002–1012. [CrossRef]
208. Missiroli, S.; Perrone, M.; Boncompagni, C.; Borghi, C.; Campagnaro, A.; Marchetti, F.; Anania, G.; Greco, P.; Fiorica, F.; Pinton, P.; et al. Targeting the NLRP3 Inflammasome as a New Therapeutic Option for Overcoming Cancer. *Cancers* **2021**, *13*, 2297. [CrossRef]
209. Guo, B.; Fu, S.; Zhang, J.; Liu, B.; Li, Z. Targeting inflammasome/IL-1 pathways for cancer immunotherapy. *Sci. Rep.* **2016**, *6*, 36107. [CrossRef] [PubMed]
210. Kaplanov, I.; Carmi, Y.; Kornetsky, R.; Shemesh, A.; Shurin, G.V.; Shurin, M.R.; Dinarello, C.A.; Voronov, E.; Apte, R.N. Blocking IL-1β reverses the immunosuppression in mouse breast cancer and synergizes with anti–PD-1 for tumor abrogation. *Proc. Natl. Acad. Sci. USA* **2018**, *116*, 1361–1369. [CrossRef]
211. Wang, Y.; Kong, H.; Zeng, X.; Liu, W.; Wang, Z.; Yan, X.; Wang, H.; Xie, W. Activation of NLRP3 inflammasome enhances the proliferation and migration of A549 lung cancer cells. *Oncol. Rep.* **2016**, *35*, 2053–2064. [CrossRef] [PubMed]
212. Okamoto, M.; Liu, W.; Luo, Y.; Tanaka, A.; Cai, X.; Norris, D.A.; Dinarello, C.A.; Fujita, M. Constitutively Active Inflammasome in Human Melanoma Cells Mediating Autoinflammation via Caspase-1 Processing and Secretion of Interleukin-1β. *J. Biol. Chem.* **2010**, *285*, 6477–6488. [CrossRef]
213. Ahmad, I.; Muneer, K.M.; Tamimi, I.A.; Chang, M.E.; Ata, M.O.; Yusuf, N. Thymoquinone suppresses metastasis of melanoma cells by inhibition of NLRP3 inflammasome. *Toxicol. Appl. Pharmacol.* **2013**, *270*, 70–76. [CrossRef] [PubMed]
214. Zaki, H.; Vogel, P.; Body-Malapel, M.; Lamkanfi, M.; Kanneganti, T.-D. IL-18 Production Downstream of the Nlrp3 Inflammasome Confers Protection against Colorectal Tumor Formation. *J. Immunol.* **2010**, *185*, 4912–4920. [CrossRef] [PubMed]
215. Huber, S.; Gagliani, N.; Zenewicz, L.; Huber, F.J.; Bosurgi, L.; Hu, B.; Hedl, M.; Zhang, W.; O'connor, W.; Murphy, A.; et al. IL-22BP is regulated by the inflammasome and modulates tumorigenesis in the intestine. *Nature* **2012**, *491*, 259–263. [CrossRef]
216. Wei, Q.; Mu, K.; Li, T.; Zhang, Y.; Yang, Z.; Jia, X.; Zhao, W.; Huai, W.; Guo, P.; Han, L. Deregulation of the NLRP3 inflammasome in hepatic parenchymal cells during liver cancer progression. *Lab. Investig.* **2013**, *94*, 52–62. [CrossRef]
217. Ayala, A.; Muñoz, M.F.; Argüelles, S. Lipid Peroxidation: Production, Metabolism, and Signaling Mechanisms of Malondialdehyde and 4-Hydroxy-2-Nonenal. *Oxid. Med. Cell. Longev.* **2014**, *2014*, 360438. [CrossRef]
218. Du, W.; Zhang, L.; Brett-Morris, A.; Aguila, B.; Kerner, J.; Hoppel, C.L.; Puchowicz, M.; Serra, D.; Herrero, L.; Rini, B.I.; et al. HIF drives lipid deposition and cancer in ccRCC via repression of fatty acid metabolism. *Nat. Commun.* **2017**, *8*, 1–12. [CrossRef]
219. Todisco, S.; Convertini, P.; Iacobazzi, V.; Infantino, V. TCA Cycle Rewiring as Emerging Metabolic Signature of Hepatocellular Carcinoma. *Cancers* **2019**, *12*, 68. [CrossRef]
220. Huang, D.; Li, T.; Li, X.; Zhang, L.; Sun, L.; He, X.; Zhong, X.; Jia, D.; Song, L.; Semenza, G.L.; et al. HIF-1-Mediated Suppression of Acyl-CoA Dehydrogenases and Fatty Acid Oxidation Is Critical for Cancer Progression. *Cell Rep.* **2014**, *8*, 1930–1942. [CrossRef]
221. Mylonis, I.; Simos, G.; Paraskeva, E. Hypoxia-Inducible Factors and the Regulation of Lipid Metabolism. *Cells* **2019**, *8*, 214. [CrossRef] [PubMed]
222. Sun, W.; Kato, H.; Kitajima, S.; Lee, K.L.; Gradin, K.; Okamoto, T.; Poellinger, L. Interaction between von Hippel-Lindau Protein and Fatty Acid Synthase Modulates Hypoxia Target Gene Expression. *Sci. Rep.* **2017**, *7*, 1–11. [CrossRef] [PubMed]
223. Han, J.S.; Lee, J.H.; Kong, J.; Ji, Y.; Kim, J.; Choe, S.S.; Kim, J.B. Hypoxia Restrains Lipid Utilization via Protein Kinase A and Adipose Triglyceride Lipase Downregulation through Hypoxia-Inducible Factor. *Mol. Cell. Biol.* **2019**, *39*. [CrossRef] [PubMed]

224. Mylonis, I.; Sembongi, H.; Befani, C.; Liakos, P.; Siniossoglou, S.; Simos, G. Hypoxia causes triglyceride accumulation via HIF-1-mediated stimulation of lipin 1 expression. *J. Cell Sci.* **2012**, *125*, 3485–3493. [CrossRef]
225. Bensaad, K.; Favaro, E.; Lewis, C.A.; Peck, B.; Lord, S.; Collins, J.M.; Pinnick, K.E.; Wigfield, S.; Buffa, F.M.; Li, J.-L.; et al. Fatty Acid Uptake and Lipid Storage Induced by HIF-1α Contribute to Cell Growth and Survival after Hypoxia-Reoxygenation. *Cell Rep.* **2014**, *9*, 349–365. [CrossRef]
226. Nejad, A.E.; Najafgholian, S.; Rostami, A.; Sistani, A.; Shojaeifar, S.; Esparvarinha, M.; Nedaeinia, R.; Javanmard, S.H.; Taherian, M.; Ahmadlou, M.; et al. The role of hypoxia in the tumor microenvironment and development of cancer stem cell: A novel approach to developing treatment. *Cancer Cell Int.* **2021**, *21*, 1–26. [CrossRef]
227. Graham, K.; Unger, E. Overcoming tumor hypoxia as a barrier to radiotherapy, chemotherapy and immunotherapy in cancer treatment. *Int. J. Nanomed.* **2018**, *13*, 6049–6058. [CrossRef]
228. Akakura, N.; Kobayashi, M.; Horiuchi, I.; Suzuki, A.; Wang, J.; Chen, J.; Niizeki, H.; Ki, K.; Hosokawa, M.; Asaka, M. Constitutive expression of hypoxia-inducible factor-1alpha renders pancreatic cancer cells resistant to apoptosis induced by hypoxia and nutrient deprivation. *Cancer Res.* **2001**, *61*, 6548–6554. [PubMed]
229. Steinbach, J.P.; Wolburg, H.; Klumpp, A.; Probst, H.; Weller, M. Hypoxia-induced cell death in human malignant glioma cells: Energy deprivation promotes decoupling of mitochondrial cytochrome c release from caspase processing and necrotic cell death. *Cell Death Differ.* **2003**, *10*, 823–832. [CrossRef]
230. Li, Z.; Jiang, L.; Chew, S.H.; Hirayama, T.; Sekido, Y.; Toyokuni, S. Carbonic anhydrase 9 confers resistance to ferroptosis/apoptosis in malignant mesothelioma under hypoxia. *Redox Biol.* **2019**, *26*, 101297. [CrossRef]
231. Zou, Y.; Palte, M.J.; Deik, A.A.; Li, H.; Eaton, J.K.; Wang, W.; Tseng, Y.-Y.; Deasy, R.; Kost-Alimova, M.; Dančík, V.; et al. A GPX4-dependent cancer cell state underlies the clear-cell morphology and confers sensitivity to ferroptosis. *Nat. Commun.* **2019**, *10*, 1–13. [CrossRef]
232. Nie, Q.; Hu, Y.; Yu, X.; Li, X.; Fang, X. Induction and application of ferroptosis in cancer therapy. *Cancer Cell Int.* **2022**, *22*, 1–19. [CrossRef]
233. Novelli, F.; Bononi, A.; Wang, Q.; Bai, F.; Patergnani, S.; Kricek, F.; Haglund, E.; Suarez, J.S.; Tanji, M.; Xu, R.; et al. BAP1 forms a trimer with HMGB1 and HDAC1 that modulates gene × environment interaction with asbestos. *Proc. Natl. Acad. Sci.* **2021**, *118*. [CrossRef]
234. Zhang, Y.; Zhuang, L.; Gan, B. BAP1 suppresses tumor development by inducing ferroptosis upon SLC7A11 repression. *Mol. Cell. Oncol.* **2019**, *6*, 1536845. [CrossRef] [PubMed]
235. Bronkhorst, I.H.G.; Ly, L.V.; Jordanova, E.S.; Vrolijk, J.; Versluis, M.; Luyten, G.P.M.; Jager, M.J. Detection of M2-Macrophages in Uveal Melanoma and Relation with Survival. *Investig. Ophthalmol. Vis. Sci.* **2011**, *52*, 643–650. [CrossRef]
236. Brouwer, N.J.; Wierenga, A.P.A.; Gezgin, G.; Marinkovic, M.; Luyten, G.P.M.; Kroes, W.G.M.; Versluis, M.; Van Der Velden, P.A.; Verdijk, R.M.; Jager, M.J. Ischemia Is Related to Tumour Genetics in Uveal Melanoma. *Cancers* **2019**, *11*, 1004. [CrossRef] [PubMed]
237. Moeller, B.J.; Cao, Y.; Li, C.Y.; Dewhirst, M.W. Radiation activates HIF-1 to regulate vascular radiosensitivity in tumors: Role of reoxygenation, free radicals, and stress granules. *Cancer Cell* **2004**, *5*, 429–441. [CrossRef]
238. Zhang, H.; Wong, C.C.L.; Wei, H.; Gilkes, D.M.; Korangath, P.; Chaturvedi, P.; Schito, L.; Chen, J.; Krishnamachary, B.; Winnard, P.T.; et al. HIF-1-dependent expression of angiopoietin-like 4 and L1CAM mediates vascular metastasis of hypoxic breast cancer cells to the lungs. *Oncogene* **2012**, *31*, 1757–1770. [CrossRef]
239. Semenza, G.L. HIF-1 Inhibitors for Cancer Therapy: From Gene Expression to Drug Discovery. *Curr. Pharm. Des.* **2009**, *15*, 3839–3843. [CrossRef]
240. Fujiwara, S.; Nakagawa, K.; Harada, H.; Nagato, S.; Furukawa, K.; Teraoka, M.; Seno, T.; Oka, K.; Iwata, S.; Ohnishi, T. Silencing hypoxia-inducible factor-1α inhibits cell migration and invasion under hypoxic environment in malignant gliomas. *Int. J. Oncol.* **2007**, *30*, 793–802. [CrossRef]
241. Greenberger, L.M.; Horak, I.D.; Filpula, D.; Sapra, P.; Westergaard, M.; Frydenlund, H.F.; Albaek, C.; Schrøder, H.D.; Ørum, H. A RNA antagonist of hypoxia-inducible factor-1α, EZN-2968, inhibits tumor cell growth. *Mol. Cancer Ther.* **2008**, *7*, 3598–3608. [CrossRef] [PubMed]
242. Koh, M.Y.; Spivak-Kroizman, T.; Venturini, S.; Welsh, S.; Williams, R.R.; Kirkpatrick, D.L.; Powis, G. Molecular mechanisms for the activity of PX-478, an antitumor inhibitor of the hypoxia-inducible factor-1α. *Mol. Cancer Ther.* **2008**, *7*, 90–100. [CrossRef]
243. Isaacs, J.S.; Jung, Y.-J.; Mimnaugh, E.G.; Martinez, A.; Cuttitta, F.; Neckers, L.M. Hsp90 Regulates a von Hippel Lindau-independent Hypoxia-inducible Factor-1α-degradative Pathway. *J. Biol. Chem.* **2002**, *277*, 29936–29944. [CrossRef]
244. Voit, R.A.; Sankaran, V.G. Stabilizing HIF to Ameliorate Anemia. *Cell* **2020**, *180*, 6. [CrossRef]
245. Bai, W.-W.; Xing, Y.-F.; Wang, B.; Lu, X.-T.; Wang, Y.-B.; Sun, Y.-Y.; Liu, X.-Q.; Guo, T.; Zhao, Y.-X. Tongxinluo Improves Cardiac Function and Ameliorates Ventricular Remodeling in Mice Model of Myocardial Infarction through Enhancing Angiogenesis. *Evidence-Based Complement. Altern. Med.* **2013**, *2013*, 1–9. [CrossRef] [PubMed]
246. Liu, X.-W.; Lu, M.-K.; Zhong, H.-T.; Wang, L.-H.; Fu, Y.-P. Panax Notoginseng Saponins Attenuate Myocardial Ischemia-Reperfusion Injury Through the HIF-1α/BNIP3 Pathway of Autophagy. *J. Cardiovasc. Pharmacol.* **2019**, *73*, 92–99. [CrossRef] [PubMed]
247. Jain, T.; Nikolopoulou, E.A.; Xu, Q.; Qu, A. Hypoxia inducible factor as a therapeutic target for atherosclerosis. *Pharmacol. Ther.* **2018**, *183*, 22–33. [CrossRef] [PubMed]

248. Ahmad, A.; Ahmad, S.; Malcolm, K.C.; Miller, S.M.; Hendry-Hofer, T.; Schaack, J.B.; White, C.W. Differential Regulation of Pulmonary Vascular Cell Growth by Hypoxia-Inducible Transcription Factor–1α and Hypoxia-Inducible Transcription Factor–2α. *Am. J. Respir. Cell Mol. Biol.* **2013**, *49*, 78–85. [CrossRef] [PubMed]
249. Dai, Z.; Zhu, M.M.; Peng, Y.; Machireddy, N.; Evans, C.; Machado, R.; Zhang, X.; Zhao, Y.-Y. Therapeutic Targeting of Vascular Remodeling and Right Heart Failure in Pulmonary Arterial Hypertension with a HIF-2α Inhibitor. *Am. J. Respir. Crit. Care Med.* **2018**, *198*, 1423–1434. [CrossRef]

biology

MDPI

Review

The Function and Therapeutic Potential of lncRNAs in Cardiac Fibrosis

Xiang Nie [1,2], Jiahui Fan [1,2] and Dao Wen Wang [1,2,*]

1 Division of Cardiology, Department of Internal Medicine, Tongji Hospital, Tongji Medical College, Huazhong University of Science and Technology, Wuhan 430030, China
2 Hubei Key Laboratory of Genetics and Molecular Mechanisms of Cardiological Disorders, Huazhong University of Science and Technology, Wuhan 430030, China
* Correspondence: dwwang@tjh.tjmu.edu.cn; Tel.: +86-27-6937-8422

Simple Summary: Cardiac fibrosis is a vital pathophysiologic change in heart disease, which eventually leads to heart failure. Several molecular mechanisms positively or negatively regulate myocardial fibrosis, among which long noncoding RNAs have gained increased attention. We summarize the contributions of lncRNAs to miRNA expression, TGF-β signaling, and ECMs synthesis, with a particular attention on the exosome-derived lncRNAs in the regulation of adverse fibrosis as well as the mode of action of lncRNAs secreted into exosomes. We also discuss how the current knowledge on lncRNAs can be applied to develop novel therapeutic strategies. This study may provide clues for the prevention and therapy of cardiac fibrosis.

Abstract: Cardiac fibrosis remains an unresolved problem in cardiovascular diseases. Fibrosis of the myocardium plays a key role in the clinical outcomes of patients with heart injuries. Moderate fibrosis is favorable for cardiac structure maintaining and contractile force transmission, whereas adverse fibrosis generally progresses to ventricular remodeling and cardiac systolic or diastolic dysfunction. The molecular mechanisms involved in these processes are multifactorial and complex. Several molecular mechanisms, such as TGF-β signaling pathway, extracellular matrix (ECM) synthesis and degradation, and non-coding RNAs, positively or negatively regulate myocardial fibrosis. Long noncoding RNAs (lncRNAs) have emerged as significant mediators in gene regulation in cardiovascular diseases. Recent studies have demonstrated that lncRNAs are crucial in genetic programming and gene expression during myocardial fibrosis. We summarize the function of lncRNAs in cardiac fibrosis and their contributions to miRNA expression, TGF-β signaling, and ECMs synthesis, with a particular attention on the exosome-derived lncRNAs in the regulation of adverse fibrosis as well as the mode of action of lncRNAs secreted into exosomes. We also discuss how the current knowledge on lncRNAs can be applied to develop novel therapeutic strategies to prevent or reverse cardiac fibrosis.

Keywords: cardiac fibrosis; lncRNAs; TGF-β; ECMs; exosome

Citation: Nie, X.; Fan, J.; Wang, D.W. The Function and Therapeutic Potential of lncRNAs in Cardiac Fibrosis. *Biology* **2023**, *12*, 154. https://doi.org/10.3390/biology12020154

Academic Editor: Gaetano Santulli

Received: 25 November 2022
Revised: 17 January 2023
Accepted: 18 January 2023
Published: 19 January 2023

1. Introduction

The essential function of the heart is to supply sufficient blood to peripheral organs and tissues during both normal and stress conditions. Normal arrangement of individual cardiomyocytes and constant transmission of contractile force are necessary to maintain cardiac structure and function. Cardiac fibrosis is initially beneficial for cardiac structure maintaining and contractility through controlling the arrangement of cardiomyocytes and maintaining normal structure of left ventricle. However, sustained fibrosis will lead to stiffness of the ventricular wall and generally progress to deterioration of both cardiac systolic and diastolic function [1,2]. The pathophysiologic mechanisms of cardiac fibrosis are diverse and complex [3]. In myocardial infarction or myocarditis, necrotic cardiomyocytes are usually replaced by scar tissue. This type of fibrosis is considered as replacement fibrosis, which reflects the regenerative capacity of the heart upon injury. In addition to cardiac structural damage, replacement fibrosis predominantly reduces systolic dysfunction as compared with diastolic function [4]. The other type of cardiac fibrosis is diffuse myocardial fibrosis, which is characterized by the excessive deposition of collagen fibers (such as collagen I, II, III, and IV) or other extracellular matrix proteins such as fibronectin (FN) and matrix metalloproteinases (MMPs). This type of fibrosis is usually present in chronic cardiovascular diseases, such as hypertension, diabetic cardiomyopathy, atrial fibrillation, and hypertrophic cardiomyopathy [1,5–7]. On the converse of replacement fibrosis, diffuse myocardial fibrosis typically reduces left ventricular diastolic function and is less pronounced on systolic function [8]. The differential pathophysiology and clinical course of cardiac fibrosis resulting from the changed structural quality and various fibrillary composition should be considered separately when exploring their pathophysiological processes [9,10].

Activated cardiac fibroblasts (CFs) as well as myofibroblasts are the main cellular effectors in the occurrence and progression of cardiac fibrosis. CFs are a subclass of interstitial cells, which produce several ECM proteins, including collagens of types I, III, IV, and V [11,12]. Under normal conditions, the ECMs not only serve as a mechanical scaffold to maintain cardiomyocyte architecture, but also control the transmission of contractile force [13]. In the injured heart, the proliferation and differentiation of fibroblasts into myofibroblasts lead to excessive ECMs synthesis and deposition, which disrupt the arrangement of myocytes and conversely attenuate the transmission of contractile force [14,15]. In addition, the increased proliferation of fibroblasts or myofibroblasts also contribute to the progress of fibrosis. Apart from the fibroblasts, other cell populations, such as myocytes, endothelial cells, and inflammatory cells, etc., may also contribute to the pathogenesis of myocardial fibrosis [16,17]. Fibroblasts dynamically interact with the various cardiac cell populations in various ways, such as mechanical, chemical, and electrophysiological, to alter gene expression and cellular behavior [18]. Endothelial cells, immune cells, and vascular smooth muscle cells also secrete molecular signals that regulate the growth or apoptosis of CFs and control fibroblast behaviors or gene expression [16]. Myocardial fibrosis is multifactorial, with various cell populations, factors, and signaling pathways involved.

Over the past few years, various noncoding RNAs, such as miRNAs, circRNAs, and lncRNAs, have been identified with significant roles in cardiac fibrosis. Long non-coding RNAs (lncRNAs) are a class of nucleic acids with lengths of more than 200 nt which have no capacity to code proteins or peptides. They play crucial roles in the progress of embryonic development, cell growth, and differentiation through regulating gene expression at the transcriptional level, post-transcriptional level, or maintaining the stability of RNAs or proteins [19,20]. To date, with development of next-generation sequencing, an increasing number of lncRNAs have been identified with significant roles in the pathogenesis of myocardial fibrosis [21–23]. In this study, we summarize the functions and mechanisms of lncRNAs in cardiac fibrosis and discuss the underlying signaling pathways, with a particular emphasis on the exosome-derived lncRNAs in regulation of fibrosis as well as the mode of lncRNAs secreted into exosomes. Better understanding of the functions of lncRNAs might lead to novel therapeutic approaches for reversing cardiac fibrosis and preventing detrimental outcomes.

2. Characteristics of lncRNAs

The ENCODE project indicates that up to 80% of the human genome do not have the function of coding proteins, which may play a significant role in regulating gene expression [24]. The transcripts of most of these genes are non-coding RNAs (ncRNAs) including rRNAs, tRNAs, circRNAs, microRNAs, and lncRNAs, etc. [25–27]. The FANTOM consortium produces a comprehensive picture of the mammalian transcriptome and publishes 34,030 lncRNAs in mice based on the cDNA sequencing [28]. The number of known human lncRNA transcripts are over 173,000 based on data from the NONCODE database [29,30]. According to the association with annotated protein-coding genes, lncRNAs are classified as intergenic transcripts, and sense or antisense transcripts that overlap other coding genes [31,32]. According to the association with subcellular structures, lncRNAs are also classified as chromatin-associated RNAs, chromatin-interlinking RNAs, nuclear bodies associated RNAs and cytoplasmic transcripts [32–34] (Figure 1). Although there are a high number of lncRNAs continually being identified, most are not validated and their functions are largely unknown.

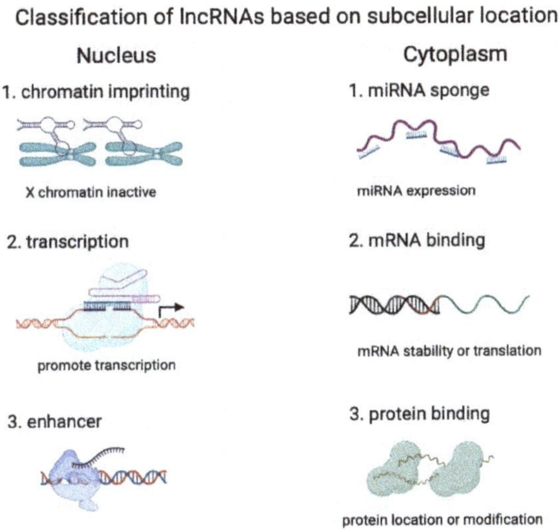

Figure 1. Classification of lncRNAs according to the subcellar localization. LncRNAs resident in the nucleus mainly function through regulating chromatin imprinting, controlling genes at transcriptional or post-transcriptional levels, or serving as enhancers. Cytoplasm-located lncRNAs through binding with proteins or RNAs to regulate miRNAs expression, mRNA stability or translation, and protein modification.

LncRNAs have a great diversity of important functions in body development, regulation of the cell cycle and apoptosis, cellular metabolism, inflammatory response, and tissue homeostasis [35–37]. They have been found with vital roles in mammalian gene regulation through various pathways, such as genomic imprinting, chromatin modification, mRNA decay, protein translation, and miRNA sponges [38–41]. Nuclear lncRNAs usually regulate gene expression through chromatin modification. For instance, the X inactive-specific transcript (Xist) gene regulates mammal X chromosome activation through producing a long noncoding RNA that modifies underlying chromatin and reduces X-linked gene expression [42–44]. LncRNA NORAD controls genomic stability through sequestering PUMILIO (pumilio-fem3-binding factor) proteins, which represses the stability and translation of mRNAs. Silencing of NORAD, PUMILIO drives chromosomal instability by inhibiting DNA repair [45,46]. Some lncRNAs also control gene expression by regulating the stability of mRNAs or modification of proteins. For example, lncRNA ZFAS1 is elevated in colorectal cancer. Knockdown of ZFAS1 decreases the RNA stabilization of SNORD12C/78 and NOP58 through binding with snoRNP to induce 2′-O-Me of 28S rRNA, which eventually inhibits the proliferation and invasion of colorectal cancer cells [47]. LncRNA LINRIS is up-regulated in colorectal cancers. LINRIS reduces IGF2BP2 mRNA expression levels by inhibiting the ubiquitin of IGF2BP2 on K139 sites to maintain its stability [48]. Another function of lncRNA is the competitive endogenous RNA (ceRNA). LncRNAs might suppress the activity of microRNA (miRNA) through serving as sponge RNAs, which lead to miRNA target gene expression increasing [49]. A new lncRNA, named MAR1 (muscle anabolic regulator 1), is significantly up-regulated in myogenesis. MAR1 enhances skeletal muscle strength by sponging miR-487b to regulate Wnt5A expression [50]. Another example of this type of lncRNA is Mirf. Silencing of Mirf promotes autophagy by reducing miR-26a expression in vivo and in vitro [51]. Although non-protein coding potential is a significant characteristic of lncRNAs, some of them might function in biological processes through producing peptides or proteins [52]. LncRNA DWORF (dwarf open reading frame) encodes a peptide of 34 amino acids. Silencing of DWORF in skeletal muscle inhibits Ca^{2+} clearance and suppresses the Ca^{2+} adenosine triphosphatase activity [53].

In the cardiovascular system, several lncRNAs, such as H19, HOTAIR, MIAT, etc., are abundantly expressed in myocardial tissues [37,54–56]. Previous studies have detected and characterized the expression and function of lncRNAs under physiological conditions or in disease states [57–59]. Several lncRNAs have been found with potential roles in heart disease or their expression levels are correlated to disease progression, especially in cardiac fibrosis [60,61]. In this review, the functions, mechanisms, and therapeutic potential of lncRNAs in regulating myocardial fibrosis are summarized and discussed in detail.

2.1. LncRNAs Serve as ceRNAs in Controlling Cardiac Fibrosis

Competing endogenous RNAs (ceRNAs) usually regulate gene expression via sponging microRNAs (miRNAs) at the post-transcriptional levels. LncRNAs might serve as ceRNAs to control miRNA expression and subsequently regulate mRNAs translation and degradation [62]. Homeostatic imbalance in lncRNA–miRNA interaction results in physiological alterations inside the cells and tissues leading to the occurrence of the heart disease [23,63]. To date, significant functions of lncRNA–miRNA interaction in the pathogenesis of myocardial fibrosis are reported by several studies (Figure 2).

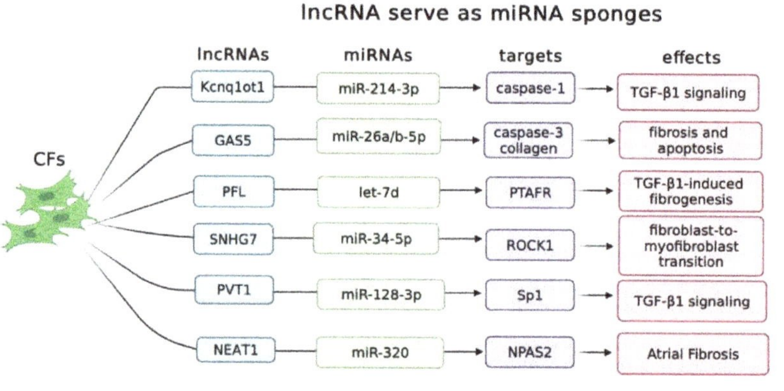

Figure 2. LncRNAs regulate cardiac fibrosis by serving as miRNA sponges. Various lncRNAs have been identified with crucial roles in cardiac fibroblasts by regulating miRNAs expression, which subsequently controls the target gene's expression, and eventually regulates cardiac fibrosis.

In diabetic cardiomyopathy (DCM) [64], lncRNA Kcnq1ot1 (KCNQ1 antisense transcript 1) is significantly up-regulated in myocardial tissues or cardiac fibroblasts treated with glucose. Silencing of Kcnq1ot1 represses the TGF-β signaling pathway via sponging miR-214-3p to suppress the target gene caspase-1 expression [65]. Another example of lncRNA–miRNA interaction in DCM is lncRNA GAS5. Knockdown of GAS5 efficiently attenuates cardiomyocyte injury and myocardial fibrosis via negatively regulating miR-26a/b-5p expression [66]. Similar molecular mechanisms are also observed in myocardial infarction (MI) [67]. Pro-fibrotic lncRNA (PFL) expression is elevated in the heart of mice induced by MI. Overexpression of PFL promotes cardiac fibrosis through increasing the viability of CFs and promoting the transition of fibroblast into myofibroblast via sponging let-7d, leading to increased platelet-activating factor receptor (PTAFR) expression [68]. LncRNA small nuclear RNA host gene 7 (SNHG7) is up-regulated in the infarcted area from left ventricle of mice after MI. Luciferase assay indicates that SNHG7 function through sponging miR-34-5p, which leads to the increased expression of ROCK1 (Rho-associated, coiled-coil domain containing protein kinases) [69]. LncRNA–miRNA interaction is also reported in atrial fibrillation (AF). Increased expression of lncRNA plasmacytoma variant translocation 1 (PVT1) is detected in AF and positively correlated with collagens expression levels. PVT1 overexpression aggravates Ang-II-induced atrial fibroblasts proliferation, collagens production, and TGF-β1 signaling activation by sponging miR-128-3p to facilitate specificity protein 1 (Sp1) expression [70,71]. LncRNA nuclear-enriched abundant transcript 1 (NEAT1) is up-regulated in atrial tissues of AF patients. NEAT1 knockdown improves Ang II-induced mouse atrial fibrosis via negatively regulating miR-320 expression leading to the up-regulation of neuronal PAS domain protein 2 (NPAS2) [72]. The above studies have revealed specific lncRNAs which when elevated function as miRNA sponges to mediate fibrotic processes associated with varied cardiac diseases.

However, the underlying mechanisms are largely undefined, and some concerns should be under consideration when exploring the role of lncRNA–miRNA interaction in myocardial fibrosis. Firstly, the physiological interactions between lncRNAs and miRNAs in normal CFs are rarely illuminated. LncRNAs, down-regulated in myocardial fibrosis, might disturb the physiological lncRNA–miRNA interaction networks, thus promoting the expression of miRNAs, and subsequently repressing the downstream mRNA translation. Secondly, the abundance of lncRNAs resident in the cells should be sufficient. Low levels of lncRNAs might fail to completely sponge high abundant miRNAs. Moreover, ceRNA activity is also influenced by other multiple factors such as the subcellular localization of ceRNA components, binding affinity of miRNAs to lncRNAs, RNA secondary structures, and RNA-binding proteins [73–75].

2.2. LncRNAs Regulate Cardiac Fibrosis through TGF-β Signaling Pathways

Transforming growth factor β (TGF-β) stimulation triggers CFs proliferation and activation, including ECM proteins synthesis and deposition as well as fibroblast-to-myofibroblast differentiation [76,77]. The TGF-β receptor is a dimeric receptor complex whose activation promotes the phosphorylation of Smad2/3 transcription factors through the canonical signaling pathway [78]. Phosphorylated Smads (Smad2, 3, and 4) transfer signal messages to the nucleus and promote gene transcription [79,80]. Fibroblast-specific silencing of TGF-β receptors markedly reduce the pressure overload-induced fibrotic response [81–83]. Knockout of Smad2/3 in fibroblasts reduces the expression of fibrosis-related genes and alleviates injury-induced cellular proliferation within the heart [84,85].

The essential interplay between lncRNA and TGF-β signaling has been widely reported [86]. TGF-β upregulates lncRNA expression in various cancers such as lung cancer, breast cancer, and hepatocellular carcinoma [87–89]. In addition to being effectors of TGF-β signaling, several lncRNAs are reported to regulate TGF-β signaling pathway through various mechanisms. For example, lncRNA LINC00941 stimulates epithelial-mesenchymal transition by directly binding with Smad4 and competing with β-trcp (beta-transducing repeat containing E3 ubiquitin) to prevent the degradation of Smad4 protein, which eventually activates the TGF-β signaling pathway [90]. However, the interaction between lncRNAs and TGF-β signaling in myocardial fibrosis is rarely discussed in detail.

In myocardial fibrosis, TGF-β stimulates several lncRNAs expression in vitro and in vivo. The increased expression of Neat1 is detected in the heart tissue from transverse aortic constriction surgery-induced mice and TGF-β1 treated cardiac fibroblasts. Neat1 recruits Ezh2 to the promoter of Smad7 resulting in decreased Smad7 expression [91]. LncRNA Safe is up-regulated in TGF-β-induced cardiac fibrosis and myocardial infarction [92]. However, the underlying mechanism of TGF-β in regulating lncRNAs expression is not currently known. Smad2/3 proteins phosphorylated by TGF-β might be involved in provoking the transcription of lncRNAs. In addition, activation of the TGF-β receptors also initiate noncanonical signaling to promote the activation of the MAPK, p38, JNK1/2, and ERK1/2 signaling pathways [93,94], which are crucial factors in regulating fibrosis-associated gene expression. More work is still required to answer these questions.

Conversely, lncRNAs also stimulate the activation of TGF-β signaling pathway. The expression of long noncoding RNA AK081284 is up-regulated in cardiac fibroblasts treated with IL-17 or high glucose. Overexpression of AK081284 in cardiac fibroblasts promotes the production of collagens and TGF-β1, while AK081284 silencing reduces collagen and TGF-β1 expression [95]. LncRNA Cfast (cardiac fibroblast-associated transcript) is significantly up-regulated during myocardial infarction. Silencing of Cfast results in reduction of fibrosis-related gene expression and the transdifferentiation of myofibroblasts into fibroblasts. Cfast inhibits the interaction between COTL1 (coactosin-Like protein 1) and TRAP1 (transforming growth factor-β receptor-associated protein 1), which eventually activates the TGF-β signaling pathway [96]. In addition, overexpression of lncRNA GAS5 suppresses TGF-β-induced fibroblast to myofibroblast differentiation. GAS5 directly binds and promotes SMAD3 binding to protein phosphatase 1A (PPM1A), and thus accelerates SMAD3 dephosphorylation in fibroblasts induced by TGF-β [97]. Increased expression of lncRNA Safe is detected in fibrotic ventricular tissues induced by myocardial infarction. Knockdown of Safe prevents TGF-β-induced fibroblast to myofibroblast transition and extracellular matrix proteins production by inhibiting neighboring gene SFRP2 (secreted frizzled-related protein 2) expression [92].

The potential mechanisms involved in these processes must also be systematically explored. LncRNAs resident in the nuclear of fibroblasts may regulate TGF-β signaling pathway-associated gene expression through epigenic or transcriptional regulation. Cytoplasmic lncRNAs may regulate gene expression by controlling the translation or metabolism of RNAs or proteins.

2.3. LncRNAs Control Cardiac Fibrosis by Regulating ECM Gene Expression

In myocardium, excessive synthesis and deposition of extracellular matrix (ECM) proteins are the significant characteristics of myocardial fibrosis. Cardiac ECMs are primarily composed of fibrillar collagens, especially for type I and III, which are the principal proteins in maintaining cardiac structure and function [98]. Cardiac ECMs also contain nonstructural matricellular glycoproteins, proteoglycans, and glycosaminoglycans. The synthesis and degradation of ECMs are predominately regulated by metalloproteinases (MMPs) [99]. The balance between ECMs synthesis and degradation is of crucial importance in cardiac structural integrity and formation of fibrosis. However, to date, few studies systematically elaborate the functions and mechanisms of lncRNAs in these processes.

In myocardial fibrosis, lncRNAs display potential functions in regulating the expression of ECM genes. Bioinformatics analysis indicates that the differentially expressed lncRNAs and extracellular matrix (ECM) protein coding genes revealed a strong association between lncRNAs and ECMs [100]. LncRNA H19 directly binds and antagonizes YB-1 (Y-Box binding protein 1) under hypoxia, which results in the de-repression of collagen 1A expression and cardiac fibrosis [101]. Myocardial infarction associated transcript (MIAT) is up-regulated in myocardial infarction heart tissues. Down-regulation of MIAT alleviates cardiac fibrosis and improves cardiac function by regulating the expression of the fibrosis-related regulators [102]. LncRNA Wisper (Wisp2 super-enhancer-associated RNA) expression is enriched in CFs and elevated in a murine model of MI. Wisper regulates cardiac fibroblasts survival and behavior by regulating lysyl hydroxylase 2 expression [70]. Silencing of lncRNA Meg3 prevents cardiac MMP-2 production, decreases cardiac fibrosis, and improves diastolic function in mice induced by transverse aortic constriction surgery [103]. Knockdown of lncRNA MALAT1 prevents fibroblast proliferation, and ECMs production in AngII-treated cardiac fibroblasts [23].

Although these studies indicate that ECM genes are dysregulated upon lncRNAs stimulation, it seems more likely that the synthesis of ECMs is a common change of myocardial fibrosis induced by lncRNAs. Whether lncRNA have a direct role in controlling ECM genes expression is still a mystery.

2.4. Exosome-Derived LncRNAs Regulate Cardiac Fibrosis

A number of studies have indicated that lncRNAs significantly regulate fibrosis by being expressed within fibroblasts and have a direct effect on ECM gene expression, TGF-β signaling pathway, and proliferation of fibroblasts or transition to myofibroblast. These effects are also proposed to be mediated by paracrine communication between donor and recipient cells, especially in cardiomyocytes and fibroblasts. Increasing evidence attaches much importance of noncoding RNAs, such as miRNAs, circRNAs, and lncRNAs, in the communication between cells by way of extracellular vesicle-mediated transfer from donor cells to the recipient cells [104–107]. This has been considered an important behavior of cardiac cells to communicate with each other and respond to cardiac injuries [108,109]. Previous studies have already confirmed the existence of exosomes in heart tissues and vessel walls using electron micrographs [110]. Fibroblasts and cardiomyocytes might interact with each other through the transfer of extracellular vesicles containing lncRNAs [111].

Involvement of lncRNAs in the crosstalk between cardiac fibroblasts and other cell populations have already been reported in recent years. Exosomes-containing lncRNA ZFAS1 induces cardiac fibrosis via the Wnt4/β-catenin signal pathway by sponging miR-4711-5p in cardiac fibroblasts [112]. LncRNA MIAT is up-regulated in serum-derived extracellular vesicles (EVs) from AF patients. MIAT aggravates the atrial remodeling and promotes AF by binding with miR-485-5p [113]. Neat1 is obviously up-regulated by P53 and HIF2A in cardiomyocytes in response to hypoxia and is enriched in cardiomyocyte-derived exosomes. Neat1 is essential for cell survival and fibroblast functions. Genetic knockout of Neat1 impairs cardiac function during myocardial infarction [111] (Figure 3 and Table 1).

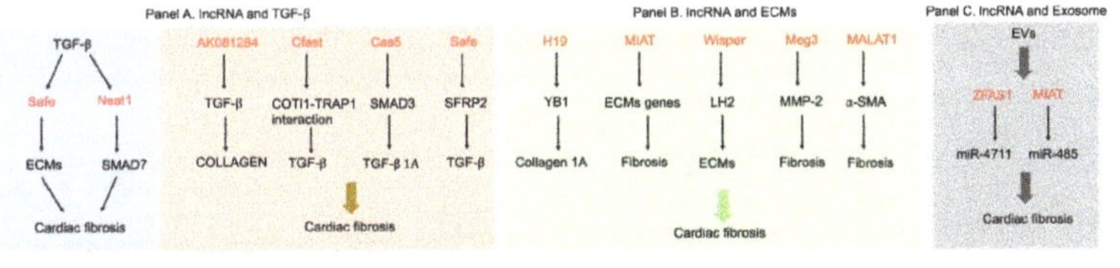

Figure 3. LncRNAs regulate cardiac fibrosis through promoting ECMs production, activating TGF-β signaling pathways or secreting into exosomes. Panel A: On the one hand, TGF-β regulates lncRNAs expression (left) and on the other hand lncRNAs also stimulate TGF-β activation (right). Panel B: LncRNAs regulates ECMs (collagens, SMA, and MMPs) production through various pathways. Panel C: Exosome-derived lncRNAs released from donor cells are accepted by fibroblasts and regulate cardiac fibrosis by targeting miRNAs.

Table 1. The functions and mechanisms of lncRNAs in cardiac fibrosis.

LncRNA Name	Expression	Experimental Model	Targeted Genes	Effects	Exosomes
Kcnq1ot1	Up-regulated	DCM	miR-214-3/Caspase1	TGF-β1 signaling	No
GAS5	Up-regulated	DCM	miR-26a/b-5p/ Caspase3	Fibrosis and apoptosis	No
PFL	Up-regulated	MI	Let-7d/PTAFR	TGF-β1-induced fibrogenesis	No
SNHG7	Up-regulated	MI	miR-34-5p/ROCK1	Fibroblast-to-myofibroblast transition	No
PVT1	Up-regulated	AF	miR-128-3p/SP1	TGF-β1 signaling	No
NEAT1	Up-regulated	AF	miR-320/NPAS2	Atrial fibrosis	No
AK081284	Up-regulated	DCM	TGF-β1	Collagen I and III production	No
CFAST	Up-regulated	MI	COTL1	Enhances TGF-β signaling	No
SAFE	Up-regulated	MI	SFRP2	Fibroblast to myofibroblast transition	No
H19	Up-regulated	MI	YB-1	Reduction of collagen 1A expression	No
Wisper	Up-regulated	MI	LH2	CF behavior and survival	No
Meg3	Down-regulated		MMP-2	Diastolic performance	No
ZFAS1	Up-regulated	DCM	miR-4711-5p	Wnt4/β-catenin signal pathway	Yes
MIAT	Up-regulated	AF	miR-485-5p	Atrial remodeling	Yes
Neat1	Up-regulated	MI	CDK1	Fibroblast and cardiomyocyte survival	Yes

DCM: Diabetic cardiomyopathy. MI: Myocardial infarction. AF: Atrial fibrillation.

2.5. The Way of LncRNAs Secreted into Exosomes

A variety of vesicles with different sizes and contents have been identified in the eukaryotic cells and tissues, including exosomes, micro-vesicles, ectosomes, apoptotic bodies, etc. [114]. Exosomes are a class of vesicles with diameters of 40 nm to 100 nm, which can be secreted by almost all cells and tissues in the body. The generation of exosomes is complex and diverse, and several molecules and physiological process are involved. In brief, vesicular endosomes uptake proteins or RNAs, followed by fusion with plasma membrane. Then, the vesicles (exosomes) are released from the donor cells by exocytosis [108]. Exosomes can fuse with live cells, transferring their cargo of lipids, glucose, proteins, and RNAs to the acceptor cells. To date, the functions of exosomal miRNAs have been well studied in various disease conditions, such as cancers, inflammation responses, and cardiovascular diseases. However, how RNAs (miRNAs and lncRNAs) are released into exosomes is still under investigation. Studies indicate that the specific motifs recognized by the RNA-binding proteins (RBPs) might be the determinant of miRNAs secreting into exosomes. For example, the 'GGAG' motif involved in miRNAs is bound by hnRNPA2B1 (a heterogeneous nuclear riboprotein), which directs miRNA trafficking to exosomes [115].

One database, named exoRBase, indicates that almost 15,500 lncRNAs have been identified in human blood exosomes [116]. How these lncRNAs are secreted into the exosome is rarely illuminated. Specific motifs present in certain lncRNAs may guide their sorting to exosomes through the interaction with specific RNA-binding proteins. Other reports find that lncRNA sorting to exosomes is regulated by changes of targeted transcript levels in the receiving cells [117]. Studies also indicate that lncRNAs with 3′ end uridylated appear over-represented in exosomes [118]. In addition, lncRNAs may also conversely regulate the secretion of exosomes. The biogenesis and secretion of exosomes involve various sorting machineries, including endosomal sorting complex required for transport (ESCRT)-dependent processes [119]. Phosphorylation of synaptosome associated protein 23 (SNAP23) increases exosome production and secretion [120]. For instance, lncRNA HOTAIR promotes the secretion of exosomes through inhibiting VAMP3 and SNAP23 colocalization to induce multivesicular bodies (MVBs) fusion with plasma membrane [121].

Although exosome-associated lncRNAs have been recently reported with potential roles in the pathogenesis of disease such as cardiac fibrosis, some concerns still need to be considered when studying the functions of lncRNAs resident in exosomes. Unlike miRNAs, lncRNAs always have a long sequence or complex secondary structure, which may limit their secretion into exosomes. LncRNAs usually bind with several proteins or miRNAs, which may hinder their secretion into exosomes. In addition, the present evidence about lncRNAs resident in exosomes is usually obtained from qPCR experiments, which might make the results unreliable.

2.6. LncRNAs Serve as Potential Therapeutic Targets

In the last few years, many efforts have been made on the application of RNA-based therapeutics in clinical practice. Chemically modified oligonucleotides and cellular RNAs are important parts of lncRNA-targeted therapeutics [122]. Antisense oligonucleotides (ASOs), a set of single-stranded DNA molecules, are complementary to target mRNA. ASOs cause mRNA degradation or pre-mRNA splicing to block protein translation [123,124]. An interesting development in biological progress is the use of natural antisense transcripts oligonucleotide. Several lncRNAs are the antisense transcripts of coding genes and are fully complementary to the target mRNA, which may be applied for natural antisense oligonucleotide. For example, over-expression of lncRNA BDNF-AS (antisense of brain-derived neurotrophic factor, BDNF) reduces BDNF expression, while silencing of this transcript increases BDNF expression and promotes neuronal differentiation [125].

MiRNA sponges are RNA molecules that contain several specific sequences which are complementary to miRNAs [126–128]. MiRNA sponges can be ideal tools for loss-of-function studies in science research. For instance, miR-181-sponge containing 10 repeated complementary miR-181 sequences significantly suppresses miR-181 expression in H9c2 cells and leads to decreased production of reactive oxygen species by upregulating target mRNA mt-COX1 expression [129]. In the heart, silencing of miR-34a using miRNA sponges is protective, whilst sustained inhibition of miR-34 may be deleterious due to its tumorigenicity [130]. Obese mice benefit from miR-122 antisense oligonucleotides treatment, as reflected by the decreased plasma cholesterol levels and liver steatosis improvement [131]. The miRNA-sponge characterization of lncRNAs indicates the potential possibility for clinical treatment.

Overall, lncRNAs are attractive approaches for disease treatment in clinical practice, such as heart failure, hypertension, and cardia fibrosis. To date, no lncRNA-associated drugs have been applied in clinical trials due to the limited efficacy and potential toxicity.

3. Conclusions

Myocardial fibrosis is the final step in cardiac remodeling in several cardiovascular diseases, such as myocardial infraction, diabetic cardiomyopathy, atrial fibrillation, and heart failure. Excessive fibrosis in heart tissue renders the myocardium stiffer mechanically and contributes to the deterioration of both systolic and diastolic function. Various molecules and signaling pathways are involved in the formation of cardiac fibrosis. TGF-β signaling pathway and ECMs produced by the activated fibroblasts or myofibroblasts play significant roles in the pathogenesis of cardiac fibrosis [132]. Modulation of TGF-β signaling and ECM gene expression is a vital contribution of lncRNAs in this process. However, details of the mechanisms which enable lncRNAs to regulate TGF-β or ECMs production remain a mystery. LncRNAs may partly function via chromatin modification, transcription regulation, and post-transcriptional modification in myocardial fibrosis [133]. Besides the direct roles of lncRNAs in fibroblasts or myofibroblasts, they also function through indirect ways by interacting with other cell types, for example, releasing into exosomes to alter cellular behaviors. Non-coding RNAs such as miRNAs, circRNAs, and lncRNAs have been identified as present in the exosomes derived from other cell populations (myocytes, endothelial cells, etc.). Exosome-containing lncRNAs released from the donor cells are ingested by the fibroblasts, which may influence the progress of fibrosis [134,135]. Our understanding of non-coding RNAs involved in cell-to-cell interactions in cardiac fibrosis is still relatively limited. A comprehensive understanding of the function and mechanisms of lncRNAs in myocardial fibrosis holds the key for disease prevention and treatment.

Since the discovery of lncRNAs as master regulators in cardiac fibrosis, their utilization for myocardium remodeling diagnosis and clinical treatment strategies is increasingly employed. However, no ongoing clinical trials are currently underway due to the dose-effect, off-target effects, and potential toxicity. The mechanistic importance as well as diagnostic and therapeutic utility of lncRNAs in cardiac fibrosis needs to be further studied.

Author Contributions: X.N. designed this study and wrote the original draft. J.F. provided help and advice on the manuscript writing. D.W.W. updated and revised the original draft. All authors contributed to editorial changes in the manuscript. All authors have read and agreed to the published version of the manuscript.

Funding: This work is supported by the National Natural Science Foundation of China (grant numbers 82100399 and 82100400) and the Project funded by the China Postdoctoral Science Foundation (grant number 2021M701315). The funders have no role in the study design, data collection and analysis, manuscript preparation, or publication decision.

Institutional Review Board Statement: Not applicable.

Informed Consent Statement: Not applicable.

Data Availability Statement: No new data are created or analyzed in this study. Data sharing is not applicable to this article.

Acknowledgments: We thank our colleagues at Wang's group for their technical assistance during this investigation.

Conflicts of Interest: The authors declare no conflict of interest.

References

1. López, B.; Ravassa, S.; Moreno, M.U.; José, G.S.; Beaumont, J.; González, A.; Díez, J. Diffuse myocardial fibrosis: Mechanisms, diagnosis and therapeutic approaches. *Nat. Rev. Cardiol.* **2021**, *18*, 479–498. [CrossRef]
2. Tallquist, M.D. Cardiac Fibroblast Diversity. *Annu. Rev. Physiol.* **2020**, *82*, 63–78. [CrossRef]
3. Weber, K.T.; Pick, R.; Jalil, J.E.; Janicki, J.S.; Carroll, E.P. Patterns of myocardial fibrosis. *J. Mol. Cell. Cardiol.* **1989**, *21* (Suppl. 5), 121–131. [CrossRef] [PubMed]
4. Kong, P.; Christia, P.; Frangogiannis, N. The pathogenesis of cardiac fibrosis. *Cell. Mol. Life Sci.* **2014**, *71*, 549–574. [CrossRef]
5. Horckmans, M.; Bianchini, M.; Santovito, D.; Megens, R.T.A.; Springael, J.Y.; Negri, I.; Vacca, M.; Di Eusanio, M.; Moschetta, A.; Weber, C.; et al. Pericardial Adipose Tissue Regulates Granulopoiesis, Fibrosis, and Cardiac Function after Myocardial Infarction. *Circulation* **2018**, *137*, 948–960. [CrossRef]
6. Prabhu, S.D.; Frangogiannis, N.G. The Biological Basis for Cardiac Repair after Myocardial Infarction: From Inflammation to Fibrosis. *Circ. Res.* **2016**, *119*, 91–112. [CrossRef]
7. Epelman, S.; Liu, P.; Mann, D. Role of innate and adaptive immune mechanisms in cardiac injury and repair. *Nat. Rev. Immunol.* **2015**, *15*, 117–129. [CrossRef]
8. Plehn, J.F.; Biederman, R. Diffuse Myocardial Fibrosis in Dilated Cardiomyopathy: Risk Marker, Risk Factor, or Does it Matter? *JACC Cardiovasc. Imaging* **2022**, *15*, 591–593. [CrossRef] [PubMed]
9. Zhang, M.; Sui, W.; Xing, Y.; Cheng, J.; Cheng, C.; Xue, F.; Zhang, J.; Wang, X.; Zhang, C.; Hao, P.; et al. Angiotensin IV attenuates diabetic cardiomyopathy via suppressing FoxO1-induced excessive autophagy, apoptosis and fibrosis. *Theranostics* **2021**, *11*, 8624–8639. [CrossRef] [PubMed]
10. Marques, F.Z.; Nelson, E.; Chu, P.Y.; Horlock, D.; Fiedler, A.; Ziemann, M.; Tan, J.K.; Kuruppu, S.; Rajapakse, N.W.; El-Osta, A.; et al. High-Fiber Diet and Acetate Supplementation Change the Gut Microbiota and Prevent the Development of Hypertension and Heart Failure in Hypertensive Mice. *Circulation* **2017**, *135*, 964–977. [CrossRef]
11. Maisch, B. Extracellular matrix and cardiac interstitium: Restriction is not a restricted phenomenon. *Herz* **1995**, *20*, 75–80.
12. Tallquist, M.D.; Molkentin, J. Redefining the identity of cardiac fibroblasts. *Nat. Rev. Cardiol.* **2017**, *14*, 484–491. [CrossRef]
13. Spinale, F.G. Matrix metalloproteinases: Regulation and dysregulation in the failing heart. *Circ. Res.* **2002**, *90*, 520–530. [CrossRef]
14. Powell, D.W.; Mifflin, R.C.; Valentich, J.D.; Crowe, S.E.; Saada, J.I.; West, A.B. Myofibroblasts. I. Paracrine cells important in health and disease. *Am. J. Physiol. Cell Physiol.* **1999**, *277*, C1–C19. [CrossRef] [PubMed]
15. Tomasek, J.J.; Gabbiani, G.; Hinz, B.; Chaponnier, C.; Brown, R.A. Myofibroblasts and mechano: Regulation of connective tissue remodelling. *Nat. Rev. Mol. Cell Biol.* **2002**, *3*, 349–363. [CrossRef] [PubMed]
16. Sun, X.; Nkennor, B.; Mastikhina, O.; Soon, K.; Nunes, S.S. Endothelium-mediated contributions to fibrosis. *Semin. Cell Dev. Biol.* **2020**, *101*, 78–86. [CrossRef]
17. Howard, C.M.; Baudino, T. Dynamic cell-cell and cell-ECM interactions in the heart. *J. Mol. Cell. Cardiol.* **2014**, *70*, 19–26. [CrossRef]
18. Deb, A. Cell-cell interaction in the heart via Wnt/β-catenin pathway after cardiac injury. *Cardiovasc. Res.* **2014**, *102*, 214–223. [CrossRef] [PubMed]
19. Tan, Y.T.; Lin, J.F.; Li, T.; Li, J.J.; Xu, R.H.; Ju, H.Q. LncRNA-mediated posttranslational modifications and reprogramming of energy metabolism in cancer. *Cancer Commun.* **2021**, *41*, 109–120. [CrossRef]
20. Nojima, T.; Proudfoot, N. Mechanisms of lncRNA biogenesis as revealed by nascent transcriptomics. *Nat. Rev. Mol. Cell Biol.* **2022**, *23*, 389–406. [CrossRef]
21. Ilieva, M.; Uchida, S. Long Non-Coding RNAs in Cardiac and Pulmonary Fibroblasts and Fibrosis. *Noncoding RNA* **2022**, *8*, 53. [CrossRef] [PubMed]
22. Yang, Z.; Jiang, S.; Shang, J.; Jiang, Y.; Dai, Y.; Xu, B.; Yu, Y.; Liang, Z.; Yang, Y. LncRNA: Shedding light on mechanisms and opportunities in fibrosis and aging. *Ageing Res. Rev.* **2019**, *52*, 17–31. [CrossRef] [PubMed]
23. Huang, S.; Zhang, L.; Song, J.; Wang, Z.; Huang, X.; Guo, Z.; Chen, F.; Zhao, X. Long noncoding RNA MALAT1 mediates cardiac fibrosis in experimental postinfarct myocardium mice model. *J. Cell Physiol.* **2019**, *234*, 2997–3006. [CrossRef] [PubMed]
24. ENCODE Project Consortium. An integrated encyclopedia of DNA elements in the human genome. *Nature* **2012**, *489*, 57–74. [CrossRef] [PubMed]
25. Elkon, R.; Agami, R. Characterization of noncoding regulatory DNA in the human genome. *Nat. Biotechnol.* **2017**, *35*, 732–746. [CrossRef] [PubMed]
26. Mattick, J.S. Non-coding RNAs: The architects of eukaryotic complexity. *EMBO Rep.* **2001**, *2*, 986–991. [CrossRef]
27. Yip, C.W.; Sivaraman, D.M.; Prabhu, A.V.; Shin, J.W. Functional annotation of lncRNA in high-throughput screening. *Essays Biochem.* **2021**, *65*, 761–773.
28. Maeda, N.; Kasukawa, T.; Oyama, R.; Gough, J.; Frith, M.; Engström, P.G.; Lenhard, B.; Aturaliya, R.N.; Batalov, S.; Beisel, K.W.; et al. Transcript annotation in FANTOM3: Mouse gene catalog based on physical cDNAs. *PLoS Genet.* **2006**, *2*, e62. [CrossRef]
29. Derrien, T.; Johnson, R.; Bussotti, G.; Tanzer, A.; Djebali, S.; Tilgner, H.; Guernec, G.; Martin, D.; Merkel, A.; Knowles, D.G.; et al. The GENCODE v7 catalog of human long noncoding RNAs: Analysis of their gene structure, evolution, and expression. *Genome Res.* **2012**, *22*, 1775–1789. [CrossRef]

30. Fang, S.; Zhang, L.; Guo, J.; Niu, Y.; Wu, Y.; Li, H.; Zhao, L.; Li, X.; Teng, X.; Sun, X.; et al. NONCODEV5: A comprehensive annotation database for long non-coding RNAs. *Nucleic Acids Res.* **2018**, *46*, D308–D314. [CrossRef]
31. Kopp, F.; Mendell, J. Functional Classification and Experimental Dissection of Long Noncoding RNAs. *Cell* **2018**, *172*, 393–407. [CrossRef] [PubMed]
32. Yousefi, H.; Maheronnaghsh, M.; Molaei, F.; Mashouri, L.; Reza Aref, A.; Momeny, M.; Alahari, S.K. Long noncoding RNAs and exosomal lncRNAs: Classification, and mechanisms in breast cancer metastasis and drug resistance. *Oncogene* **2020**, *39*, 953–974. [CrossRef] [PubMed]
33. Mondal, T.; Rasmussen, M.; Pandey, G.K.; Isaksson, A.; Kanduri, C. Characterization of the RNA content of chromatin. *Genome Res.* **2010**, *20*, 899–907. [CrossRef]
34. Mao, Y.S.; Zhang, B.; Spector, D. Biogenesis and function of nuclear bodies. *Trends Genet.* **2011**, *27*, 295–306. [CrossRef] [PubMed]
35. Wang, Y.; Sun, X. The functions of LncRNA in the heart. *Diabetes Res. Clin. Pract.* **2020**, *168*, 108249. [CrossRef]
36. Bhan, A.; Soleimani, M.; Mandal, S. Long Noncoding RNA and Cancer: A New Paradigm. *Cancer Res.* **2017**, *77*, 3965–3981. [CrossRef] [PubMed]
37. Liang, Y.; Rong, X.; Luo, Y.; Li, P.; Han, Q.; Wei, L.; Wang, E. A novel long non-coding RNA LINC00355 promotes proliferation of lung adenocarcinoma cells by down-regulating miR-195 and up-regulating the expression of CCNE1. *Cell Signal.* **2020**, *66*, 109462. [CrossRef]
38. Qian, X.; Zhao, J.; Yeung, P.Y.; Zhang, Q.C.; Kwok, C.K. Revealing lncRNA Structures and Interactions by Sequencing-Based Approaches. *Trends Biochem. Sci.* **2019**, *44*, 33–52. [CrossRef]
39. Schmitz, S.U.; Grote, P.; Herrmann, B. Mechanisms of long noncoding RNA function in development and disease. *Cell. Mol. Life Sci.* **2016**, *73*, 2491–2509. [CrossRef]
40. Chen, L.L. Linking Long Noncoding RNA Localization and Function. *Trends Biochem. Sci.* **2016**, *41*, 761–772. [CrossRef]
41. Johnson, S.J.; Cooper, T. Overlapping mechanisms of lncRNA and expanded microsatellite RNA. *Wiley Interdiscip. Rev. RNA* **2021**, *12*, e1634. [CrossRef] [PubMed]
42. Brockdorff, N.; Bowness, J.; Wei, G. Progress toward understanding chromosome silencing by Xist RNA. *Genes Dev.* **2020**, *34*, 733–744. [CrossRef]
43. Yu, B.; Qi, Y.; Li, R.; Shi, Q.; Satpathy, A.T.; Chang, H.Y. B cell-specific XIST complex enforces X-inactivation and restrains atypical B cells. *Cell* **2021**, *184*, 1790–1803. [CrossRef] [PubMed]
44. Engreitz, J.M.; Pandya-Jones, A.; McDonel, P.; Shishkin, A.; Sirokman, K.; Surka, C.; Kadri, S.; Xing, J.; Goren, A.; Lander, E.S.; et al. The Xist lncRNA exploits three-dimensional genome architecture to spread across the X chromosome. *Science* **2013**, *341*, 1237973. [CrossRef]
45. Lee, S.; Kopp, F.; Chang, T.C.; Sataluri, A.; Chen, B.; Sivakumar, S.; Yu, H.; Xie, Y. Noncoding RNA NORAD Regulates Genomic Stability by Sequestering PUMILIO Proteins. *Cell* **2016**, *164*, 69–80. [CrossRef] [PubMed]
46. Elguindy, M.M.; Mendell, J. NORAD-induced Pumilio phase separation is required for genome stability. *Nature* **2021**, *595*, 303–308. [CrossRef]
47. Wu, H.; Qin, W.; Lu, S.; Wang, X.; Zhang, J.; Sun, T.; Hu, X.; Li, Y.; Chen, Q.; Wang, Y.; et al. Long noncoding RNA ZFAS1 promoting small nucleolar RNA-mediated 2′-O-methylation via NOP58 recruitment in colorectal cancer. *Mol. Cancer* **2020**, *19*, 95. [CrossRef]
48. Wang, Y.; Lu, J.H.; Wu, Q.N.; Jin, Y.; Wang, D.S.; Chen, Y.X.; Liu, J.; Luo, X.J.; Meng, Q.; Pu, H.Y.; et al. LncRNA LINRIS stabilizes IGF2BP2 and promotes the aerobic glycolysis in colorectal cancer. *Mol. Cancer* **2019**, *18*, 174. [CrossRef]
49. Thomson, D.W.; Dinger, M. Endogenous microRNA sponges: Evidence and controversy. *Nat. Rev. Genet.* **2016**, *17*, 272–283. [CrossRef]
50. Zhang, Z.K.; Li, J.; Guan, D.; Liang, C.; Zhuo, Z.; Liu, J.; Lu, A.; Zhang, G.; Zhang, B.T. A newly identified lncRNA MAR1 acts as a miR-487b sponge to promote skeletal muscle differentiation and regeneration. *J. Cachexia Sarcopenia Muscle* **2018**, *9*, 613–626. [CrossRef]
51. Liang, H.; Su, X.; Wu, Q.; Shan, H.; Lv, L.; Yu, T.; Zhao, X.; Sun, J.; Yang, R.; Zhang, L.; et al. LncRNA 2810403D21Rik/Mirf promotes ischemic myocardial injury by regulating autophagy through targeting Mir26a. *Autophagy* **2020**, *16*, 1077–1091. [CrossRef] [PubMed]
52. Quinn, J.J.; Chang, H. Unique features of long non-coding RNA biogenesis and function. *Nat. Rev. Genet.* **2016**, *17*, 47–62. [CrossRef] [PubMed]
53. Nelson, B.R.; Makarewich, C.A.; Anderson, D.M.; Winders, B.R.; Troupes, C.D.; Wu, F.; Reese, A.L.; McAnally, J.R.; Chen, X.; Kavalali, E.T.; et al. A peptide encoded by a transcript annotated as long noncoding RNA enhances SERCA activity in muscle. *Science* **2016**, *351*, 271–275. [CrossRef]
54. Lai, Y.; He, S.; Ma, L.; Lin, H.; Ren, B.; Ma, J.; Zhu, X.; Zhuang, S. HOTAIR functions as a competing endogenous RNA to regulate PTEN expression by inhibiting miR-19 in cardiac hypertrophy. *Mol. Cell. Biochem.* **2017**, *432*, 179–187. [CrossRef] [PubMed]
55. Zhu, X.H.; Yuan, Y.X.; Rao, S.L.; Wang, P. LncRNA MIAT enhances cardiac hypertrophy partly through sponging miR-150. *Eur. Rev. Med. Pharmacol. Sci.* **2016**, *20*, 3653–3660. [PubMed]

56. Liu, L.; An, X.; Li, Z.; Song, Y.; Li, L.; Zuo, S.; Liu, N.; Yang, G.; Wang, H.; Cheng, X.; et al. The H19 long noncoding RNA is a novel negative regulator of cardiomyocyte hypertrophy. *Cardiovasc. Res.* **2016**, *111*, 56–65. [CrossRef] [PubMed]
57. Greco, S.; Salgado Somoza, A.; Devaux, Y.; Martelli, F. Long Noncoding RNAs and Cardiac Disease. *Antioxid. Redox Signal.* **2018**, *29*, 880–901. [CrossRef]
58. Lozano-Vidal, N.; Bink, D.; Boon, R. Long noncoding RNA in cardiac aging and disease. *J. Mol. Cell Biol.* **2019**, *11*, 860–867. [CrossRef]
59. Scheuermann, J.C.; Boyer, L. Getting to the heart of the matter: Long non-coding RNAs in cardiac development and disease. *EMBO J.* **2013**, *32*, 1805–1816. [CrossRef]
60. Li, H.; Trager, L.E.; Liu, X.; Hastings, M.H.; Xiao, C.; Guerra, J.; To, S.; Li, G.; Yeri, A.; Rodosthenous, R.; et al. lncExACT1 and DCHS2 Regulate Physiological and Pathological Cardiac Growth. *Circulation* **2022**, *145*, 1218–1233. [CrossRef]
61. Viereck, J.; Bührke, A.; Foinquinos, A.; Chatterjee, S.; Kleeberger, J.A.; Xiao, K.; Janssen-Peters, H.; Batkai, S.; Ramanujam, D.; Kraft, T.; et al. Targeting muscle-enriched long non-coding RNA H19 reverses pathological cardiac hypertrophy. *Eur. Heart J.* **2020**, *41*, 3462–3474. [CrossRef] [PubMed]
62. Wang, H.; Huo, X.; Yang, X.R.; He, J.; Cheng, L.; Wang, N.; Deng, X.; Jin, H.; Wang, N.; Wang, C.; et al. STAT3-mediated upregulation of lncRNA HOXD-AS1 as a ceRNA facilitates liver cancer metastasis by regulating SOX4. *Mol. Cancer* **2017**, *16*, 136. [CrossRef] [PubMed]
63. Ghafouri-Fard, S.; Abak, A.; Talebi, S.F.; Shoorei, H.; Branicki, W.; Taheri, M.; Akbari Dilmaghani, N. Role of miRNA and lncRNAs in organ fibrosis and aging. *Biomed. Pharmacother.* **2021**, *143*, 112132. [CrossRef]
64. Tanwar, V.S.; Reddy, M.A.; Natarajan, R. Emerging Role of Long Non-Coding RNAs in Diabetic Vascular Complications. *Front. Endocrinol.* **2021**, *12*, 665811. [CrossRef] [PubMed]
65. Yang, F.; Qin, Y.; Lv, J.; Wang, Y.; Che, H.; Chen, X.; Jiang, Y.; Li, A.; Sun, X.; Yue, E.; et al. Silencing long non-coding RNA Kcnq1ot1 alleviates pyroptosis and fibrosis in diabetic cardiomyopathy. *Cell Death Dis.* **2018**, *9*, 1000. [CrossRef] [PubMed]
66. Zhu, C.; Zhang, H.; Wei, D.; Sun, Z. Silencing lncRNA GAS5 alleviates apoptosis and fibrosis in diabetic cardiomyopathy by targeting miR-26a/b-5p. *Acta Diabetol.* **2021**, *58*, 1491–1501. [CrossRef]
67. Varzideh, F.; Kansakar, U.; Donkor, K.; Wilson, S.; Jankauskas, S.S.; Mone, P.; Wang, X.; Lombardi, A.; Santulli, G. Cardiac Remodeling After Myocardial Infarction: Functional Contribution of microRNAs to Inflammation and Fibrosis. *Front. Cardiovasc. Med.* **2022**, *9*, 863238. [CrossRef]
68. Liang, H.; Pan, Z.; Zhao, X.; Liu, L.; Sun, J.; Su, X.; Xu, C.; Zhou, Y.; Zhao, D.; Xu, B.; et al. LncRNA PFL contributes to cardiac fibrosis by acting as a competing endogenous RNA of let-7d. *Theranostics* **2018**, *8*, 1180–1194. [CrossRef]
69. Wang, J.; Zhang, S.; Li, X.; Gong, M. LncRNA SNHG7 promotes cardiac remodeling by upregulating ROCK1 via sponging miR-34-5p. *Aging* **2020**, *12*, 10441–10456. [CrossRef]
70. Micheletti, R.; Plaisance, I.; Abraham, B.J.; Sarre, A.; Ting, C.C.; Alexanian, M.; Maric, D.; Maison, D.; Nemir, M.; Young, R.A.; et al. The long noncoding RNA Wisper controls cardiac fibrosis and remodeling. *Sci. Transl. Med.* **2017**, *9*, eaai9118. [CrossRef]
71. Cao, F.; Li, Z.; Ding, W.M.; Yan, L.; Zhao, Q.Y. LncRNA PVT1 regulates atrial fibrosis via miR-128-3p-SP1-TGF-β1-Smad axis in atrial fibrillation. *Mol Med.* **2019**, *25*, 7. [CrossRef] [PubMed]
72. Dai, H.; Zhao, N.; Liu, H.; Zheng, Y.; Zhao, L. LncRNA Nuclear-Enriched Abundant Transcript 1 Regulates Atrial Fibrosis via the miR-320/NPAS2 Axis in Atrial Fibrillation. *Front. Pharmacol.* **2021**, *12*, 647124. [CrossRef] [PubMed]
73. Qi, X.; Zhang, D.H.; Wu, N.; Xiao, J.H.; Wang, X.; Ma, W. ceRNA in cancer: Possible functions and clinical implications. *J. Med. Genet.* **2015**, *52*, 710–718. [CrossRef] [PubMed]
74. Xu, J.; Xu, J.; Liu, X.; Jiang, J. The role of lncRNA-mediated ceRNA regulatory networks in pancreatic cancer. *Cell Death Discov.* **2022**, *8*, 287. [CrossRef]
75. Chiu, H.S.; Martínez, M.R.; Komissarova, E.V.; Llobet-Navas, D.; Bansal, M.; Paull, E.O.; Silva, J.; Yang, X.; Sumazin, P.; Califano, A. The number of titrated microRNA species dictates ceRNA regulation. *Nucleic Acids Res.* **2018**, *46*, 4354–4369. [CrossRef]
76. Saadat, S.; Noureddini, M.; Mahjoubin-Tehran, M.; Nazemi, S.; Shojaie, L.; Aschner, M.; Maleki, B.; Abbasi-Kolli, M.; Rajabi Moghadam, H.; Alani, B.; et al. Pivotal Role of TGF-β/Smad Signaling in Cardiac Fibrosis: Non-coding RNAs as Effectual Players. Front. *Cardiovasc. Med.* **2020**, *7*, 588347. [CrossRef]
77. Hu, H.H.; Chen, D.Q.; Wang, Y.N.; Feng, Y.L.; Cao, G.; Vaziri, N.D.; Zhao, Y.Y. New insights into TGF-β/Smad signaling in tissue fibrosis. *Chem. Biol. Interact.* **2018**, *292*, 76–83. [CrossRef]
78. Khalil, H.; Kanisicak, O.; Prasad, V.; Correll, R.N.; Fu, X.; Schips, T.; Vagnozzi, R.J.; Liu, R.; Huynh, T.; Lee, S.J.; et al. Fibroblast-specific TGF-β-Smad2/3 signaling underlies cardiac fibrosis. *J. Clin. Investig.* **2017**, *127*, 3770–3783. [CrossRef]
79. Wang, L.; Wang, H.L.; Liu, T.T.; Lan, H.Y. TGF-Beta as a Master Regulator of Diabetic Nephropathy. *Int. J. Mol. Sci.* **2021**, *22*, 7881. [CrossRef]
80. Yoshida, K.; Matsuzaki, K.; Murata, M.; Yamaguchi, T.; Suwa, K.; Okazaki, K. Clinico-Pathological Importance of TGF-β/Phospho-Smad Signaling during Human Hepatic Fibrocarcinogenesis. *Cancers* **2018**, *10*, 183. [CrossRef]
81. Györfi, A.H.; Matei, A.; Distler, J. Targeting TGF-β signaling for the treatment of fibrosis. *Matrix Biol.* **2018**, *68*, 8–27. [CrossRef] [PubMed]
82. Wei, Y.; Kim, T.J.; Peng, D.H.; Duan, D.; Gibbons, D.L.; Yamauchi, M.; Jackson, J.R.; Le Saux, C.J.; Calhoun, C.; Peters, J.; et al. Fibroblast-specific inhibition of TGF-β1 signaling attenuates lung and tumor fibrosis. *J. Clin Investig.* **2017**, *127*, 3675–3688. [CrossRef] [PubMed]

83. Wang, Q.; Yu, Y.; Zhang, P.; Chen, Y.; Li, C.; Chen, J.; Wang, Y.; Li, Y. The crucial role of activin A/ALK4 pathway in the pathogenesis of Ang-II-induced atrial fibrosis and vulnerability to atrial fibrillation. *Basic Res. Cardiol.* **2017**, *112*, 47. [CrossRef]
84. Chen, X.; Long, L.; Cheng, Y.; Chu, J.; Shen, Z.; Liu, L.; Li, J.; Xie, Q.; Liu, H.; Wu, M.; et al. Qingda granule attenuates cardiac fibrosis via suppression of the TGF-β1/Smad2/3 signaling pathway in vitro and in vivo. *Biomed. Pharmacother.* **2021**, *137*, 111318. [CrossRef] [PubMed]
85. Shen, Z.; Shen, A.; Chen, X.; Wu, X.; Chu, J.; Cheng, Y.; Peng, M.; Chen, Y.; Weygant, N.; Wu, M.; et al. Huoxin pill attenuates myocardial infarction-induced apoptosis and fibrosis via suppression of p53 and TGF-β1/Smad2/3 pathways. *Biomed. Pharmacother.* **2020**, *130*, 110618. [CrossRef]
86. Yousefi, F.; Shabaninejad, Z.; Vakili, S.; Derakhshan, M.; Movahedpour, A.; Dabiri, H.; Ghasemi, Y.; Mahjoubin-Tehran, M.; Nikoozadeh, A.; Savardashtaki, A.; et al. TGF-β and WNT signaling pathways in cardiac fibrosis: Non-coding RNAs come into focus. *Cell Commun Signal.* **2020**, *18*, 87. [CrossRef]
87. Yuan, J.H.; Yang, F.; Wang, F.; Ma, J.Z.; Guo, Y.J.; Tao, Q.F.; Liu, F.; Pan, W.; Wang, T.T.; Zhou, C.C.; et al. A long noncoding RNA activated by TGF-β promotes the invasion-metastasis cascade in hepatocellular carcinoma. *Cancer Cell* **2014**, *25*, 666–681. [CrossRef]
88. Pachera, E.; Assassi, S.; Salazar, G.A.; Stellato, M.; Renoux, F.; Wunderlin, A.; Blyszczuk, P.; Lafyatis, R.; Kurreeman, F.; de Vries-Bouwstra, J.; et al. Long noncoding RNA H19X is a key mediator of TGF-β-driven fibrosis. *J. Clin. Investig.* **2020**, *130*, 4888–4905. [CrossRef]
89. Janakiraman, H.; House, R.P.; Gangaraju, V.K.; Diehl, J.A.; Howe, P.H.; Palanisamy, V. The Long (lncRNA) and Short (miRNA) of It: TGFβ-Mediated Control of RNA-Binding Proteins and Noncoding RNAs. *Mol. Cancer Res.* **2018**, *16*, 567–579. [CrossRef]
90. Meng, X.M.; Nikolic-Paterson, D.; Lan, H. TGF-β: The master regulator of fibrosis. *Nat. Rev. Nephrol.* **2016**, *12*, 325–338. [CrossRef]
91. Ge, Z.; Yin, C.; Li, Y.; Tian, D.; Xiang, Y.; Li, Q.; Tang, Y.; Zhang, Y. Long noncoding RNA NEAT1 promotes cardiac fibrosis in heart failure through increased recruitment of EZH2 to the Smad7 promoter region. *J. Transl. Med.* **2022**, *20*, 7. [CrossRef] [PubMed]
92. Hao, K.; Lei, W.; Wu, H.; Wu, J.; Yang, Z.; Yan, S.; Lu, X.A.; Li, J.; Xia, X.; Han, X.; et al. LncRNA-Safe contributes to cardiac fibrosis through Safe-Sfrp2-HuR complex in mouse myocardial infarction. *Theranostics* **2019**, *9*, 7282–7297. [CrossRef]
93. Derynck, R.; Zhang, Y. Smad-dependent and Smad-independent pathways in TGF-beta family signalling. *Nature* **2003**, *425*, 577–584. [CrossRef] [PubMed]
94. Weng, J.; Tu, M.; Wang, P.; Zhou, X.; Wang, C.; Wan, X.; Zhou, Z.; Wang, L.; Zheng, X.; Li, J.; et al. Amiodarone induces cell proliferation and myofibroblast differentiation via ERK1/2 and p38 MAPK signaling in fibroblasts. *Biomed. Pharmacother.* **2019**, *115*, 108889. [CrossRef]
95. Zhang, Y.; Zhang, Y.Y.; Li, T.T.; Wang, J.; Jiang, Y.; Zhao, Y.; Jin, X.X.; Xue, G.L.; Yang, Y.; Zhang, X.F.; et al. Ablation of interleukin-17 alleviated cardiac interstitial fibrosis and improved cardiac function via inhibiting long non-coding RNA-AK081284 in diabetic mice. *J. Mol. Cell. Cardiol.* **2018**, *115*, 64–72. [CrossRef]
96. Zhang, F.; Fu, X.; Kataoka, M.; Liu, N.; Wang, Y.; Gao, F.; Liang, T.; Dong, X.; Pei, J.; Hu, X.; et al. Long noncoding RNA Cfast regulates cardiac fibrosis. *Mol. Ther. Nucleic Acids* **2021**, *23*, 377–392. [CrossRef]
97. Tang, R.; Wang, Y.C.; Mei, X.; Shi, N.; Sun, C.; Ran, R.; Zhang, G.; Li, W.; Staveley-O'Carroll, K.F.; Li, G.; et al. LncRNA GAS5 attenuates fibroblast activation through inhibiting Smad3 signaling. *Am. J. Physiol. Cell Physiol.* **2020**, *319*, C105–C115. [CrossRef]
98. Li, L.; Zhao, Q.; Kong, W. Extracellular matrix remodeling and cardiac fibrosis. *Matrix Biol.* **2018**, *68*, 490–506. [CrossRef] [PubMed]
99. Chaudhuri, O.; Cooper-White, J.; Janmey, P.A.; Mooney, D.J.; Shenoy, V.B. Effects of extracellular matrix viscoelasticity on cellular behaviour. *Nature* **2020**, *584*, 535–546. [CrossRef]
100. Huang, Z.P.; Ding, Y.; Chen, J.; Wu, G.; Kataoka, M.; Hu, Y.; Yang, J.H.; Liu, J.; Drakos, S.G.; Selzman, C.H.; et al. Long non-coding RNAs link extracellular matrix gene expression to ischemic cardiomyopathy. *Cardiovasc. Res.* **2016**, *112*, 543–554. [CrossRef]
101. Choong, O.K.; Chen, C.Y.; Zhang, J.; Lin, J.H.; Lin, P.J.; Ruan, S.C.; Kamp, T.J.; Hsieh, P.C.H. Hypoxia-induced H19/YB-1 cascade modulates cardiac remodeling after infarction. *Theranostics* **2019**, *9*, 6550–6567. [CrossRef] [PubMed]
102. Qu, X.; Du, Y.; Shu, Y.; Gao, M.; Sun, F.; Luo, S.; Yang, T.; Zhan, L.; Yuan, Y.; Chu, W.; et al. MIAT Is a Pro-fibrotic Long Non-coding RNA Governing Cardiac Fibrosis in Post-Infarct Myocardium. *Sci. Rep.* **2017**, *7*, 42657. [CrossRef]
103. Piccoli, M.T.; Gupta, S.K.; Viereck, J.; Foinquinos, A.; Samolovac, S.; Kramer, F.L.; Garg, A.; Remke, J.; Zimmer, K.; Batkai, S.; et al. Inhibition of the Cardiac Fibroblast-Enriched lncRNA Meg3 Prevents Cardiac Fibrosis and Diastolic Dysfunction. *Circ. Res.* **2017**, *121*, 575–583. [CrossRef] [PubMed]
104. Barile, L.; Moccetti, T.; Marban, E.; Vassalli, G. Roles of exosomes in cardioprotection. *Eur. Heart J.* **2017**, *38*, 1372–1379. [CrossRef]
105. Mathiyalagan, P.; Sahoo, S. Exosomes-Based Gene Therapy for MicroRNA Delivery. *Methods Mol. Biol.* **2017**, *1521*, 139–152.
106. Ruan, Y.; Lin, N.; Ma, Q.; Chen, R.; Zhang, Z.; Wen, W.; Chen, H.; Sun, J. Circulating LncRNAs Analysis in Patients with Type 2 Diabetes Reveals Novel Genes Influencing Glucose Metabolism and Islet β-Cell Function. *Cell. Physiol. Biochem.* **2018**, *46*, 335–350. [CrossRef]
107. Tao, S.C.; Rui, B.Y.; Wang, Q.Y.; Zhou, D.; Zhang, Y.; Guo, S.C. Extracellular vesicle-mimetic nanovesicles transport LncRNA-H19 as competing endogenous RNA for the treatment of diabetic wounds. *Drug Deliv.* **2018**, *25*, 241–255. [CrossRef] [PubMed]

108. Mathieu, M.; Martin-Jaular, L.; Lavieu, G.; Thery, C. Specificities of secretion and uptake of exosomes and other extracellular vesicles for cell-to-cell communication. *Nat. Cell Biol.* **2019**, *21*, 9–17. [CrossRef]
109. Valadi, H.; Ekstrom, K.; Bossios, A.; Sjostrand, M.; Lee, J.J.; Lotvall, J.O. Exosome-mediated transfer of mRNAs and microRNAs is a novel mechanism of genetic exchange between cells. *Nat. Cell Biol.* **2007**, *9*, 654–659. [CrossRef]
110. Chaturvedi, P.; Kalani, A.; Medina, I.; Familtseva, A.; Tyagi, S.C. Cardiosome mediated regulation of MMP9 in diabetic heart: Role of mir29b and mir455 in exercise. *J. Cell. Mol. Med.* **2015**, *19*, 2153–2161. [CrossRef]
111. Kenneweg, F.; Bang, C.; Xiao, K.; Boulanger, C.M.; Loyer, X.; Mazlan, S.; Schroen, B.; Hermans-Beijnsberger, S.; Foinquinos, A.; Hirt, M.N.; et al. Long Noncoding RNA-Enriched Vesicles Secreted by Hypoxic Cardiomyocytes Drive Cardiac Fibrosis. *Mol. Ther. Nucleic Acids* **2019**, *18*, 363–374. [CrossRef]
112. Wang, Y.; Cao, X.; Yan, L.; Zheng, Y.; Yu, J.; Sun, F.; Lian, Z.; Sun, L. Exosome-derived long non-coding RNA ZFAS1 controls cardiac fibrosis in chronic kidney disease. *Aging* **2021**, *13*. [CrossRef]
113. Chen, Y.; Chen, X.; Li, H.; Li, Y.; Cheng, D.; Tang, Y.; Sang, H. Serum extracellular vesicles containing MIAT induces atrial fibrosis, inflammation and oxidative stress to promote atrial remodeling and atrial fibrillation via blockade of miR-485-5p-mediated CXCL10 inhibition. *Clin. Transl. Med.* **2021**, *11*, e482. [CrossRef] [PubMed]
114. Raposo, G.; Stahl, P. Extracellular vesicles: A new communication paradigm? *Nat. Rev. Mol. Cell Biol.* **2019**, *20*, 509–510. [CrossRef]
115. Villarroya-Beltri, C.; Gutierrez-Vazquez, C.; Sanchez-Cabo, F.; Perez-Hernandez, D.; Vazquez, J.; Martin-Cofreces, N.; Martinez-Herrera, D.J.; Pascual-Montano, A.; Mittelbrunn, M.; Sanchez-Madrid, F. Sumoylated hnRNPA2B1 controls the sorting of miRNAs into exosomes through binding to specific motifs. *Nat. Commun.* **2013**, *4*, 2980. [CrossRef] [PubMed]
116. Li, S.; Li, Y.; Chen, B.; Zhao, J.; Yu, S.; Tang, Y.; Zheng, Q.; Li, Y.; Wang, P.; He, X.; et al. exoRBase: A database of circRNA, lncRNA and mRNA in human blood exosomes. *Nucleic Acids Res.* **2018**, *46*, D106–D112. [CrossRef] [PubMed]
117. Squadrito, M.L.; Baer, C.; Burdet, F.; Maderna, C.; Gilfillan, G.D.; Lyle, R.; Ibberson, M.; De Palma, M. Endogenous RNAs modulate microRNA sorting to exosomes and transfer to acceptor cells. *Cell Rep.* **2014**, *8*, 1432–1446. [CrossRef] [PubMed]
118. Koppers-Lalic, D.; Hackenberg, M.; Bijnsdorp, I.V.; van Eijndhoven, M.A.J.; Sadek, P.; Sie, D.; Zini, N.; Middeldorp, J.M.; Ylstra, B.; de Menezes, R.X.; et al. Nontemplated nucleotide additions distinguish the small RNA composition in cells from exosomes. *Cell Rep.* **2014**, *8*, 1649–1658. [CrossRef]
119. Hurley, J.H.; Hanson, P. Membrane budding and scission by the ESCRT machinery: It's all in the neck. *Nat. Rev. Mol. Cell Biol.* **2010**, *11*, 556–566. [CrossRef]
120. Puri, N.; Roche, P. Ternary SNARE complexes are enriched in lipid rafts during mast cell exocytosis. *Traffic* **2006**, *7*, 1482–1494. [CrossRef]
121. Yang, L.; Peng, X.; Li, Y.; Zhang, X.; Ma, Y.; Wu, C.; Fan, Q.; Wei, S.; Li, H.; Liu, J. Long non-coding RNA HOTAIR promotes exosome secretion by regulating RAB35 and SNAP23 in hepatocellular carcinoma. *Mol. Cancer* **2019**, *18*, 78. [CrossRef] [PubMed]
122. Crooke, S.T.; Witztum, J.L.; Bennett, C.F.; Baker, B.F. RNA-Targeted Therapeutics. *Cell Metab.* **2018**, *27*, 714–739. [CrossRef] [PubMed]
123. Crooke, S.T. Molecular Mechanisms of Antisense Oligonucleotides. *Nucleic Acid Ther.* **2017**, *27*, 70–77. [CrossRef] [PubMed]
124. Lim, K.H.; Han, Z.; Jeon, H.Y.; Kach, J.; Jing, E.; Weyn-Vanhentenryck, S.; Downs, M.; Corrionero, A.; Oh, R.; Scharner, J.; et al. Antisense oligonucleotide modulation of non-productive alternative splicing upregulates gene expression. *Nat. Commun.* **2020**, *11*, 3501. [CrossRef]
125. Modarresi, F.; Faghihi, M.A.; Lopez-Toledano, M.A.; Fatemi, R.P.; Magistri, M.; Brothers, S.P.; van der Brug, M.P.; Wahlestedt, C. Inhibition of natural antisense transcripts in vivo results in gene-specific transcriptional upregulation. *Nat. Biotechnol.* **2012**, *30*, 453–459. [CrossRef]
126. Lima, J.F.; Cerqueira, L.; Figueiredo, C.; Oliveira, C.; Azevedo, N.F. Anti-miRNA oligonucleotides: A comprehensive guide for design. *RNA Biol.* **2018**, *15*, 338–352. [CrossRef]
127. Kluiver, J.; Slezak-Prochazka, I.; Smigielska-Czepiel, K.; Halsema, N.; Kroesen, B.J.; van den Berg, A. Generation of miRNA sponge constructs. *Methods* **2012**, *58*, 113–117. [CrossRef]
128. Wronska, A. The Role of microRNA in the Development, Diagnosis, and Treatment of Cardiovascular Disease: Recent Developments. *J. Pharmacol. Exp. Ther.* **2023**, *384*, 123–132. [CrossRef]
129. Das, S.; Kohr, M.; Dunkerly-Eyring, B.; Lee, D.I.; Bedja, D.; Kent, O.A.; Leung, A.K.; Henao-Mejia, J.; Flavell, R.A.; Steenbergen, C. Divergent Effects of miR-181 Family Members on Myocardial Function Through Protective Cytosolic and Detrimental Mitochondrial microRNA Targets. *J. Am. Heart Assoc.* **2017**, *6*, e004694. [CrossRef]
130. Bernardo, B.C.; Gregorevic, P.; Ritchie, R.H.; McMullen, J.R. Generation of MicroRNA-34 Sponges and Tough Decoys for the Heart: Developments and Challenges. *Front. Pharmacol.* **2018**, *9*, 1090. [CrossRef]
131. Esau, C.; Davis, S.; Murray, S.F.; Yu, X.X.; Pandey, S.K.; Pear, M.; Watts, L.; Booten, S.L.; Graham, M.; McKay, R.; et al. miR-122 regulation of lipid metabolism revealed by in vivo antisense targeting. *Cell Metab.* **2006**, *3*, 87–98. [CrossRef] [PubMed]
132. Rockey, D.C.; Bell, P.; Hill, J. Fibrosis—A common pathway to organ injury and failure. *N. Engl. J. Med.* **2015**, *372*, 1138–1149. [CrossRef] [PubMed]
133. Zhang, K.; Han, X.; Zhang, Z.; Zheng, L.; Hu, Z.; Yao, Q.; Cui, H.; Shu, G.; Si, M.; Li, C.; et al. The liver-enriched lnc-LFAR1 promotes liver fibrosis by activating TGFβ and Notch pathways. *Nat. Commun.* **2017**, *8*, 144. [CrossRef]

134. Bang, C.; Antoniades, C.; Antonopoulos, A.S.; Eriksson, U.; Franssen, C.; Hamdani, N.; Lehmann, L.; Moessinger, C.; Mongillo, M.; Muhl, L.; et al. Intercellular communication lessons in heart failure. *Eur. J. Heart Fail.* **2015**, *17*, 1091–1103. [CrossRef] [PubMed]
135. Zanotti, S.; Gibertini, S.; Blasevich, F.; Bragato, C.; Ruggieri, A.; Saredi, S.; Fabbri, M.; Bernasconi, P.; Maggi, L.; Mantegazza, R.; et al. Exosomes and exosomal miRNAs from muscle-derived fibroblasts promote skeletal muscle fibrosis. *Matrix Biol.* **2018**, *74*, 77–100. [CrossRef]

biology

MDPI

Article

Tissue and Serum Biomarkers in Degenerative Aortic Stenosis-Insights into Pathogenesis, Prevention and Therapy

Alkistis Kapelouzou [1], Styliani Geronikolou [1], Irene Lidoriki [2], Christos Kontogiannis [3], Loukas Kaklamanis [4], Loukas Tsourelis [4,†] and Dennis V. Cokkinos [1,*]

1 Center for Clinical, Experimental Surgery and Translational Research, Biomedical Research Foundation of the Academy of Athens, 11527 Athens, Greece
2 1st Department of Surgery, School of Medicine, National and Kapodistrian University of Athens, Laiko Hospital, 11527 Athens, Greece
3 Department of Clinical Therapeutics, National and Kapodistrian University of Athens Alexandra Hospital, 11527 Athens, Greece
4 Department of Pathology, Onassis Cardiac Surgery Center, 17674 Athens, Greece
* Correspondence: dcokkinos@bioacademy.gr
† Deceased.

Simple Summary: We found higher levels of six biomarkers significantly involved in cardiovascular pathology, i.e., irisin, periostin, osteoglycin, interleukin 18, high mobility group box 1 and proprotein convertase subtilisin/kexin type 9 in the serum at the protein level, and in the tissue at both the protein and mRNA levels of patients with AS (N = 60). Higher levels of all factors were found in DAS patients' serum than in normal C (N = 22). All biomarkers were seen in the aortic valve cusps with DAS, but no trace of PCR mRNA was found in the five transplantation valves.

Abstract: Background and Aim. Degenerative Aortic Stenosis (DAS) is a common disease that causes substantial morbidity and mortality worldwide, especially in the older population. Our aim was to further investigate novel serum and tissue biomarkers to elucidate biological processes involved in this entity. Material and Methods. We evaluated the expression of six biomarkers significantly involved in cardiovascular pathology, i.e., irisin, periostin, osteoglycin, interleukin 18, high mobility group box 1 and proprotein convertase subtilisin/kexin type 9 in the serum at the protein level, and in the tissue at both the protein and mRNA levels of patients with AS (N = 60). Five normal valves obtained after transplantation from hearts of patients with idiopathic dilated cardiomyopathy were also studied. Serum measurements were also performed in 22 individuals without valvular disease who served as controls (C). Results. Higher levels of all factors were found in DAS patients' serum than in normal C. IHC and PCR mRNA tissue analysis showed the presence of all biomarkers in the aortic valve cusps with DAS, but no trace of PCR mRNA was found in the five transplantation valves. Moreover, periostin serum levels correlated significantly with IHC and mRNA tissue levels in AS patients. Conclusion. We showed that six widely prevalent biomarkers affecting the atherosclerotic process were also involved in DAS, suggesting a strong osteogenic and pro-inflammatory profile, indicating that aortic valve calcification is a multifactorial biological process.

Keywords: valvular heart disease; degenerative aortic valve stenosis; biomarkers; irisin; periostin; osteoglycin; IL-18; HMGB-1; PCSK9

Citation: Kapelouzou, A.; Geronikolou, S.; Lidoriki, I.; Kontogiannis, C.; Kaklamanis, L.; Tsourelis, L.; Cokkinos, D.V. Tissue and Serum Biomarkers in Degenerative Aortic Stenosis-Insights into Pathogenesis, Prevention and Therapy. *Biology* **2023**, *12*, 347. https://doi.org/10.3390/biology12030347

Academic Editors: Gaetano Santulli and Guo-Chang Fan

Received: 7 December 2022
Revised: 3 February 2023
Accepted: 9 February 2023
Published: 22 February 2023

1. Introduction

The interest in degenerative aortic stenosis (DAS) remains unabated, since it is an increasingly frequent cause of cardiac morbidity and mortality [1,2]. In a U.S. population study, its prevalence was 2.8% for peoples aged > 75 years [3]. It is correlated to atherosclerotic vascular disease, with many factors being prevalent in both conditions [1,2]. However, Ortlepp et al. [4] point out, that the predictive power of risk factors is not equal for the two entities. It is currently being accepted that DAS is not a passive process, but involves many mechanisms, such as lipoprotein profile derangement, oxidation, inflammation and valvular cell apoptosis [5–7]. All these are compounded by hemodynamic factors, since the initial valve sclerosis causes flow turbulence and nonlinear flow promoting further progress of the sclerosis/calcification process [1,6].

Livia Passos et al. [8] argue that cardiovascular calcification is an inflammatory disease, through crosstalk between innate and adaptive immune cell components.

A great number of biomarkers has been studied in DAS.

The purpose of our study is to identify additional serum and tissue biomarkers involved in patients with DAS, and also to investigate the correlation between their tissue and serum levels. Since 2015 we have been investigating tissue and serum biomarkers in this entity. We have shown that patients with DAS express more pro-inflammatory, calcification, fibrosis, proliferation and apoptosis biomarkers. We have also shown that Toll-like receptors and interleukin-37 are differentially expressed in aortic compared to mitral valves, indicating a higher pro-calcific and pro-inflammatory process in the aortic valve [7], in addition to the hemodynamic factors and the turbulent aortic flow.

Trying to obtain further insight in the calcification process we measured six biomarkers in the tissue of stenotic aortic valves excised at surgery for aortic valve replacement and compared them to normal aortic valves obtained at cardiac transplantation.

These six biomarkers were also measured in the serum of patients and controls without any cardiovascular disease; the reasons for their inclusion are explained below.

1.1. Irisin Levels

The myokine irisin, which is cleaved from the plasma membrane FNDC5, is more highly expressed in cardiac muscle than in skeletal [9]. It is also highly elevated in patients with heart failure and preserved ejection fraction (EF) more than in those with reduced EF; in the former, it correlated with total antioxidant capacity [10]. Irisin is increased in acute heart failure and is an independent factor for 1 year all-cause mortality [11]. However, it is decreased in myocardial infarction, both clinical and experimental [12,13]. It is correlated negatively to coronary artery disease severity [14]. Patients in the higher SYNTAX score levels were older and had lower irisin levels than younger ones. However, it has also been reported to be cardioprotective [15].

Irisin has been found to promote osteoblast proliferation and differentiation via the MAP kinase pathways by Qiao et al. [16] who postulated its possible use in osteopenia. It increases sclerostin expression in osteocytes to induce bone resorption. It mediates this effect via αV integrin receptors [17]. In our previous study we have found sclerostin to be increased in DAS [5]. Irisin administration has been proposed for treatment of osteoporosis by Colaioanni et al. [18].

1.2. Periostin

Hakuno et al. [19] found PN to be increased in degenerative or rheumatically affected heart valves. The same authors also found that in wild type mice a high fat diet markedly increased its expression in both AV and MV together with the fibrotic markers MMP2 and MMP13. It is associated with myocardial fibrosis in human heart failure [20]. It was increased together with MMP2 activity. It is increased in hypertrophic mice hearts together with interstitial fibrosis [21]. It decreases together with a reduction in myocardial fibrosis in hearts unloaded both clinically (LVAD) and experimentally in mice (aortic arch de-banding) [22]. It is also a potential marker biomarker for coronary artery disease with acute heart failure [23]. Furthermore patients with CAD had higher levels than controls at similar ages (around 63 y).

1.3. Osteoglycin

The same pattern for PN was found for OGN, which is implicated in matrix homeostasis. It modulates fibrosis and inflammation. Deckx et al. [24] found that its levels were higher in patients with AS with less severe myocardial fibrosis, in whom it was negatively correlated with collagen content in the myocardium, but they did not measure it in the valves. Van Aelst et al. [25] found that it prevents cardiac dilation and dysfunction after myocardial infarction through infarct collagen strengthening. Zuo et al. [26] have found that osteoglycin attenuates cardiac fibrosis; it could be an antifibrotic, but is also pro-calcific by suppressing myofibroblast proliferation. Circulating osteoglycin and NGAL/MMP9 complex concentrations predict 1 y MACE after coronary arteriography [27]. It was statistically slightly higher in CAD patients aged 70 vs. 65 y [28] Tanaka et al. have stated that it is a humoral bone anabolic factor [28]

1.4. Interleukin 18

IL-18 is a dominant pro-inflammatory cytokine. In the heart, it is produced by infiltrating neutrophils, resident macrophages, endothelial cells, smooth muscle cells and cardiomyocytes [29]. In the non-rheumatic aortic valve, increased tissue levels have been found as compared to controls [30], and are correlated to advanced clinical severity. It promotes myofibroblast activation of porcine valvular interstitial cells [31]. Interestingly, the administration of increased doses of IL-18 upregulated the expression of osteopontin [32] which we have found to be pro-osteogenic [5]. It is prospectively and independently associated with CVD risk [33] in patients of similar ages (52.5 y).

1.5. HMGB1

The high mobility group box 1 (HMGB1) which is also a pro-inflammatory factor has been implicated in the pathogenesis of DAS. It has been found to be increased in the calcific regions; the same was found in regard to TLR by Shen et al. [34], who postulate that TLR4 may function as an essential mediator of HMGB1-induced calcification and in the activation of p38 and NFkB. Wang et al. [35] found that HMGB1 protein and TLR4 are upregulated in vitro by HMGB1 in aortic valvular interstitial cells. We have described in detail the role of TLRs in aortic and mitral valve stenosis [7]. Its increase in the serum is related to severity of coronary artery stenosis [36].

1.6. PCSK9

Recent experimental studies have demonstrated that PCSK9 might directly promote inflammation, apoptotic cell death and endothelial dysfunction in the atherosclerotic process and plaque formation [37] and is associated with the severity of CAD [38] in hypercholesterolemia, cardiovascular inflammation and diabetes [39–41].

Wang et al. [42] and Poggio et al. [43] found lower calcium content in aortic valves of $PCSK9^{-}/^{-}$ mice than wild type animals. The former were resistant to production of aortic calcification with two pro-calcification diets (β-glycero–phosphate and ascorbic acid). The same authors found that human calcified aortic valves expressed higher PCSK9 than non-calcified ones. Interestingly, Salaun et al. [44] found that elevated plasma levels of PCSK9 were a risk factor for hemodynamic deterioration of surgically implanted bioprosthetic aortic valves. The same held true for LP-PLA2 and HoMA. PCSK9, apart from its action on LDL receptors, can promote apoptotic cell death of neurons [45].

2. Material and Methods

2.1. Ethics Statement

This study was approved by the Ethics Committees of the Onassis Cardiac Surgery Center (OCSC) and conformed to the principles outlined in the Declaration of Helsinki. Written informed consent was obtained from all patients.

2.2. Study Population

The present, prospective, open-label study, extends the results of our previously published studies [5,7]. The age of the 60 DAS patients was 66.1 ± 12.5 years, 50% were women. Echocardiography was obtained in all patients. Aortic valve area was 0.9 to 0.5 cm [2].

Patients were taking antihypertensive (75%) and antilipidemic (80%) drugs during the last two years. However, none had significant coronary artery disease necessitating concomitant aortocoronary bypass surgery. Patients with rheumatic cardiac disease, bicuspid aortic valve or connective tissue disorders were excluded. The excised valve cusps were harvested during surgery. Blood was collected from AS patients before valve surgery and from twenty-two (22) healthy subjects without any chronic cardiovascular or metabolic disease, and not receiving any long-term medication, who served as healthy control group (C), for comparison of serum biomarkers; mean age was 34.4 ± 7.5 years, 50% were women. None were steadily employing radioactive drugs.

We also obtained 5 aortic valves from patients undergoing cardiac transplantation, 3 men and 2 women (min age 48.4), because of terminal heart failure due to idiopathic dilated cardiomyopathy, who did not show any sclerotic or calcific aortic valve changes.

2.3. Blood Analysis

Blood samples were collected by venipuncture after subjects were fasted overnight. Serum was collected and stored at −80 °C, analysis was performed in duplicate; dilution was assessed as per protocol. We used commercially available enzyme immunoassay kits for irisin (FNDC5) (EK-067-29, Phoenix Pharmaceuticals, California, CA, USA); periostin (PN) (EHPOSTN, Thermoscientific, Germany); osteoglycin (OGN) (CSB EL016314HU, Cusabio, Houston, TX, USA); IL-18 (DY318-05, R&D, Minneapolis, MN, USA); high-mobility group box 1, HMGB1 (LS-F11641, Lifespan Ltd., Seattle, DC, USA); PCSK9 (DPC900, R&D, Minneapolis, MN, USA) and quantified each protein according to the protocol of the manufacturer with an ELISA reader system (Spectramax 190; Molecular Devices, Sunnyvale, Calif, CA, USA).

2.4. Valve Cusp Immunohistochemistry and Quantitative Morphometrical Analysis

Aortic valve cusps were excised and one part of each valve tissue was placed in a container for immunohistochemistry (IHC) analysis at the pathology department of the OCSC and the Biomedical Research Foundation of Academy of Athens according to our previous protocol. The protocol of IHC has been described and validated in our lab [46]; FNDC-5 (PA5-62368, 5 μg/mL, Invitrogen, CA, USA); PN (PA5-82458, 5 μg/mL, Invitrogen, CA, USA); OGN (PA5-48255, 5 μg/mL, Invitrogen, CA, USA); IL-18 (PA5-79479, Invitrogen, CA, USA) HMGB1 (PA5-80691, 5 μg/mL, Invitrogen, CA, USA); PCSK9 (PA5-79789, 5 μg/mL, Invitrogen, CA, USA) were used for IHC. IHC was performed according to the manufacturer's protocol by using the development kit (Zytochem Plus; Zytomed system, Germany). Appropriate isotype negative controls were performed at the same concentrations as the primary antibodies. Microscopic investigation of the IHC sections was performed with stereology upright Leica DMRA2 camera, and were analyzed by stereo-investigator 10 program (version 10.1, MBF Bioscience, Microbrightfield. Inc., Willinston, VT, USA) in order to quantify the extent of the tissue covered by each antibody.

2.5. RNA Isolation and qRT PCR Analysis

Total RNA was extracted using Trizol reagent (Sigma, Saint Louis, MO, USA) according to the manufacturer's instructions [46]. The RNA quality was assessed with agarose gel electrophoresis and quantitated spectrophotometrically. cDNA was synthesized by RT (MMLV, reverse transcriptase; Sigma), and real-time quantitative polymerase chain reaction was performed by using SYBR Green (Invitrogen, Life Technologies, New York, NY, USA). The primers synthesized by Origine (Herford, Germany) were used as documented in Table 1. The thermal cycling protocol was performed according to our lab protocol [5,7].

Table 1. Primers sequences. Abbreviations: Irisin (FNDC5); periostin (PN); osteoglycin (OGN); interleukin 18 (IL-18); high-mobility group box 1 (HMGB1); proprotein convertase subtilisin kexin 9 (PCSK9).

Primers	Forward	Reverse	bp	NM
b actin	CACCATTGGCAATGAGCGGTTC	AGGTCTTTGCGGATGTCCACGT	101	1101
FNDC5	AGCGAGCCTGTGCTCTTCAAGA	GAACAGGACCACGACGATGATC	120	1171940
PN	CAGCAAACCACCTTCACGGATC	TTAAGGAGGCGCTGAACCATGC	118	1135934
OGN	CCATAATGCCCTGGAATCCGTG	CAGGCGTATCTCTTCAATGCGG	125	14057
IL-18	GATAGCCAGCCTAGAGGTATGG	CCTTGATGTTATCAGGAGGATTCA	120	1243211
HMGB-1	GCGAAGAAACTGGGAGAGATGTG	GCATCAGGCTTTCCTTTAGCTCG	114	1313892
PCKS-9	GACACCAGCATACAGAGTGACC	GTGCCATGACTGTCACACTTGC	118	174936

We also measured the 6 biomarkers already discussed in the 5 aortic valves explanted from patients who had undergone cardiac transplantation by RNA isolation and qRT analysis.

All had idiopathic dilated cardiomyopathy without coronary artery disease at arteriography: 2 were women aged 28 and 56 years, and were 3 men aged 48, 54 and 58 years.

2.6. Statistical Analysis

Shapiro–Wilks test for normality showed that none of the variables had normal distribution. Thus, Univariate and Multivariate analysis, one-way ANOVA or *t*-test were inappropriate tests for the analysis. We performed non-parametric tests instead. The Mann–Whitney test for evaluating the patients versus control serum biomarkers showed significant differences between all serum markers. All correlations were performed with non-parametric Spearman's rho. Alpha was set at 0.05. Statistics were performed with SPSS28.

3. Results

3.1. Serum Findings

(a) The non-parametric Mann–Whitney test showed significant differences in all serum biomarkers between patients and the control group. (Figure 1, Table 2).

Table 2. Statistical analysis of serum biomarkers in all study groups. Mann–Whitney test for all markers measured in the serum of both groups (control and AS); alpha was set at 0.05 C and AS. Abbreviations: C (control); AS (aortic stenosis patients); Irisin (FNDC5); periostin (PN); osteoglycin (OGN); interleukin 18 (IL-18); high-mobility group box 1 (HMGB1); proprotein convertase subtilisin kexin 9 (PCSK9).

Serum Markers	Median Control [95% CI]	Median AS [95% CI]	Mann–Whitney U	*p*-Value
IL-18	2.46 [1.26, 3.02]	3.36 [2.95, 3.56]	259.5	0.004
HMGB1	0.59 [0.44, 0.76]	2.98 [2.4, 13.86]	0.00	<0.001
PCSK9	2.05 [1.2, 82.86]	262.00 [1.57, 224.286]	0.00	<0.001
OGN	4.7 [3.5, 29.78]	29.5 [27.64, 30.9]	0.00	<0.001
PN	25.00 [23.28, 45.75]	240.00 [227.54, 265.36]	0.00	<0.001
FNDC5	1.63 [1.1, 43.74]	2.955 [2.2, 3.73]	31.00	<0.001

(b) Positive correlations were found in HMGB1 with PCSK9 and PN vs. PSCK9. Negative correlations were observed in OGN with PN and IL-18.

Any other correlation was found insignificant ($p > 0.05$).

(c) Positive correlations were found in HMGB1 with PCSK9 and PN with PSCK9. Negative correlations were observed in OGN with PN and IL-18.

(d) Any other correlation was found insignificant ($p > 0.05$).

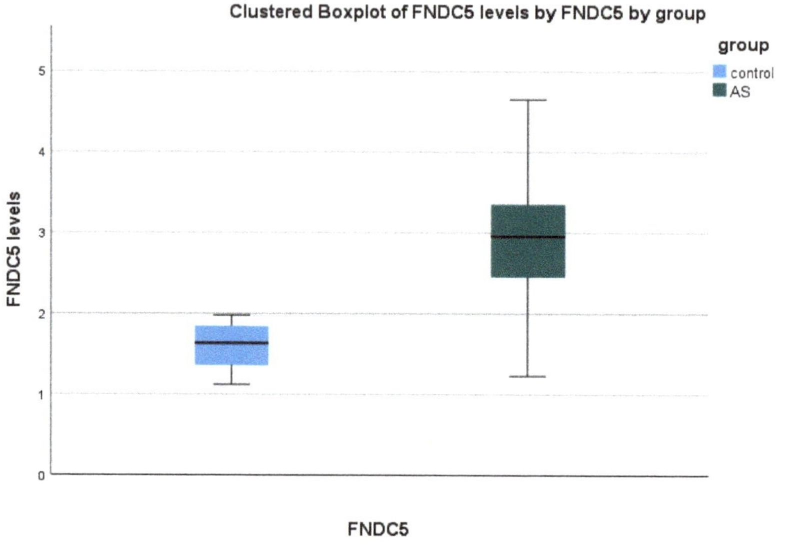

(a)

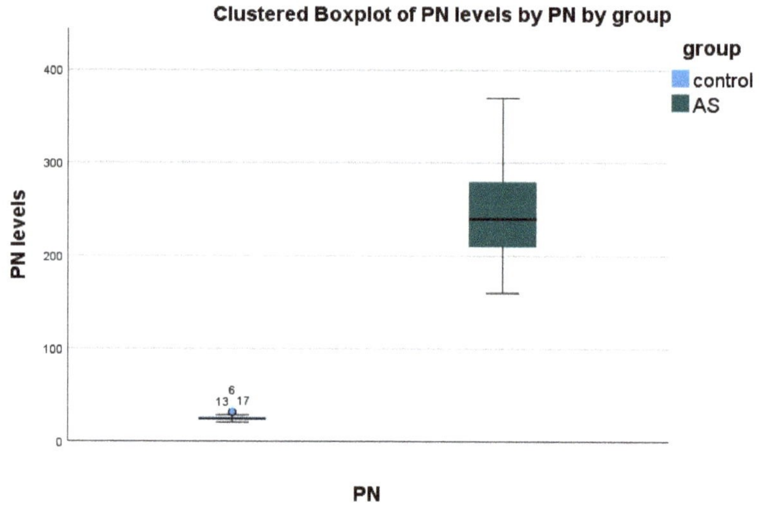

(b)

Figure 1. *Cont.*

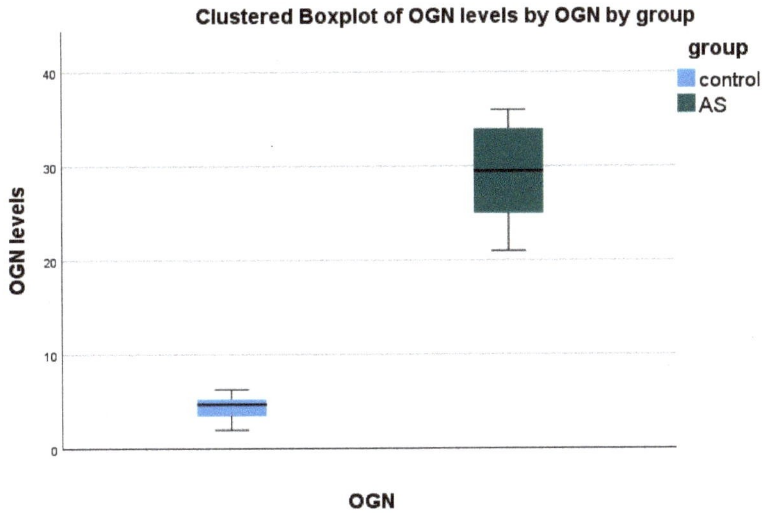

(c)

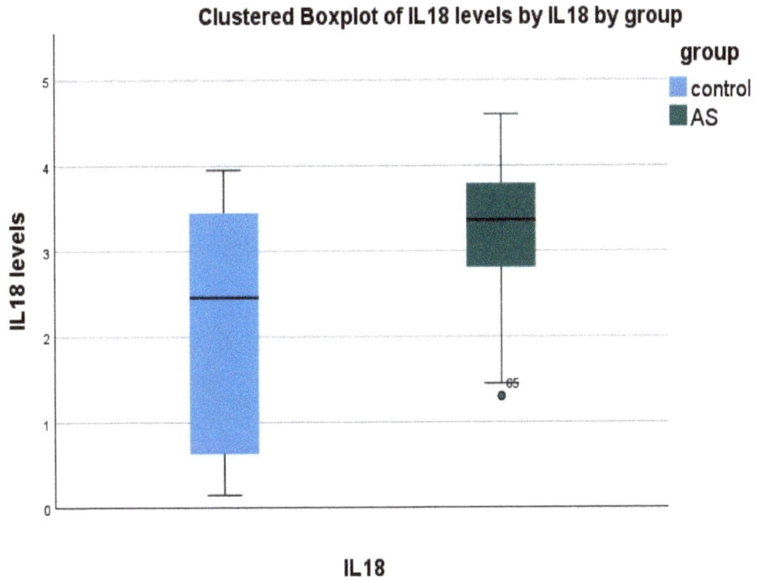

(d)

Figure 1. *Cont.*

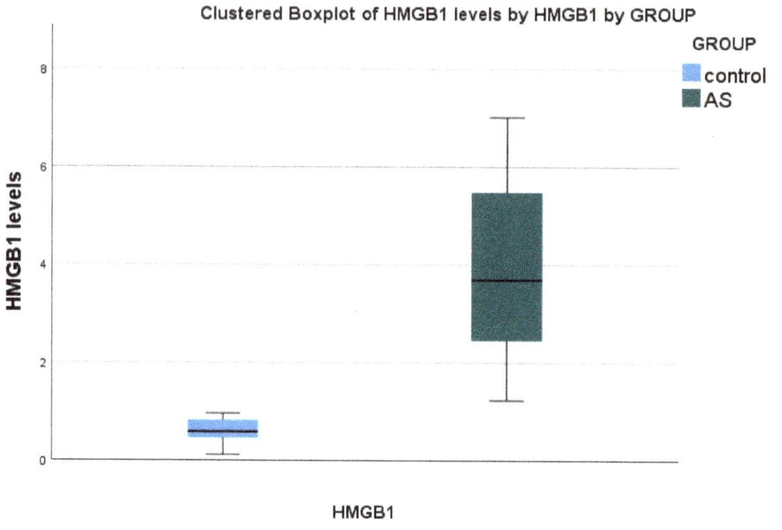

(e)

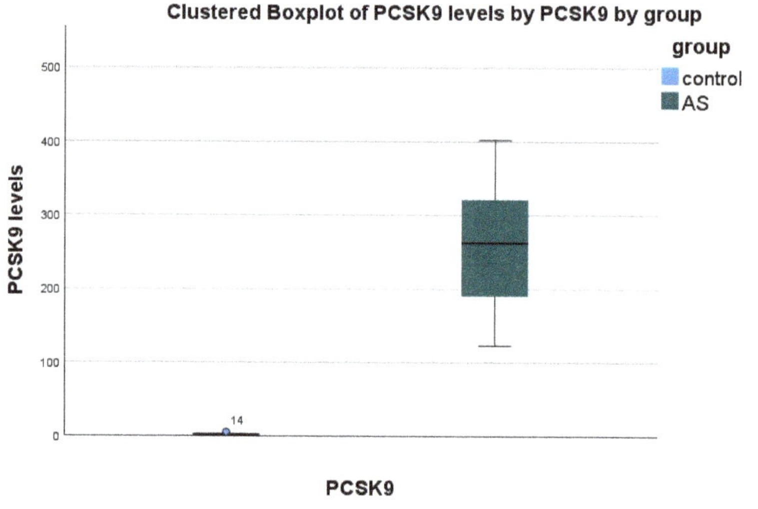

(f)

Figure 1. Serum biomarkers' concentration in all study groups. Data presented as boxplots: Blue color, Control; Green color, (AS) aortic stenosis patients. Abbreviations: (**a**) irisin (FNDC5) (**b**) periostin (PN); (**c**) osteoglycin (OGN); (**d**) interleukin 18 (IL-18); (**e**) high-mobility group box 1 (HMGB1); (**f**) proprotein convertase subtilisin kexin 9 (PCSK9).

3.2. Tissue vs. Serum

(a) Of all biomarkers found significantly correlated only PN tissue levels correlate with the same marker's levels detected in serum.

(b) For all other tissue markers described in Table 3, the correlations were found in non-identical markers: (positive) tissue HMGB1 with serum OGN, tissue PCSK9 with serum OGN, (negative) tissue IL-18 with serum PCSK9 and OCN, tissue HMGB1 with serum IL-18, tissue PCSK9 with serum HMGB1 and IL-18. (Table 3)

(c) The insignificant correlations were not included in Table 3.

Table 3. The Spearman's rho coefficients, *p*-value and 95% Confidence Intervals of every significant correlation between and within the tissues under investigation.

	Biomarkers Compared	Spearman Rho	*p*-Value	95% Confidence Intervals
Serum vs. serum	OGN with PN	−0.359	0.005	[−0.567, 0.108]
	IL-18 with OGN	−0.441	<0.001	[−0.630, 0.203]
	HMGB1 with PCSK9	0.445	0.001	[0.173, 0.610]
	PN vs PSCK9	0.286	0.027	[0.027, 0.509]
Tissue vs. tissue	IL-18 with HMGB1	−0.425	<0.001	[−0.618, −0.185]
	IL-18 with PCSK9	−0.528	<0.001	[−0.694, −0.310]
	HMGB1 with PCSK9	0.420	0.001	[0.178, 0.613]
Tissue vs. serum	IL-18 with FNDC5	−0.291	0.024	[−0.513, −0.033]
	IL-18 with OGN	−0.419	0.001	[−0.613, −0.177]
	HMGB1 with IL-18	−0.326	0.011	[−0.541, −0.071]
	HMGB1 with OGN	0.282	0.029	[0.022, 0.506]
	PCSK9 with IL-18	−0.405	0.001	[−0.602, −0.161]
	PCSK9 with HMGB1	−0.326	0.011	[−0.541, −0.070]
	PCSK9 with OGN	0.604	<0.001	[0.407, 0.747]
	PN with PN	0.289	0.025	[0.030, 0.511]

3.3. Immunohistochemistry Biomarkers in Aortic Valve Cusps

Immunohistochemistry staining was performed (Figure 2). Biomarker tissue presence was confirmed in all AVC samples.

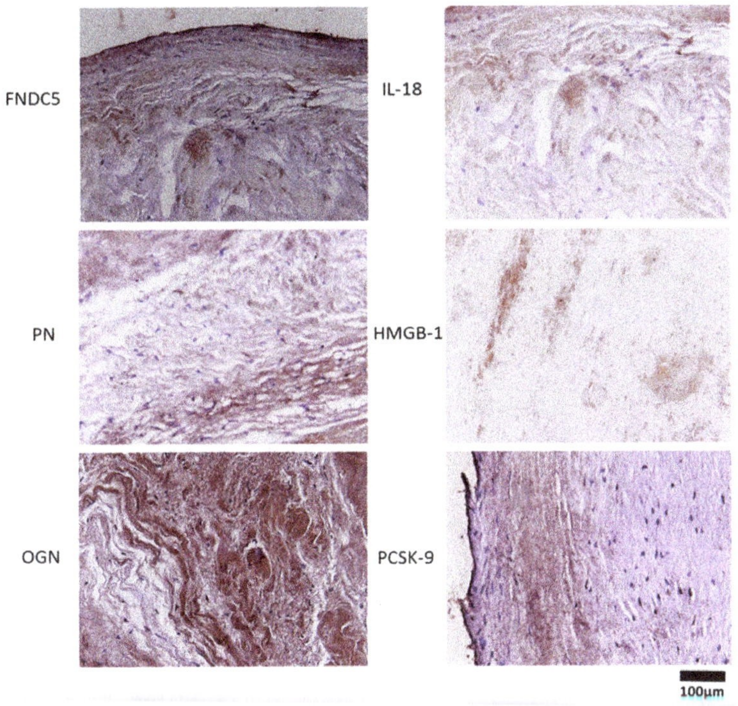

Figure 2. Immunohistochemistry staining in the AS group. Representative photos of tissue biomarkers from aortic valve cusps. Nuclei stained with celestine blue are shown in blue, expression of biomarkers

shown in brown. 40× magnification. Abbreviations: irisin (FNDC5); periostin (PN); osteoglycin (OGN); interleukin 18 (IL-18); high-mobility group box 1 (HMGB1); proprotein convertase subtilisin kexin 9 (PCSK9).

3.4. mRNA Expression of Inflammation and Calcification Biomarkers in AS Patients

Tissue mRNA levels of all biomarkers was present in AVC (Figure 3).

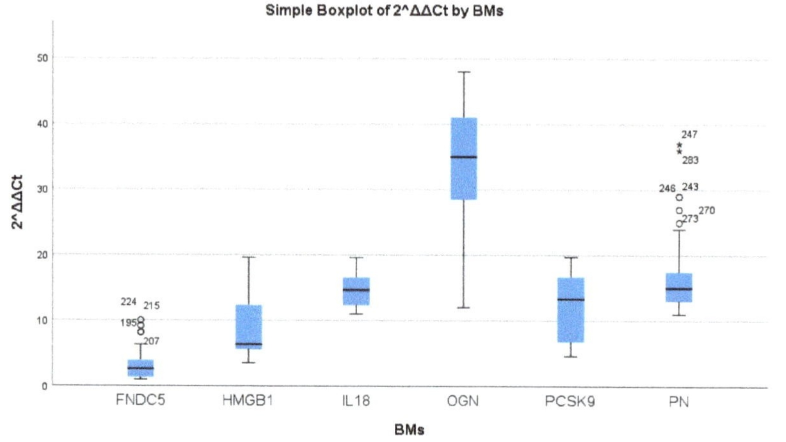

Figure 3. mRNA expression of tissue biomarker concentrations in AS patients. Data presented as boxplots; Abbreviations: irisin (FNDC5); periostin (PN); osteoglycin (OGN); interleukin 18 (IL-18); high-mobility group box 1 (HMGB1); proprotein convertase subtilisin kexin 9 (PCSK9). Non-normality confirmed by outliers imposed non-parametric evaluation opted herein.

The highest expression was found for osteoglycin, while FNDC5 was only mildly elevated None of the valves from patients with cardiac transplantation had any expression of mRNA of the above biomarkers.

The five valves from the cardiac transplantation hearts showed no expression at all.

3.5. Tissue Biomarkers Correlations

We found significant correlations between most tissue biomarkers.

4. Discussion

In this study we continue the investigation of the calcification process through the novel body of biomarkers examined in DAS [5,7]. We investigated six factors, both in serum and valve tissue, of which some have been very scantily studied. All of them have also been involved in CAD. Thus, we did not consider it practical to add a comparison Group with CAD only and without DAS.

We obtained valves from five patients undergoing cardiac transplantation, which did not show any expression of RNA of these biomarkers. With regard to the serum biomarkers, it must be realized that the control individuals were younger. However, this difference reflects a real-life situation, that individuals develop DAS later in life. It must be realized that DAS and CAD co-exist in around 50% of patients, as presented by Ortlep et al. [4]. However, none of our patients needed any concomitant surgery for CAD, thus excluding significant disease.

Our study may offer another biomarker, periostin, which may in the future provide prognostic information. Up to now, NT-proBNP [47] and BNP [48] have provided data with regard to intervention time because they reflect myocardial stress which would prompt information. A valve-specific marker such as periostin may provide means of following the course of sclerosis to stress.

Again, we must stress that we did not perform a population study so as to assess the importance of these biomarkers for predicting the presence of DAS.

New technologies to measure early calcification and inflammation are available. Dweck et al. have provided data from position emission tomography in vivo [49] in patents with DAS.

It must be noted that we did not find any microscopic evidence of changes in the normal aortic valves obtained at cardiac transplantation, nor by mRNA PCR, which is more reliable than immunohistochemistry [50].

Our findings confirm the association of PCSK9 with valve tissue calcification, and supports the postulations that PCSK9 inhibitors or drugs preventing its production by the liver could be a consideration for prevention of the course towards AS once aortic valve sclerosis has been diagnosed. In fact, in the FOURIER study, PCSK9 inhibition was associated with a lower hazard of new or worsening stenosis AS [51]. This is especially relevant since initial trials with statins have not been successful in preventing AS [52,53]. While there was some promise in a rosuvastatin trial with echo measurements [54], this was not validated in the ASTRONOMER TRIAL [55]. A reason might be that statins do lower cholesterol but increase PCSK9 levels [56]. Thus, well-controlled therapeutic trials with PCSK9 inhibitors are warranted.

In the coronary arteries statins actually promote calcification, while decreasing atherosclerotic plaque burden and fibrosis [57]. This action may explain their lack of influence on the progression of aortic calcification. We found that the PCSK9 receptor is strongly expressed in the aortic valve both by IHC and mRNA analysis and increased in the serum as compared to controls. This supports the population study of Perrot et al. [58] who found that AS was less frequent in carriers of the PCSK9 R461 loss-of-function variants. They also found increased expression of PCSK9 in calcified human valves.

The high interrelation between PCSK9 and aortic valve calcification has already been stressed [59]. Additionally, this family of drugs antagonizes apoptosis; [45,56] endothelial apoptosis [60] is found in endothelial cells in DAS.

PCSK9 inhibitors alleviate oxidation and inflammation [61].

Another candidate drug family would be SGLT2 inhibitors. Interestingly, they have been found to attenuate the secretion of IL1β and IL-18 [59] by repressing the HMGB1-TLR4 receptor axis [62]. They also antagonize many inflammatory interleukins [63–65]. They are also considered powerful antioxidants [66]. Possibly, if these families of drugs for PCSK9 and SGLT2 are given together they could exert a synergistic effect.

Another practical aspect of our finding is if serum levels could predict or follow the course of aortic valve sclerosis towards actual stenosis, by a set of easily measured biomarkers. This held true for periostin in our patients.

We tried to address the mechanisms of DAS. The fact that none of these biomarkers are found in the tissue of normal aortic valves in control patients undergoing cardiac transplantation of a similar age to those with AVR suggests that this is not a problem regulated only by age. As it regards serum levels, all our control patients had much lower levels than those with DAS, which is not surprising. Furthermore, biomarkers in DAS are legion and concern all aspects of inflammation, oxidative stress, pro-calcification effects and lipid metabolism.

It should also be stressed that they all have been associated with CAD as well, since approximately half of the patients with DAS have CVD as well. As already stated, coronary artery disease is strongly associated with DAS in many studies considering the age of our patients; its prevalence could be estimated between 41% and 51% [4,67].

Indeed, non-obstructive aortic valve calcification has become a window into significant coronary artery disease [68].

However, since as already stated, many biomarkers are being studied, we did not address any other correlations in the tissue which would create confusion, since it would be difficult to find practical explanations and promote far-fetched postulation.

We believe that our findings add impetus to the efforts towards preventing the progression of aortic sclerosis to frank DAS by drugs affecting causative factors, such as hyperlipidemia, inflammation, oxidation and endothelial apoptosis. This is a realistic goal in parallel to CAD, where, although interventional and surgical therapies have attained excellent results, efforts at prevention continue unabated.

5. Study Limitations

The lower serum levels of most markers in our controls may be because they are of younger age and there is an absence of comorbidities. However, we do not consider that we have a new diagnostic and pragmatic biomarker, but we have investigated the pathobiology of this entity.

6. Conclusions

Patients with stenotic aortic valves express higher pro-inflammatory, calcification, fibrosis, proliferation and apoptosis-expressing markers in their serum than normal controls. They are all strongly expressed in calcified, but not normal, valves. We found that PN concentration is an important finding that can lead us to a consideration of the prognostic role of serum biomarkers in the course of this pathological process. Our findings point toward a higher pro-calcification and pro-inflammatory profile in DAS patients. We believe that our findings provide interesting data for the diagnosis and prevention of aortic sclerosis, and possibly treatment of mild AS.

Author Contributions: Data curation, and management activities to annotate (produce metadata), scrub and maintain research data (including software code, where it is necessary for interpreting the data itself) for initial use and later re-use: A.K., S.G., I.L. and C.K.; Formal analysis and application of statistical, mathematical, computational, or other formal techniques to analyze or synthesize study data: A.K. and S.G.; Funding acquisition, acquisition of the financial support for the project leading to this publication: D.V.C.; Investigation, conducting the research and investigation process, specifically performing the experiments, or data/evidence collection: A.K., L.K., L.T. and D.V.C.; Methodology, development or design of methodology, creation of models: A.K. and D.V.C.; Project administration, management and coordination responsibility for the research activity planning and execution: A.K., L.T. and D.V.C.; Resources, provision of study materials, reagents, materials, patients, laboratory samples, animals, instrumentation, computing resources, or other analysis tools: A.K., S.G., I.L. and L.T.; Software, programming, software development, designing computer programs, implementation of the computer code and supporting algorithms, testing of existing code components: A.K., S.G., I.L. and C.K.; Supervision, oversight and leadership responsibility for the research activity planning and execution, including mentorship external to the core team: A.K. and D.V.C.; Validation, verification, whether as a part of the activity or separate, of the overall replication/reproducibility of results/experiments and other research outputs: A.K., S.G., I.L., C.K., L.K. and D.V.C.; Visualization, preparation, creation and/or presentation of the published work, specifically visualization/data presentation: A.K. and D.V.C.; Writing—original draft, preparation, creation and/or presentation of the published work, specifically writing the initial draft (including substantive translation): A.K., S.G., I.L., C.K. and D.V.C.; Writing—review and editing, preparation, creation and/or presentation of the published work by those from the original research group, specifically critical review, commentary or revision—including pre- or post-publication stages: A.K. and D.V.C. All authors have read and agreed to the published version of the manuscript.

Funding: This research did not receive any specific additional grant from funding agencies in the public, commercial or not-for-profit sectors.

Institutional Review Board Statement: This was approved by the University of Athens on 12 February 2014, Protocol number 2257.

Informed Consent Statement: Informed consent was obtained from all subjects involved in the study.

Data Availability Statement: The data is not available due to privacy or ethical restrictions.

Conflicts of Interest: The authors declare that they have no known competing financial interests or personal relationships that could have appeared to influence the work reported in this paper.

Abbreviations

AS, aortic stenosis; AVC, aortic valve cusp; CAD, coronary artery disease; CVD, cardiovascular disease; FNDC5, irisin; PN, periostin; OGN, osteoglycin; IL-18, interleukin 18; HMGB-1, high mobility group box 1; IHC, immunochemistry; PCSK9, proprotein convertase subtilisin/kexin type 9; qRT-PCR, quantitative reverse transcription polymerase chain reaction.

References

1. Otto, C.M.; Prendergast, B. Aortic-Valve Stenosis—From Patients at Risk to Severe Valve Obstruction. *N. Engl. J. Med.* **2014**, *371*, 744–756. [CrossRef] [PubMed]
2. Benamer, H.; Auffret, V.; Cayla, G.; Chevalier, B.; Commeau, P.; Dupouy, P.; Eltchaninoff, H.; Gilard, M.; Guerin, P.; Lung, B.; et al. Position papier français (GACI) pour l'implantation de valve aortique percutanée (TAVI). *Arch. Mal. Coeur Vaiss. Prat.* **2018**, *2018*, 32–40. [CrossRef]
3. Nkomo, V.T.; Gardin, J.M.; Skelton, T.N.; Gottdiener, J.S.; Scott, C.G.; Enriquez-Sarano, M. Burden of valvular heart diseases: A population-based study. *Lancet* **2006**, *368*, 1005–1011. [CrossRef]
4. Ortlepp, J.R.; Schmitz, F.; Bozoglu, T.; Hanrath, P.; Hoffmann, R. Cardiovascular risk factors in patients with aortic stenosis predict prevalence of coronary artery disease but not of aortic stenosis: An angiographic pair matched case-control study. *Heart* **2003**, *89*, 1019–1022. [CrossRef]
5. Kapelouzou, A.; Tsourelis, L.; Kaklamanis, L.; Degiannis, D.; Kogerakis, N.; Cokkinos, D.V. Serum and tissue biomarkers in aortic stenosis. *Glob. Cardiol. Sci. Pract.* **2015**, *2015*, 49. [CrossRef]
6. Helske, S.; Kupari, M.; Lindstedt, K.A.; Kovanen, P.T. Aortic valve stenosis: An active atheroinflammatory process. *Curr. Opin. Infect. Dis.* **2007**, *18*, 483–491. [CrossRef] [PubMed]
7. Kapelouzou, A.; Kontogiannis, C.; Tsilimigras, D.I.; Georgiopoulos, G.; Kaklamanis, L.; Tsourelis, L.; Cokkinos, D.V. Differential expression patterns of Toll Like Receptors and Interleukin-37 between calcific aortic and mitral valve cusps in humans. *Cytokine* **2019**, *116*, 150–160. [CrossRef]
8. Passos, L.S.; Lupieri, A.; Becker-Greene, D.; Aikawa, E. Innate and adaptive immunity in cardiovascular calcification. *Atherosclerosis* **2020**, *306*, 59–67. [CrossRef] [PubMed]
9. Aydin, S.; Kuloglu, T.; Aydin, S.; Eren, M.N.; Celik, A.; Yilmaz, M.; Kalayci, M.; Sahin, I.; Gungor, O.; Gurel, A.; et al. Cardiac, skeletal muscle and serum irisin responses to with or without water exercise in young and old male rats: Cardiac muscle produces more irisin than skeletal muscle. *Peptides* **2014**, *52*, 68–73. [CrossRef]
10. Silvestrini, A.; Bruno, C.; Vergani, E.; Venuti, A.; Favuzzi, A.M.R.; Guidi, F.; Nicolotti, N.; Meucci, E.; Mordente, A.; Mancini, A. Circulating irisin levels in heart failure with preserved or reduced ejection fraction: A pilot study. *PLoS ONE* **2019**, *14*, e0210320. [CrossRef] [PubMed]
11. Shen, S.; Gao, R.; Bei, Y.; Li, J.; Zhang, H.; Zhou, Y.; Yao, W.; Xu, D.; Zhou, F.; Jin, M.; et al. Serum Irisin Predicts Mortality Risk in Acute Heart Failure Patients. *Cell. Physiol. Biochem.* **2017**, *42*, 615–622. [CrossRef]
12. Emanuele, E.; Minoretti, P.; Pareja-Galeano, H.; Sanchis-Gomar, F.; Garatachea, N.; Lucia, A. Serum Irisin Levels, Precocious Myocardial Infarction, and Healthy Exceptional Longevity. *Am. J. Med.* **2014**, *127*, 888–890. [CrossRef]
13. Kuloglu, T.; Aydin, S.; Eren, M.N.; Yilmaz, M.; Sahin, I.; Kalayci, M.; Sarman, E.; Kaya, N.; Yilmaz, O.F.; Turk, A.; et al. Irisin: A potentially candidate marker for myocardial infarction. *Peptides* **2014**, *55*, 85–91. [CrossRef]
14. Efe, T.H.; Açar, B.; Ertem, A.G.; Yayla, K.G.; Algül, E.; Yayla, K.G.; Ünal, S.; Bilgin, M.; Çimen, T.; Kirbaş, Ö.; et al. Serum Irisin Level Can Predict the Severity of Coronary Artery Disease in Patients with Stable Angina. *Korean Circ. J.* **2017**, *47*, 44–49. [CrossRef]
15. Zhao, Y.T.; Wang, H.; Zhang, S.; Du, J.; Zhuang, S.; Zhao, T.C. Irisin Ameliorates Hypoxia/Reoxygenation-Induced Injury through Modulation of Histone Deacetylase 4. *PLoS ONE* **2016**, *11*, e0166182. [CrossRef] [PubMed]
16. Qiao, X.; Nie, Y.; Ma, Y.; Chen, Y.; Cheng, R.; Yin, W.; Hu, Y.; Xu, W.; Xu, L. Irisin promotes osteoblast proliferation and differentiation via activating the MAP kinase signaling pathways. *Sci. Rep.* **2016**, *6*, 18732. [CrossRef]
17. Kim, H.; Wrann, C.D.; Jedrychowski, M.; Vidoni, S.; Kitase, Y.; Nagano, K.; Zhou, C.; Chou, J.; Parkman, V.J.A.; Novick, S.J.; et al. Irisin mediates effects on bone and fat via αV Integrin receptors. *Cell* **2018**, *175*, 1756–1768. [CrossRef]
18. Colaianni, G.; Cinti, S.; Colucci, S.; Grano, M. Irisin and musculoskeletal health. *Ann. N. Y. Acad. Sci.* **2017**, *1402*, 5–9. [CrossRef] [PubMed]
19. Hakuno, D.; Kimura, N.; Yoshioka, M.; Mukai, M.; Kimura, T.; Okada, Y.; Yozu, R.; Shukunami, C.; Hiraki, Y.; Kudo, A.; et al. Periostin advances atherosclerotic and rheumatic cardiac valve degeneration by inducing angiogenesis and MMP production in humans and rodents. *J. Clin. Investig.* **2010**, *120*, 2292–2306. [CrossRef] [PubMed]
20. Zhao, S.; Wu, H.; Xia, W.; Chen, X.; Zhu, S.; Zhang, S.; Shao, Y.; Ma, W.; Yang, D.; Zhang, J. Periostin expression is upregulated and associated with myocardial fibrosis in human failing hearts. *J. Cardiol.* **2014**, *63*, 373–378. [CrossRef]
21. Oka, T.; Xu, J.; Kaiser, R.A.; Melendez, J.; Hambleton, M.; Sargent, M.A.; Lorts, A.; Brunskill, E.W.; Dorn, G.W.; Conway, S.J.; et al. Genetic manipulation of periostin expression reveals a role in cardiac hypertrophy and ventricular remodeling. *Circ. Res.* **2007**, *101*, 313–321. [CrossRef] [PubMed]

22. Stansfield, W.E.; Andersen, N.M.; Tang, R.-H.; Selzman, C.H. Periostin Is a Novel Factor in Cardiac Remodeling After Experimental and Clinical Unloading of the Failing Heart. *Ann. Thorac. Surg.* **2009**, *88*, 1916–1921. [CrossRef]
23. Qiu, X.; Ma, F.; Zhang, H. Circulating Levels of IL-13, TGF-β1, and Periostin as Potential Biomarker for Coronary Artery Disease with Acute Heart Failure. *Evid. Based Complement. Altern. Med.* **2021**, *2021*, 1690421. [CrossRef] [PubMed]
24. Deckx, S.; Heggermont, W.; Carai, P.; Rienks, M.; Dresselaers, T.; Himmelreich, U.; van Leeuwen, R.; Lommen, W.; van der Velden, J.; Gonzalez, A.; et al. Osteoglycin prevents the development of age-related diastolic dysfunction during pressure overload by reducing cardiac fibrosis and inflammation. *Matrix Biol.* **2018**, *66*, 110–124. [CrossRef] [PubMed]
25. Van Aelst, L.N.; Voss, S.; Carai, P.; Van Leeuwen, R.; Vanhoutte, D.; Sanders-van Wijk, S.; Eurlings, L.; Swinnen, M.; Verheyen, F.K.; Verbeken, E.; et al. Osteoglycin prevents cardiac dilatation and dysfunction after myocardial infarction through infarct collagen strengthening. *Circ. Res.* **2015**, *116*, 425–436. [CrossRef]
26. Zuo, C.; Li, X.; Huang, J.; Chen, D.; Ji, K.; Yang, Y.; Xu, T.; Zhu, D.; Yan, C.; Gao, P. Osteoglycin attenuates cardiac fibrosis by suppressing cardiac myofibroblast proliferation and migration through antagonizing lysophosphatidic acid 3/matrix metalloproteinase 2/epidermal growth factor receptor signalling. *Cardiovasc. Res.* **2018**, *114*, 703–712. [CrossRef] [PubMed]
27. Cheng, J.M.; Akkerhuis, K.M.; Meilhac, O.; Oemrawsingh, R.M.; Garcia-Garcia, H.M.; van Geuns, R.-J.; Piquer, D.; Merle, D.; du Paty, E.; Galéa, P.; et al. Circulating Osteoglycin and NGAL/MMP9 Complex Concentrations Predict 1-Year Major Adverse Cardiovascular Events after Coronary Angiography. *Arter. Thromb. Vasc. Biol.* **2014**, *34*, 1078–1084. [CrossRef]
28. Tanaka, K.-I.; Matsumoto, E.; Higashimaki, Y.; Katagiri, T.; Sugimoto, T.; Seino, S.; Kaji, H. Role of Osteoglycin in the Linkage between Muscle and Bone. *J. Biol. Chem.* **2012**, *287*, 11616–11628. [CrossRef]
29. Wang, M.; Markel, T.A.; Meldrum, D.R. Interleukin 18 in the heart. *Shock* **2008**, *30*, 3–10. [CrossRef]
30. Naito, Y.; Tsujino, T.; Wakabayashi, K.; Matsumoto, M.; Ohyanagi, M.; Mitsuno, M.; Miyamoto, Y.; Hao, H.; Hirota, S.; Okamura, H.; et al. Increased interleukin-18 expression in nonrheumatic aortic valve stenosis. *Int. J. Cardiol.* **2010**, *144*, 260–263. [CrossRef]
31. Zhou, J.; Zhu, J.; Jiang, L.; Zhang, B.; Zhu, D.; Wu, Y. Interleukin 18 promotes myofibroblast activation of valvular interstitial cells. *Int. J. Cardiol.* **2016**, *221*, 998–1003. [CrossRef] [PubMed]
32. Yu, Q.; Vazquez, R.; Khojeini, E.V.; Patel, C.; Venkataramani, R.; Larson, D.F. IL-18 induction of osteopontin mediates cardiac fibrosis and diastolic dysfunction in mice. *Am. J. Physiol. Heart Circ. Physiol.* **2009**, *297*, H76–H85. [CrossRef] [PubMed]
33. Jefferis, B.J.; Papacosta, O.; Owen, C.G.; Wannamethee, S.G.; Humphries, S.E.; Woodward, M.; Lennon, L.T.; Thomson, A.; Welsh, P.; Rumley, A.; et al. Interleukin 18 and coronary heart disease: Prospective study and systematic review. *Atherosclerosis* **2011**, *217*, 227–233. [CrossRef] [PubMed]
34. Shen, W.; Zhou, J.; Wang, C.; Xu, G.; Wu, Y.; Hu, Z. High mobility group box 1 induces calcification of aortic valve interstitial cells via toll-like receptor 4. *Mol. Med. Rep.* **2017**, *15*, 2530–2536. [CrossRef]
35. Wang, B.; Li, F.; Zhang, C.; Wei, G.; Liao, P.; Dong, N. High-mobility group box-1 protein induces osteogenic phenotype changes in aortic valve interstitial cells. *J. Thorac. Cardiovasc. Surg.* **2016**, *151*, 255–262. [CrossRef] [PubMed]
36. Hu, X.; Jiang, H.; Bai, Q.; Zhou, X.; Xu, C.; Lu, Z.; Cui, B.; Wen, H. Increased serum HMGB1 is related to the severity of coronary artery stenosis. *Clin. Chim. Acta* **2009**, *406*, 139–142. [CrossRef]
37. Cheng, J.M.; Oemrawsingh, R.M.; Garcia-Garcia, H.M.; Boersma, E.; van Geuns, R.-J.; Serruys, P.W.; Kardys, I.; Akkerhuis, K.M. PCSK9 in relation to coronary plaque inflammation: Results of the ATHEROREMO-IVUS study. *Atherosclerosis* **2016**, *248*, 117–122. [CrossRef] [PubMed]
38. Peng, J.; Xing, C.Y.; Zhao, K.; Deng, J.; Olmedo, D.A.; Ma, Z.; Zhang, M.; Wang, Y. Associations of pro-protein convertase subtilisin-like kexin type 9, soluble low-density lipoprotein receptor and coronary artery disease: A case-control study. *Int. J. Cardiol.* **2022**, *350*, 9–15. [CrossRef]
39. Donato, L.J.; Saenger, A.K.; Train, L.J.; Kotzer, K.E.; Lagerstedt, S.A.; Hornseth, J.M.; Basu, A.; Winters, J.L.; Baudhuin, L.M. Genetic and biochemical analyses in dyslipidemic patients undergoing LDL apheresis. *J. Clin. Apher.* **2014**, *29*, 256–265. [CrossRef]
40. Liberale, L.; Montecucco, F.; Camici, G.G.; Dallegri, F.; Vecchie, A.; Carbone, F.; Bonaventura, A. Treatment with Proprotein Convertase Subtilisin/Kexin Type 9 (PCSK9) Inhibitors to Reduce Cardiovascular Inflammation and Outcomes. *Curr. Med. Chem.* **2017**, *24*, 1403–1416. [CrossRef]
41. Vavlukis, M.; Kedev, S. Effects of High Intensity Statin Therapy in the Treatment of Diabetic Dyslipidemia in Patients with Coronary Artery Disease. *Curr. Pharm. Des.* **2018**, *24*, 427–441. [CrossRef]
42. Wang, W.-G.; He, Y.-F.; Chen, Y.-L.; Zhao, F.-M.; Song, Y.-Q.; Zhang, H.; Ma, Y.-H.; Guan, X.; Zhang, W.-Y.; Chen, X.-L.; et al. Proprotein convertase subtilisin/kexin type 9 levels and aortic valve calcification: A prospective, cross sectional study. *J. Int. Med. Res.* **2016**, *44*, 865–874. [CrossRef]
43. Poggio, P.; Songia, P.; Cavallotti, L.; Barbieri, S.S.; Zanotti, I.; Arsenault, B.J.; Valerio, V.; Ferri, N.; Capoulade, R.; Camera, M. PCSK9 Involvement in Aortic Valve Calcification. *J. Am. Coll. Cardiol.* **2018**, *72*, 3225–3227. [CrossRef] [PubMed]
44. Salaun, E.; Mahjoub, H.; Dahou, A.; Mathieu, P.; Larose, É.; Després, J.-P.; Rodés-Cabau, J.; Arsenault, B.J.; Puri, R.; Clavel, M.-A.; et al. Hemodynamic Deterioration of Surgically Implanted Bioprosthetic Aortic Valves. *J. Am. Coll. Cardiol.* **2018**, *72*, 241–251. [CrossRef]
45. Kysenius, K.; Muggalla, P.; Mätlik, K.; Arumäe, U.; Huttunen, H.J. PCSK9 regulates neuronal apoptosis by adjusting ApoER2 levels and signaling. *Cell. Mol. Life Sci.* **2012**, *69*, 1903–1916. [CrossRef] [PubMed]
46. Chomczynski, P.; Sacchi, N. Single-step method of RNA isolation by acid guanidinium thiocyanate-phenol-chloroform extraction. *Anal. Biochem.* **1987**, *162*, 156–159. [CrossRef]

47. Weber, M.; Arnold, R.; Rau, M.; Brandt, R.; Berkovitsch, A.; Mitrovic, V.; Hamm, C. Relation of N-Terminal Pro–B-Type Natriuretic Peptide to Severity of Valvular Aortic Stenosis. *Am. J. Cardiol.* **2004**, *94*, 740–745. [CrossRef] [PubMed]
48. Everett, R.J.; Clavel, M.-A.; Pibarot, P.; Dweck, M.R. Timing of intervention in aortic stenosis: A review of current and future strategies. *Heart* **2018**, *104*, 2067–2076. [CrossRef]
49. Dweck, M.R.; Jones, C.; Joshi, N.V.; Fletcher, A.M.; Richardson, H.; White, A.; Marsden, M.; Pessotto, R.; Clark, J.C.; Wallace, W.A.; et al. Assessment of Valvular Calcification and Inflammation by Positron Emission Tomography in Patients with Aortic Stenosis. *Circulation* **2012**, *125*, 76–86. [CrossRef]
50. Awan, M.S.; Irfan, B.; Zahid, I.; Mirza, Y.; Ali, S.A. Comparison of Polymerase Chain Reaction and Immunohistochemistry Assays for Analysing Human Papillomavirus Infection in Oral Squamous Cell Carcinoma. *J. Clin. Diagn. Res.* **2017**, *11*, XC10–XC13. [CrossRef]
51. Bergmark, B.A.; O'Donoghue, M.L.; Murphy, S.A.; Kuder, J.F.; Ezhov, M.V.; Češka, R.; Gouni-Berthold, I.; Jensen, H.K.; Tokgozoglu, S.L.; Mach, F.; et al. An Exploratory Analysis of Proprotein Convertase Subtilisin/Kexin Type 9 Inhibition and Aortic Stenosis in the FOURIER Trial. *JAMA Cardiol.* **2020**, *5*, 709–713. [CrossRef]
52. Cowell, S.J.; Newby, D.E.; Prescott, R.J.; Bloomfield, P.; Reid, J.; Northridge, D.B.; Boon, N.A. A Randomized Trial of Intensive Lipid-Lowering Therapy in Calcific Aortic Stenosis. *N. Engl. J. Med.* **2005**, *352*, 2389–2397. [CrossRef]
53. Rossebø, A.B.; Pedersen, T.R.; Boman, K.; Brudi, P.; Chambers, J.B.; Egstrup, K.; Gerdts, E.; Gohlke-Bärwolf, C.; Holme, I.; Kesäniemi, Y.A.; et al. Intensive Lipid Lowering with Simvastatin and Ezetimibe in Aortic Stenosis. *N. Engl. J. Med.* **2008**, *359*, 1343–1356. [CrossRef]
54. Moura, L.M.; Ramos, S.F.; Zamorano, J.L.; Barros, I.M.; Azevedo, L.F.; Rocha-Gonçalves, F.; Rajamannan, N.M. Rosuvastatin affecting aortic valve endothelium to slow the progression of aortic stenosis. *J. Am. Coll. Cardiol.* **2007**, *49*, 554–561. [CrossRef]
55. Chan, K.L.; Teo, K.; Dumesnil, J.G.; Ni, A.; Tam, J.; ASTRONOMER Investigators. Effect of Lipid lowering with rosuvastatin on progression of aortic stenosis: Results of the aortic stenosis progression observation: Measuring effects of rosuvastatin (ASTRONOMER) trial. *Circulation* **2010**, *121*, 306–314. [CrossRef]
56. Cokkinos, D.V.; Cokkinos, P.; Kolovou, G. Proprotein convertase subtilisin/kexin type 9 inhibitors: New insights into cardiovascular atherosclerotic pathophysiology with therapeutic implications. *Arch. Cardiovasc. Dis.* **2019**, *112*, 455–458. [CrossRef]
57. Ferencik, M.; Chatzizisis, Y.S. Statins and the coronary plaque calcium "paradox": Insights from non-invasive and invasive imaging. *Atherosclerosis* **2015**, *241*, 783–785. [CrossRef]
58. Perrot, N.; Valerio, V.; Moschetta, D.; Boekholdt, S.M.; Dina, C.; Chen, H.Y.; Abner, E.; Martinsson, A.; Manikpurage, H.D.; Rigade, S.; et al. Genetic and In Vitro Inhibition of PCSK9 and Calcific Aortic Valve Stenosis. *JACC Basic Transl. Sci.* **2020**, *5*, 649–661. [CrossRef]
59. Sundararaman, S.S.; Döring, Y.; van Der Vorst, E.P. PCSK9: A multi-faceted protein that is involved in cardiovascular biology. *Biomedicines* **2021**, *9*, 793. [CrossRef]
60. Lee, Y.S.; Chou, Y.Y. Pathogenetic mechanism of senile calcific aortic stenosis: The role of apoptosis. *Chin. Med. J.* **1998**, *111*, 934–939.
61. Punch, E.; Klein, J.; Diaba-Nuhoho, P.; Morawietz, H.; Garelnabi, M. Effects of PCSK9 Targeting: Alleviating Oxidation, Inflammation, and Atherosclerosis. *J. Am. Heart Assoc.* **2022**, *11*, e023328. [CrossRef] [PubMed]
62. Jigheh, Z.A.; Haghjo, A.G.; Argani, H.; Roshangar, L.; Rashtchizadeh, N.; Sanajou, D.; Ahmad, S.N.S.; Rashedi, J.; Dastmalchi, S.; Abbasi, M.M. Empagliflozin alleviates renal inflammation and oxidative stress in streptozotocin-induced diabetic rats partly by repressing HMGB1-TLR4 receptor axis. *Iran. J. Basic Med. Sci.* **2019**, *22*, 384–390. [CrossRef]
63. Sukhanov, S.; Higashi, Y.; Yoshida, T.; Mummidi, S.; Aroor, A.R.; Russell, J.J.; Bender, S.B.; DeMarco, V.G.; Chandrasekar, B. The SGLT2 inhibitor Empagliflozin attenuates interleukin-17A-induced human aortic smooth muscle cell proliferation and migration by targeting TRAF3IP2/ROS/NLRP3/Caspase-1-dependent IL-1β and IL-18 secretion. *Cell. Signal.* **2021**, *77*, 109825. [CrossRef]
64. Uthman, L.; Homayr, A.; Hollmann, M.W.; Zuurbier, C.J.; Weber, N.C. Administration of SGLT2 inhibitor empagliflozin against TNF-α induced endothelial dysfunction in human venous and arterial endothelial cells. *FASEB J.* **2018**, *32*, 569.4. [CrossRef]
65. Pirklbauer, M.; Sallaberger, S.; Staudinger, P.; Corazza, U.; Leierer, J.; Mayer, G.; Schramek, H. Empagliflozin Inhibits IL-1β-Mediated Inflammatory Response in Human Proximal Tubular Cells. *Int. J. Mol. Sci.* **2021**, *22*, 5089. [CrossRef]
66. Tsai, K.-F.; Chen, Y.-L.; Chiou, T.; Chu, T.-H.; Li, L.-C.; Ng, H.-Y.; Lee, W.-C.; Lee, C.-T. Emergence of SGLT2 Inhibitors as Powerful Antioxidants in Human Diseases. *Antioxidants* **2021**, *10*, 1166. [CrossRef]
67. Kvidal, P.; Bergström, P.R.; Hörte, L.G.; Ståhle, E. Observed and relative survival after aortic valve replacement. *J. Am. Coll. Cardiol.* **2000**, *35*, 747–756. [CrossRef]
68. Adler, Y.; Vaturi, M.; Herz, I.; Iakobishvili, Z.; Toaf, J.; Fink, N.; Battler, A.; Sagie, A. Nonobstructive aortic valve calcification: A window to significant coronary artery disease. *Atherosclerosis* **2002**, *161*, 193–197. [CrossRef] [PubMed]

biology

MDPI

Article

Identification and Verification of Biomarkers and Immune Infiltration in Obesity-Related Atrial Fibrillation

Zhonghui Xie [1,2,†], Chuanbin Liu [3,†], Xu Lu [1,2,†], Zhijie Chen [4], Nan Zhang [1,2], Xinyan Wang [1,2], Xiaoqian Li [4] and Yang Li [1,2,*]

1 Medical School of Chinese PLA, Beijing 100853, China
2 Senior Department of Cardiology, The Sixth Medical Centre of Chinese PLA General Hospital, Beijing 100048, China
3 Western Medical Branch of the Chinese PLA General Hospital, Beijing 100853, China
4 Department of Cardiology, Fujian Provincial Hospital, Provincial Clinical Medicine College of Fujian Medical University, Fuzhou 350001, China
* Correspondence: liyangbsh@163.com
† These authors contributed equally as co-first authors.

Simple Summary: Obesity is an independent risk factor for atrial fibrillation, which, in the ensuing decades, will probably increase the global burden. Previous studies have indicated that inflammation is a central mediator between obesity and atrial fibrillation. However, the mechanisms underlying this crosstalk are still being uncovered, and there are insufficient specific biomarkers. We co-analyzed the atrial fibrillation and obesity microarrays to investigate the possible molecular mechanism of obesity-related atrial fibrillation. We found that *MNDA*, *CYBB*, *CD86*, *FCGR2C*, *NCF2*, *LCP2*, *TLR8*, *HLA-DRA*, *LCP1*, and *PTPN22* were only elevated in blood samples of obese atrial-fibrillation patients. Atrial-fibrillation patients' left atrial appendage had increased infiltration of naïve B cells and decreased infiltration of memory B cells. Ten validated hub genes were related positively to naïve B cells and negatively to memory B cells. Ten validated genes identified by bioinformatics analysis, specifically correlated with obesity-related atrial fibrillation, may serve as biomarkers for obesity-related atrial fibrillation. These findings may also aid in comprehending pathophysiological mechanisms and identifying possible treatment targets for obesity-related atrial fibrillation.

Abstract: Obesity is an independent risk factor for atrial fibrillation (AF). However, the mechanisms underlying this crosstalk are still being uncovered. Co-differentially expressed genes (co-DEGs) of AF and obesity microarrays were identified by bioinformatics analysis. Subsequently, functional enrichment, cell-type enrichment, and protein–protein interaction network analyses of co-DEGs were carried out. Then, we validated the hub genes by qRT-PCR of patients' blood samples. Finally, CIBERSORT was utilized to evaluate the AF microarray to determine immune infiltration and the correlation between validated hub genes and immune cells. A total of 23 co-up-regulated DEGs in AF and obesity microarrays were identified, and these genes were enriched in inflammation- and immune-related function. The enriched cells were whole blood, CD33+ myeloid, and CD14+ monocytes. The hub genes were identified as *MNDA*, *CYBB*, *CD86*, *FCGR2C*, *NCF2*, *LCP2*, *TLR8*, *HLA-DRA*, *LCP1*, and *PTPN22*. All hub genes were only elevated in blood samples of obese-AF patients. The CIBERSORT analysis revealed that the AF patients' left atrial appendage had increased infiltration of naïve B cells and decreased infiltration of memory B cells. The hub genes were related positively to naïve B cells and negatively to memory B cells. Ten hub genes may serve as biomarkers for obesity-related AF. These findings may also aid in comprehending pathophysiological mechanisms for obesity-related AF.

Keywords: atrial fibrillation; obesity; immune infiltration; inflammation; biomarkers

Citation: Xie, Z.; Liu, C.; Lu, X.; Chen, Z.; Zhang, N.; Wang, X.; Li, X.; Li, Y. Identification and Verification of Biomarkers and Immune Infiltration in Obesity-Related Atrial Fibrillation. *Biology* **2023**, *12*, 121. https://doi.org/10.3390/biology12010121

Academic Editor: Gaetano Santulli

Received: 29 November 2022
Revised: 6 January 2023
Accepted: 10 January 2023
Published: 12 January 2023

1. Introduction

Atrial fibrillation (AF) is one of the most common cardiac arrhythmias, while a study published in 2020 based on the prospective urban and rural epidemiological (PURE) study demonstrated that the incidence of AF is 270–360 cases per 100,000 [1]. Obesity was first identified as an independent risk factor for AF in the Framingham cohort study [2]. In recent years, an umbrella review and meta-analysis of observational and Mendelian randomization studies suggested that an increase of one unit in body mass index (BMI) was associated with an increased risk of AF [3]. Given that obesity has spread widely across the globe, it is likely to increase the global burden of AF in the coming decades. More seriously, obese patients have a lower success rate for catheter ablation and are more likely to relapse after surgery, while both BMI [4] and epicardial fat [5,6] are associated with the risk of AF progression and its recurrence. Therefore, it is necessary to investigate the potential pathogenesis of AF associated with obesity.

At present, the mechanisms involved in obesity-related AF remain unclear. Obesity may cause AF through local effects, such as abnormal secretion of adipokines, inflammation, hypoxia, and adipose and fibrous tissue infiltration [7,8], which may also increase the susceptibility to AF through systemic effects mechanisms such as insulin resistance, metabolism disorders, inflammatory states, and hemodynamic changes [9,10]. The mechanism appears diverse, but many studies proved that inflammation might be a pivotal mediator between obesity and AF [11]. However, there is still a lack of effective prediction methods for obesity-related AF, and the conditions and prognoses of these patients are not easily predictable. Therefore, it is necessary to search for specific biomarkers to identify the possible mechanism of obesity-related AF.

CIBERSORT is a widely used analysis tool, which can carry out deconvolution analysis of gene expression profiles and predict fractions of immune cell types in a mixed cell population [12]. Leukocyte signature matrix (LM22) can be used as the reference gene matrix for estimating 22 different types of immune cells. CIBERSORT is effective against numerous non-tumor diseases.

In this study, we used a microarray study as an effective method to explore possible pathogenesis and to find specific markers. Due to the lack of a microarray study of obesity-related AF patients, we analyzed the AF and obesity microarrays to investigate the co-DEGs and possible biomarkers of obesity-related AF. Then, we validated the hub genes in obesity-related AF patients' blood samples. Finally, CIBERSORT was used to compare the immune infiltration of left atrial appendages (LAAs) between patients with sinus rhythm (SR) and AF. We investigated the relationship between hub genes and immune cell types further. Consequently, this study has important implications for identifying specific biomarkers and potential pathogenic mechanisms.

2. Materials and Methods

2.1. Materials and Methods

Figure 1 displays the study's methodology. GSE79768 and GSE94752 were retrieved from Gene Expression Omnibus (GEO) database (http://www.ncbi.nlm.nih.gov/geo/, accessed on 26 April 2022). GSE79768 (platform: GPL570) has paired LAA and right atrial appendages (RAA) from 7 AF patients and 6 SR patients. GSE94752 (platform: GLP11532) has subcutaneous white adipose tissue (WAT) specimens from 9 lean and 39 obese patients. LAA data of the GSE79768 dataset were used to identify differentially expressed genes (DEGs) between patients with AF and SR. Data from GSE94752 was used to identify mRNA associated with obesity susceptibility. We first calculated the DEGs of these two microarrays and found that the up-regulated and down-regulated DEGs from these two microarrays intersected. By analyzing the co-DEGs of the two gene datasets listed above, we determined the potential relationship between AF and obesity. As described elsewhere, the AF patients in this study had a disease course lasting longer than one month, whereas the SR patients had no clinical evidence of AF [13]. We only analyzed the data of the left atrium because there is still evidence that the left atrium and right atrium have different gene expressions, and the left atrium plays a more critical role in the pathogenesis of AF [14]. The lean patients had a BMI of less than 25 kg/m^2, while the BMI of obese patients was over 30 kg/m^2, as indicated elsewhere [15].

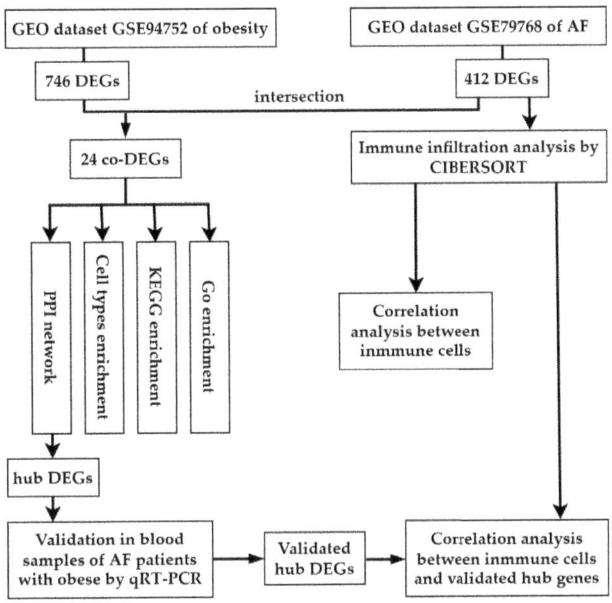

Figure 1. Workflow of the bioinformatics analysis methods in the present study. GEO, Gene Expression Omnibus; AF, arial fibrillation; DEGs, differentially expressed genes.

2.2. Data Processing and Identification of DEGs and Co-DEGs

The raw datasets of GSE79768 and GSE94752 were accessed and quality controlled by R (Version 4.1.2) packages of "affy", "affyPLM" and "limma". Robust multiarray average (RMA) was used to normalize the data. The adjusted p-value < 0.05 and $|\log_2 FC| < 0.585$ were used as thresholds to filter out DEGs. Hierarchical cluster heat maps were used to show distinguished mRNA expression levels of DEGs by the "pheatmap" package in R. Volcano plots were generated to represent all mRNA expression levels and the corresponding statistical inference value by "ggplot2" and "ggrepel" packages. Further, we calculated and made Venn diagrams for co-DEGs of AF- and obesity-DEGs by the "VennDiagram" package.

2.3. Gene Ontology, Pathways, and Cell-Type Enrichment Analyses of Co-DEGs

Gene ontology (GO) and Kyoto Encyclopedia of Genes and Genomes (KEGG) pathway enrichment analyses (http://www.kegg.jp/ or http://www.genome.jp/kegg/, accessed on 15 May 2022) of co-DEGs were carried out using the "clusterProfiler" and "pathview" packages of R software. Bar graphs were used to display enrichment results. The KEGG pathway graph was used to display the gene enrichment pathway. The q value (adjusted p-value) threshold was set to 0.05. Further, Enrichr (https://maayanlab.cloud/Enrichr/, accessed on 15 May 2022) and the Human Gene Atlas database were used to evaluate co-DEG cell types with enriched co-DEGs.

2.4. Protein–Protein Interaction Network Integration of Co-DEGs

Protein–protein interaction network integration (PPI) of co-DEGs was analyzed by the search tool for the retrieval of interacting genes (STRING database, Version 11.5; http://string-db.org/ accessed on 15 May 2022), which is suitable for identifying interactions between input genes. Subsequently, the analytic results of the STRING database were fed into Cytoscape software (Version 3.7.2). The biological networks and node degrees were analyzed and represented graphically, and the top ten hub genes were identified.

2.5. Validation for the Potential Role of Hub Genes

The expression levels of the top ten hub genes in obesity-related AF were confirmed by quantitative real-time polymerase chain reaction (qRT-PCR). This study included patients from the Chinese PLA General Hospital. To further investigate the function of hub genes, we obtained blood samples from eight lean-AF, eight obese-AF, seven lean-SR, and eight obese-SR patients. Patients' clinical information can be seen in Table S1. Total RNA from blood was prepared using a high-performance blood total RNA extraction kit (DP443, TIANGEN). Reverse RNA transcription was performed by a reverse-transcription system (K1622, Thermo Fisher, Waltham, MA, USA). Obtained cDNA was amplified using SYBR premix kit (A25742, Thermo Fisher) by BIO-RAD CFX96 (Bio-Rad Laboratories, Hercules, CA, USA). Predesigned gene-specific primers are listed in Table S2. β-actin RNA levels were used as an internal control. This study protocol involving human subjects had been reviewed and approved by the Chinese People's Liberation Army General Hospital Ethics Committee. The patients/participants provided their written informed consent to participate in the study. The AF patients were diagnosed with an electrocardiogram (ECG), whereas the SR patients did not exhibit any symptoms of AF (such as palpitations, dizziness, chest pain, or pressure) or ECG evidence. The lean patients' BMI were no more than 25 kg/m^2, and they did not have any metabolic disorders, for example, hyperlipidemia or diabetes mellitus. The patients who had a BMI greater than 30 kg/m^2 were regarded as obese patients.

2.6. Immune Infiltration by CIBERSORT Analysis of AF-DEGs

We used the CIBERSORT algorithm to analyze the normalized data generated from raw data, and the leukocyte signature matrix (LM22), which contains 22 types of immune cells, was used as the reference gene matrix [12]. LM22 includes macrophages M0, M1, and M2, memory B cells, B naïve cells, plasma cells, CD8 T cells, gamma-delta T cells, follicular helper T cells, CD4 memory-activated T cells, CD4 naïve T cells, CD4 memory resting T cells, regulatory T cells (Tregs), neutrophils, monocytes, activated NK cells, resting NK cells, resting dendritic cells, activated dendritic cells, eosinophils, activated mast cells, and resting mast cells. With a p-value less than 0.05, the samples were considered successful for deconvolution analysis and further subjected to differential analysis of proportions of immune cells. Further, "ggplot2", "pheatmap", and "vioplot" packages were utilized to visualize the percentage and difference of immune cell infiltration. The relationships between two immune cells and between immune cells and validated hub genes were calculated by the "corrplot" package.

2.7. Statistical Analysis

R (4.1.2) was utilized to conduct the bioinformatics analysis. This study's qRT-PCR data were processed using GraphPad Prism (Version 9). Using F-tests, the expression levels of hub genes were analyzed. Adjusted p-value (q value) and p-value less than 0.05 were deemed statistically significant.

3. Results

3.1. Identification of DEGs in Atrial Fibrillation and Obesity

We identified 54,675 probes corresponding to 20,267 genes in the GSE79768 dataset and 33,298 probes corresponding to 19,409 genes in the GSE94752. There are 412 DEGs between AF patients and SR patients in GSE79768 LA specimens, including 292 up-regulated genes and 120 down-regulated genes. There are 746 DEGs between obese and lean patients in GSE94752 WAT specimens, including 547 up-regulated genes and 199 down-regulated genes. Following this, we discovered one co-down-regulated gene and twenty-three co-up-regulated genes in the aforementioned two datasets. The above data are available in

Table S3. Heat maps of the top 50 up-regulated and top 50 down-regulated genes of AF-DEGs and obese-DEGs are presented, and these figures can be found in Figures S1 and S2. Figure 2a illustrates the number of co-expressed genes of AF- and obese-DEGs, and these genes are labeled in Figure 2b, which displays the distribution of the differences in gene expressions. Further, Figure 2c,d demonstrate hierarchical clustering analysis of AF- and obese-related DEGs, respectively.

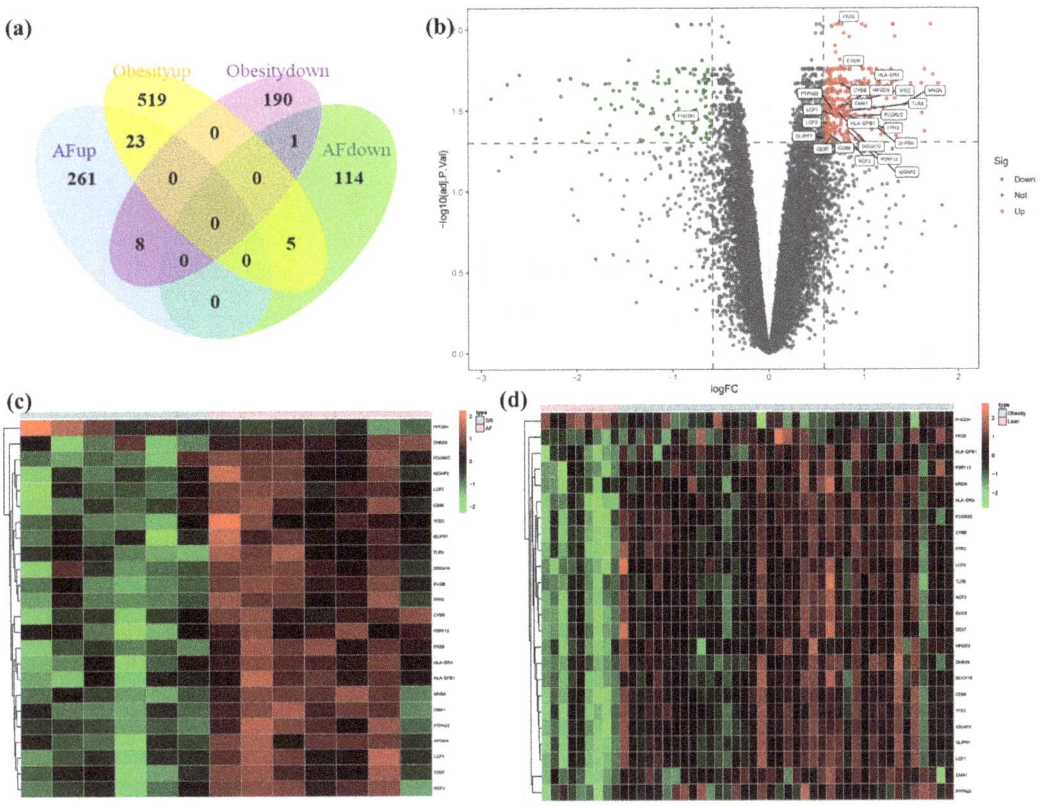

Figure 2. Venn diagram, volcano plot, and heat map of AF- and obese-related DEGs. (**a**) Venn diagram of DEGs related to AF and obesity. (**b**) Volcano plot of DEGs related to AF. Gray points represent the adjusted *p*-value > 0.05; red points represent adjusted *p*-value > 0.05 and up-regulated genes; green points represent adjusted *p*-value < 0.05 and down-regulated genes. (**c**) Heat map of GSE79768 with co-DEGs related to AF and obesity. The vertical axis represents samples, and the horizontal axis represents co-DEGs. Pink, AF samples; blue, SR samples; red, greater expression; green, less expression. (**d**) Heat map of GSE94752 with co-DEGs related to AF and obesity. Pink, obese samples; blue, lean samples; red, greater expression; green, less expression.

3.2. GO terms, KEGG Pathway, and Cell-Type Enrichment in Co-DEGs

Given that there is only one down-regulated gene in co-DEGs, only the up-regulated genes were utilized for enrichment analysis. The GO-terms enrichment results are shown in Figure 3a and Table S4. The top ten GO terms related to biological processes (BPs) among these genes were: positive regulation of cytokine production, immune response-regulating signaling pathway, immune response-activating cell-surface-receptor signaling pathway, immune response-activating signal transduction, immune response-regulating cell-surface-receptor-signaling pathway, activation of the immune response, lymphocyte activation involved in immune response, positive regulation of α-β T cell activation, positive regulation of interferon-gamma (IFN-γ) production, and positive regulation of T cell activation. In terms of cellular components (CCs), there was a significant correlation in NADPH oxidase complex, MHC class II protein complex, MHC protein complex, an integral component of the lumenal side of endoplasmic reticulum membrane, the lumenal side of endoplasmic reticulum membrane, the endocytic vesicle, the lumenal side of the membrane, an integral component of endoplasmic reticulum membrane, an intrinsic component of endoplasmic reticulum membrane, and the lysosomal membrane. The terms related to molecular function (MF) mainly include superoxide-generating NAD(P)H oxidase activity, oxidoreductase activity, acting on NAD(P)H, oxygen as acceptor, MHC class II protein-complex binding, MHC protein-complex binding, peptide antigen binding, GTPase binding, oxidoreductase activity, acting on NAD(P)H, peptide-binding, electron-transfer activity, amide-binding.

The KEGG pathway enrichment results are shown in Figure 3b and Table S2. In the KEGG analysis, the up-regulated genes were mainly enriched in leishmaniasis, phagosome, staphylococcus aureus infection, allograft rejection, graft-versus-host disease, type I diabetes mellitus, the intestinal immune network for IgA production, autoimmune thyroid disease, viral myocarditis, rheumatoid arthritis, hematopoietic cell lineage, asthma, osteoclast differentiation, systemic lupus erythematosus, cell adhesion molecules, tuberculosis, diabetic cardiomyopathy, inflammatory bowel disease, and antigen processing and presentation.

Enrichment analysis of cell types indicated that up-regulated genes in co-DEGs are more likely to identify whole blood, CD33+ myeloid, and CD14+ monocytes (Figure 3c and Table S4) The interaction of enriched cell types is shown in Figure 3d, which shows an interaction among whole blood, CD33+ myeloid, and CD14+ monocytes. Further, the relationship between cell-type enrichment and genes is shown in Figure 3e. Figure 3f depicts the pathway map of the phagosome, which, according to the aforementioned enrichment analysis, may play a role in obesity-related AF.

3.3. PPI Network Analysis in Co-DEGs

As 21 co-DEGs filtered into the PPI network, we identified 16 nodes and 43 edges, and these data appear in Figure 4a. The hub genes calculated by Cytoscape software are *MNDA* (score 320), *CYBB* (score 314), CD86 (score 312), *FCGR2C* (score 242), *NCF2* (score 240), *LCP2* (score 182), *TLR8* (score 148), *HLA-DRA* (score 54), *LCP1* (score 30), and *PTPN22* (score 8), which are considered to be associated with obesity-related AF. (Figure 4b) Among these genes, *MNDA*, *CYBB*, and *CD86* possessed the top three scores, calculated by the maximal clique centrality (MCC) algorithm.

3.4. Validation for the Potential Role of Hub Genes

We confirmed hub-gene expression levels in lean-SR, obese-SR, lean-AF, and obese-AF patients' blood samples. As depicted in Figure 5, all ten hub genes were uniquely expressed at a higher level in obese-AF patients compared with lean-SR, obese-SR, and lean-AF patients.

3.5. Immune Infiltration Analyses

Due to the co-DEGs being enriched in whole blood, CD33+ myeloid, and CD14+ monocytes, we further analyzed the DEGs in LAA samples between SR and AF patients by the CIBERSORT algorithm. The overall differential expressions of immune fractions between SR and AF patients are depicted in Figure 6a. More specifically, in Figure 6b, the LAA of AF patients exhibited a higher infiltration of naïve B cells and a lower infiltration of memory B cells compared to SR patients. As shown in Figure 6c, naïve B cells were negatively associated with memory B cells, regulatory T cells (Tregs), resting NK cells, M2 macrophages, and activated mast cells, whereas they were positively associated with gamma-delta T cells, M1 macrophages, and resting mast cells. Memory B cells were negatively associated with naïve B cells, CD4 memory resting T cells, gamma-delta T cells,

M1 macrophages, resting mast cells, and neutrophils, while positively associated with Tregs, resting NK cells, M2 macrophages, activated dendritic cells, and activated mast cells. However, no correlation was found between monocytes, memory B cells, and naïve B cells. Therefore, we further analyzed the correlation between hub genes and immune cells. As illustrated in Figure 6d, naïve B cells were positively associated with *LCP1*, *HLA-DRA*, *TLR8*, *CD86*, and *CYBB*. Meanwhile, memory B cells were negatively associated with *PTPN22*, *LCP1*, *HLA-DRA*, *TLR8*, *LCP2*, *NCF2*, *FCGR2C*, *CD86*, *CYBB*, and *MNDA*.

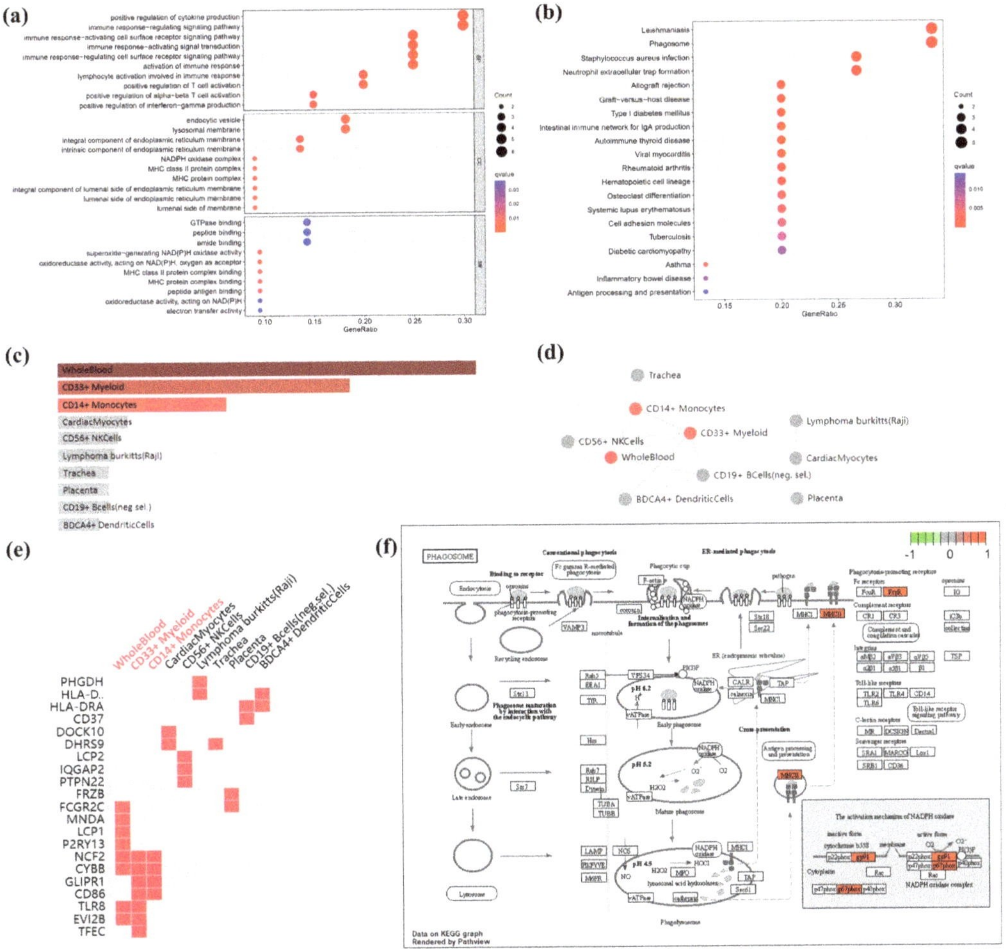

Figure 3. GO terms, KEGG pathway, and cell-type enrichment. (**a**) GO term enrichment for AF- and obese-related co-DEGs. Dot sizes, counts of enriched DEGs; dot colors, q value. (**b**) KEGG pathway of AF- and obese-related co DEGs. Dot sizes, counts of enriched DEGs; dot colors, q value. (**c**) Bar graph showing *p*-value for Enrichr cell-type enrichment, sorted by *p*-value ranking. Red nodes, adjust *p*-value < 0.05; gray nodes, adjust *p*-value > 0.05. (**d**) Network showing the interaction of enriched cell types. Red nodes, adjust *p*-value < 0.05; gray nodes, adjust *p*-value > 0.05. (**e**) Clustergram shows the relationship between cell-type enrichment and genes, sorted by *p* value ranking. Enriched terms are the columns, input genes are the rows, and cells in the matrix indicate if a gene is associated with a term. (**f**) KEGG pathway map of the phagosome, which is enrichment in co DEGs: *CYBB* (gp91), *NCF2* (p67phox), *HLA-DRA* (MHCII), and *FCGR2C* (FcγR). Red, up-regulated genes; green, down-regulated genes.

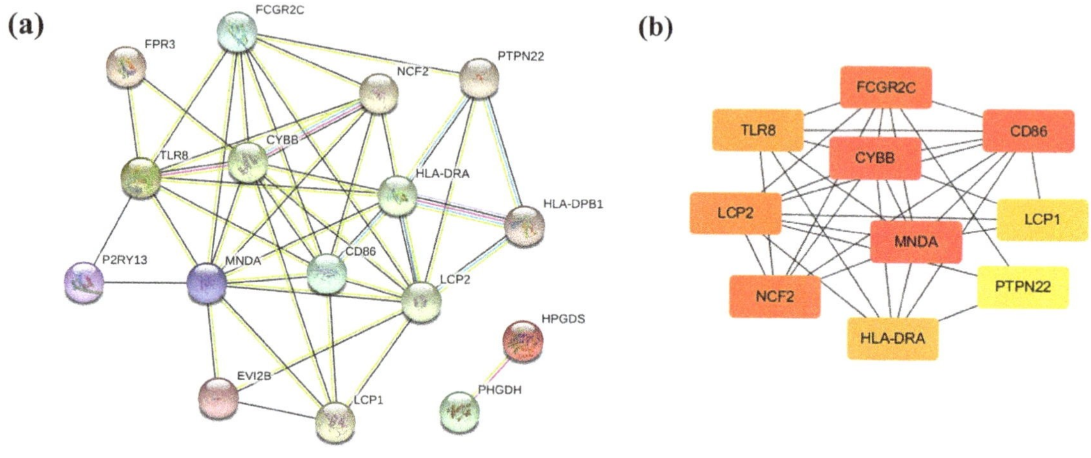

Figure 4. PPI network and hub genes. (**a**) PPI networks constructed by STRING database for co-DEGs (threshold > 0.4). (**b**) Hub genes were calculated by Cytoscape software. Red, higher MCC score; yellow, lower MCC score.

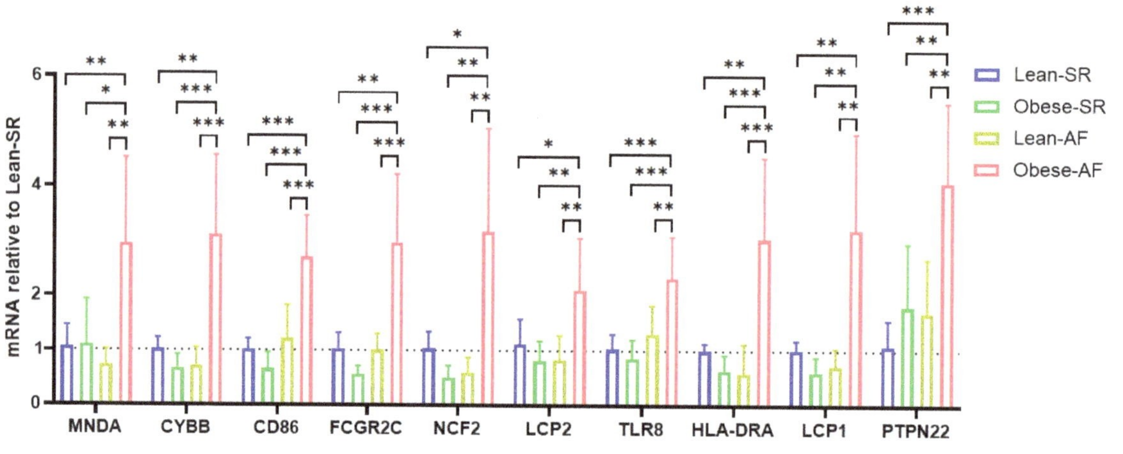

Figure 5. Validation of hub genes. Validation of hub genes in patients' blood cells by qPCR, each of genes is relative to lean-SR group. * $p < 0.05$, ** $p < 0.01$, *** $p < 0.001$.

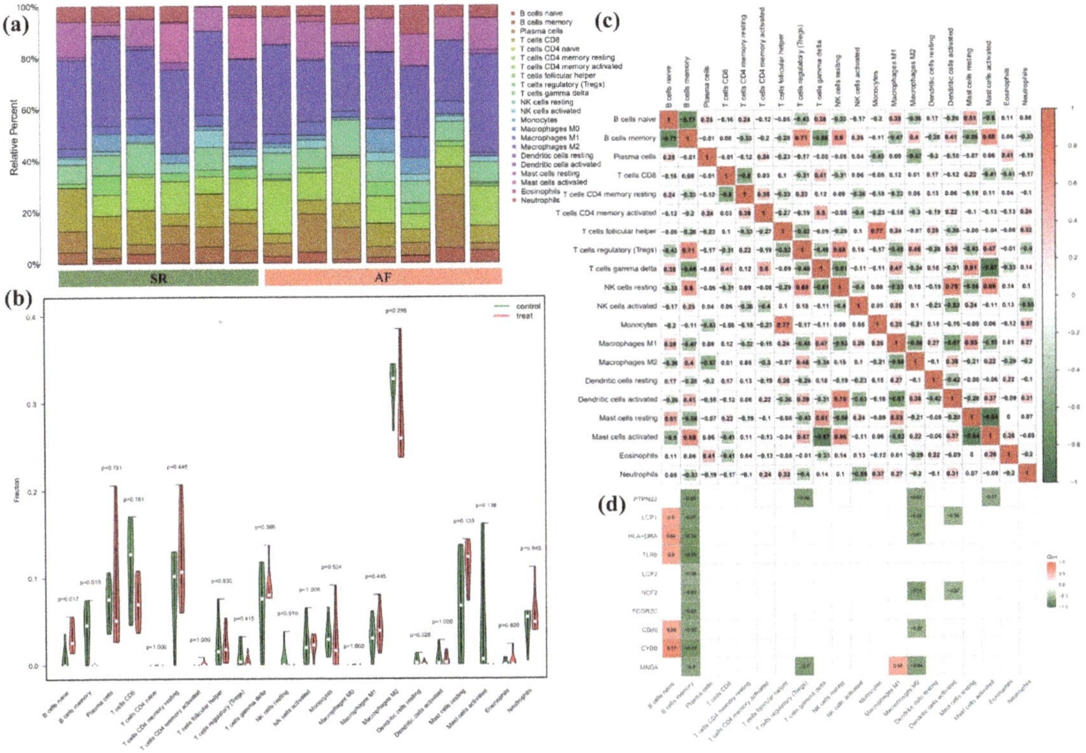

Figure 6. Results of CIBERSORT analysis of GSE79768 (AF-related array). (**a**) Cumulative histogram of immune cell infiltration. (**b**) Violin diagram of immune cell proportions in two groups. Green and fusiform fractions on the left, SR group; red and fusiform fractions on the right, AF group. (**c**) Correlation matrix of infiltration degree of immune cells. Red, positive correlation between two immune cells; blue, negative correlation between two immune cells. The bigger size of the numbers statistics data, the more positive or negative correlation. (**d**) Correlation matrix of infiltration degree of immune cells and hub genes. Red, positive correlation between two immune cells; blue, negative correlation between two immune cells.

4. Discussion

In recent years, the incidences of atrial fibrillation (AF) and obesity have increased simultaneously, garnering more attention [16]. In obese patients, the adipose tissue (AT) is in a state of inflammation, which can cause AF [17], and recent studies have revealed that AF can be reduced by intervening in the "quality" of AT [18]. EAT is a highly active visceral tissue that produces many different pro-inflammatory adipokines that can diffuse directly to the myocardium and cause immune-cell infiltration and inflammation that may lead to AF [19]. The current study has identified and verified ten biomarkers for obesity-related AF by analyzing co-DEGs of AF microarray and obese adipose tissue microarray. In addition, we demonstrated a positive correlation between these biomarkers and the ratio of naïve B cells to memory B cells in the LAA of AF patients, which aids in the identification of initiating factors and potential therapeutic targets in obesity-related AF.

We found 23 co-DEGs in the two microarrays, 22 of which were up-regulated and 1 was down-regulated. First, based on a complete study of GO, KEGG, and cell-type enrichment of these co-up-regulated genes, we hypothesize that AT of obese patients and LAA of AF patients shares certain physiological processes. The blood's CD33+ myeloid cells and CD14+ monocytes adhere to and infiltrate the tissues. The superoxide-generating activity of

NADPH oxidase complex, phagocytosis activity, and MHCII antigen-presentation function of monocyte-derived macrophages was enhanced. In addition, the neutrophil extracellular trap formation was strengthened. Further, lymphocytes were activated in the tissues, and the secretion of cytokines(including IFN-γ) was enhanced. Second, the hub genes in co-up-regulated DEGs were identified as *MNDA, CYBB, CD86, FCGR2C, NCF2, LCP2, TLR8, HLA-DRA, LCP1,* and *PTPN22* by the PPI network. These hub genes were validated in blood samples of obesity-related AF. Third, the CIBERSORT analysis of the LAA of AF patients suggested increased infiltration of B cell naïve and decreased infiltration of memory B cells. The ratio of naïve/memory B cells had a positive association with CD4 memory resting T cells, gamma-delta T cells, M1 macrophages, resting mast cells, and neutrophils; and a negative association with Tregs, resting NK cells, M2 macrophages, activated dendritic cells, and activated mast cells. The validated hub genes were positively associated with the naïve/memory B cell ratio.

In obese patients, adipose tissue can release danger-associated molecular patterns (DAMPs) and then cause immune cell infiltration by Toll-like receptors (TLR) [20]. In 2008, Chen et al. first identified that AF patients have more inflammatory cells identified as CD45+ infiltrating the atria than normal controls [21]. Then, more research revealed that inflammation is an essential common pathway for AF caused by obesity [22]. This is consistent with our research. Co-DEGs were primarily enriched in immune function in our study. The ten hub genes pertain to the immune system. All ten hub genes were only raised in the blood of obese-AF patients, but not in the blood of lean-SR, lean-AF, or obese-AF patients. At present, few studies have found specific markers for obesity-related AF, while our research found that blood samples qRT-PCR of these genes may become a novel convenient method for predicting obesity-related AF.

CYBB, FCGR2C, NCF2, and *HLA-DRA* of validated hub genes were enriched in the phagosome pathway. *LCP1*, LCP2, *CD86*, and *HLA-DRA* were enriched in lymphocyte activation and immune response. *MNDA, CYBB, CD86, TLR8,* and *PTPN22* were enriched in positive regulation of cytokine production. Research published in 2023 in Science also suggests that a history of obesity reprograms mononuclear phagocytes, leading to transcription of pro-inflammatory cytokines [23]. Some studies have suggested that the inflammatory increase of macrophages and inflammatory factors can lead to AF through ion channel remodeling and structural remodeling [24,25]. The inflammatory-related genes are generally increased in classically activated macrophages (M1). Macrophages with increased NOX2 and enhanced phagocytosis can stimulate neutrophil extracellular trap formation [26]. In addition, this type of macrophage can not only secrete inflammatory cytokines, and enhance MHCII antigen presentation, but also stimulate the activation of CD4+ T cells and B cells [27], which is consistent with our study's findings.

CYBB and *NCF2*, both NADPH oxidase (NOX), are present in phagocytic cells and participate in the "respiratory burst" to produce a large number of reactive oxygen species (ROS) and contribute to T cell activation [28]. The ROS can be released from the phagocytic cells and act on other cells [29]. ROS is one of the primary mechanisms causing obesity-related AF [22], and there exists a mechanism called "ROS-induced ROS release" (RIRR) [30]. Therefore, so finding the trigger source of ROS may play a critical role in obesity-related AF prevention. There has been some controversy about the role of NOX2 in the pathogenesis of AF. A 2015 study suggested intermittent hypoxia may induce AF by activating NOX2, thereby decreasing atrial Cx40 and Cx43 [31]. A study in 2020 suggested that NOX2 may increase the incidence of AF in obese mice by reducing I_{Na}, I_{Kur}, and atrial action potential duration (APD), and mitochondrial antioxidants could prevent the occurrence of AF [32]. However, another study in 2021 suggested that in atrial NOX2 overexpression mice, NOX2 overexpression could only slightly increase the induction rate of AF. However, no effect on electrophysiology and structural remodeling was found, and ROS induced by overexpression of human NOX2 in mouse myocardium is not the primary cause of the rising incidence of AF [33]. Atrial ROS main sources vary in AF courses and atrial matrix [34]. In our study, the increase of NOX2 in obesity was mainly concentrated in monocytes, so NOX2 and ROS causing AF may be mainly related to the increase of inflammatory-monocyte-derived macrophages rather than cardiomyocytes. Currently, studies on the co-effects of *NCF2* and *CYBB* on ROS are mainly seen in autoimmune, infectious, and ischemic diseases [35–37]. The mechanism of *NCF2* in AF needs to be further studied.

HLA-DRA and *CD86* are both biomarkers of M1 macrophages. A study based on human cardiomyocytes suggests that M1 macrophages with high expression of *CD86* in atrial fibrillation may promote extracellular matrix remodeling of atrial fibroblasts and participate in the occurrence of atrial fibrillation [38]. *TLR8* (toll-like receptor 8) recognizes pathogen-associated molecular patterns (PAMPs) and mediates the production of cytokines necessary for the development of effective immunity. A study reported that *TLR8* was correlated with levels of IL6, IL1β, and a greater inflammatory response, which is an important mechanism that causes atrial fibrillation [30]. However, the mechanism of *NCF2, HLA-DRA, TLR8, FCGR2C, MNDA, LCP1, LCP2*, and *PTPN22* in AF has not been studied. A comprehensive examination of these genes could aid in developing targeted protective measures to limit the damage caused by abnormal immunity.

In 2010, a study revealed the infiltration of CD3+ T cells and a small number of CD20+ B cells in the atria of patients with AF by the immunohistochemistry method [39], but the exact mechanism of lymphocyte involvement in AF has not been investigated fully. Our findings suggest a possibility that the activated lymphocytes can secrete lymphokines such as IFN-γ, further enhancing the phagocytosis and respiratory burst of monocytes [40]. It has been reported that in systemic sclerosis, the activation and apoptosis of the memory subset of B cells stimulate the proliferation of the naïve subset of B cells, resulting in an increased proportion of naïve/memory B cells [41], so a higher percentage of naïve/memory B cells may be associated with a pro-inflammatory state. It has been suggested that obesity, pro-inflammatory factors, and phagocytosis-derived Fc ligand can cause the depletion of memory B cells [42,43], which may explain the increase in the ratio of naïve/memory B cells in this study. Furthermore, in this study, the pro-inflammatory immune cells like, CD4 memory resting T cells, gamma-delta T cells, M1 macrophages, resting mast cells, and neutrophils were positively correlated with the ratio of naïve/memory B cells. In addition, the expression of the validated hub genes is directly proportional to the naïve/memory B cells ratio and may be involved in regulating the infiltration and activation of B cells.

Still, our study has several limitations. First, we have tentatively verified the correlation among these hub genes, obesity, and AF in small-sample subjects. Therefore, the results of this study need to be further investigated in a cohort study with larger sample sizes. Second, this study only focused on one AF dataset due to limited datasets with intra-dataset distinction and correctable inter-dataset batch effect, so greater sample sizes of LAA samples from obese-AF patients should be further verified. Finally, more in vivo and in vitro studies containing protein level verifications are warranted to elucidate the underlying mechanisms among these validated genes, infiltrating immune cells and obesity-related AF.

5. Conclusions

In summary, we identified ten genes by bioinformatics analysis, which were specifically correlated with obesity-related AF, and these genes could be potential biomarkers for obesity-related AF. Our findings may potentially contribute to a deeper understanding of the pathophysiological mechanism and the identification of possible treatment targets in obesity-related AF.

Supplementary Materials: The following supporting information can be downloaded at: https://www.mdpi.com/article/10.3390/biology12010121/s1. Table S1: Patients' baseline characteristics; Table S2 Primers used in this study for RT-qPCR experiments; Figure S1 Hierarchical clustering analysis of AF-related top 50 up-regulated and top 50 down-regulated expressed genes and Figure S2. Hierarchical clustering analysis of obese-related top 50 up-regulated and top 50 down-regulated expressed genes; Table S3: Gene expression matrix; Table S4: Enrichment results.

Author Contributions: Z.X., conceptualization, methodology, software usage, writing, visualization; X.L. (Xu Lu), resources, validation, and investigation; Z.C., N.Z., X.W. and X.L. (Xiaoqian Li), resources, data curation; C.L. and Y.L., supervision, review, and funding acquisition. All authors have read and agreed to the published version of the manuscript.

Funding: This research was funded by the National Natural Science Foundation of China, grant number 82070328 to Yang Li and 82200366 to Chuanbin Liu.

Institutional Review Board Statement: The study was conducted in accordance with the Declaration of Helsinki, and approved by the Ethics Committee of the Chinese People's Liberation Army General Hospital (protocol code S2021-612-03 and date of approval 20220224).

Informed Consent Statement: Informed consent was obtained from all subjects involved in the study. Written informed consent has been obtained from the patient(s) to publish this paper.

Data Availability Statement: Data are available in a publicly accessible repository that does not issue DOIs. Publicly available datasets were analyzed in this study. This data can be found here: http://www.ncbi.nlm.nih.gov/geo/ (GSE79768 and GSE94752F, accessed on 26 April 2022).

Conflicts of Interest: The authors declare no conflict of interest.

References

1. Joseph, P.G.; Healey, J.S.; Raina, P.; Connolly, S.J.; Ibrahim, Q.; Gupta, R.; Avezum, A.; Dans, A.L.; Lopez-Jaramillo, P.; Yeates, K.; et al. Global variations in the prevalence, treatment, and impact of atrial fibrillation in a multi-national cohort of 153 152 middle-aged individuals. *Cardiovasc. Res.* **2021**, *117*, 1523–1531. [CrossRef]
2. Wang, T.J.; Parise, H.; Levy, D.; D'Agostino, R.B., Sr.; Wolf, P.A.; Vasan, R.S.; Benjamin, E.J. Obesity and the risk of new-onset atrial fibrillation. *JAMA* **2004**, *292*, 2471–2477. [CrossRef]
3. Kim, M.S.; Kim, W.J.; Khera, A.V.; Kim, J.Y.; Yon, D.K.; Lee, S.W.; Shin, J.I.; Won, H.H. Association between adiposity and cardiovascular outcomes: An umbrella review and meta-analysis of observational and Mendelian randomization studies. *Eur. Heart J.* **2021**, *42*, 3388–3403. [CrossRef]
4. Providencia, R.; Adragao, P.; de Asmundis, C.; Chun, J.; Chierchia, G.; Defaye, P.; Anselme, F.; Creta, A.; Lambiase, P.D.; Schmidt, B.; et al. Impact of Body Mass Index on the Outcomes of Catheter Ablation of Atrial Fibrillation: A European Observational Multicenter Study. *J. Am. Heart Assoc.* **2019**, *8*, e012253. [CrossRef] [PubMed]
5. Hu, X.; Jiang, W.; Wu, S.; Xu, K.; Zhang, D.; Zhang, Y.; Liu, X.; Qin, M. Extra-pulmonary vein driver mapping and ablation for persistent atrial fibrillation in obese patients. *Europace* **2021**, *23*, 701–709. [CrossRef]
6. Powell-Wiley, T.M.; Poirier, P.; Burke, L.E.; Després, J.P.; Gordon-Larsen, P.; Lavie, C.J.; Lear, S.A.; Ndumele, C.E.; Neeland, I.J.; Sanders, P.; et al. Obesity and Cardiovascular Disease: A Scientific Statement From the American Heart Association. *Circulation* **2021**, *143*, e984–e1010. [CrossRef]
7. Poggi, A.L.; Gaborit, B.; Schindler, T.H.; Liberale, L.; Montecucco, F.; Carbone, F. Epicardial fat and atrial fibrillation: The perils of atrial failure. *Europace* **2022**, *24*, 1201–1212. [CrossRef] [PubMed]
8. Konwerski, M.; Gąsecka, A.; Opolski, G.; Grabowski, M.; Mazurek, T. Role of Epicardial Adipose Tissue in Cardiovascular Diseases: A Review. *Biology* **2022**, *11*, 355. [CrossRef]
9. Costa, R.M.; Neves, K.B.; Tostes, R.C.; Lobato, N.S. Perivascular Adipose Tissue as a Relevant Fat Depot for Cardiovascular Risk in Obesity. *Front. Physiol.* **2018**, *9*, 253. [CrossRef] [PubMed]
10. Goudis, C.A.; Korantzopoulos, P.; Ntalas, I.V.; Kallergis, E.M.; Ketikoglou, D.G. Obesity and atrial fibrillation: A comprehensive review of the pathophysiological mechanisms and links. *J. Cardiol.* **2015**, *66*, 361–369. [CrossRef]
11. Vyas, V.; Hunter, R.J.; Longhi, M.P.; Finlay, M.C. Inflammation and adiposity: New frontiers in atrial fibrillation. *Europace* **2020**, *22*, 1609–1618. [CrossRef] [PubMed]
12. Newman, A.M.; Liu, C.L.; Green, M.R.; Gentles, A.J.; Feng, W.; Xu, Y.; Hoang, C.D.; Diehn, M.; Alizadeh, A.A. Robust enumeration of cell subsets from tissue expression profiles. *Nat. Methods* **2015**, *12*, 453–457. [CrossRef] [PubMed]
13. Tsai, F.C.; Lin, Y.C.; Chang, S.H.; Chang, G.J.; Hsu, Y.J.; Lin, Y.M.; Lee, Y.S.; Wang, C.L.; Yeh, Y.H. Differential left-to-right atria gene expression ratio in human sinus rhythm and atrial fibrillation: Implications for arrhythmogenesis and thrombogenesis. *Int. J. Cardiol.* **2016**, *222*, 104–112. [CrossRef] [PubMed]
14. Thomas, A.M.; Cabrera, C.P.; Finlay, M.; Lall, K.; Nobles, M.; Schilling, R.J.; Wood, K.; Mein, C.A.; Barnes, M.R.; Munroe, P.B.; et al. Differentially expressed genes for atrial fibrillation identified by RNA sequencing from paired human left and right atrial appendages. *Physiol. Genomics.* **2019**, *51*, 323–332. [CrossRef]
15. Kulyté, A.; Ehrlund, A.; Arner, P.; Dahlman, I. Global transcriptome profiling identifies KLF15 and SLC25A10 as modifiers of adipocytes insulin sensitivity in obese women. *PLoS ONE* **2017**, *12*, e0178485. [CrossRef]
16. Ardissino, M.; Reddy, R.K.; Slob, E.A.W.; Patel, K.H.K.; Ryan, D.K.; Gill, D.; Ng, F.S. Sleep Disordered Breathing, Obesity and Atrial Fibrillation: A Mendelian Randomisation Study. *Genes* **2022**, *13*, 104. [CrossRef]

17. Haemers, P.; Hamdi, H.; Guedj, K.; Suffee, N.; Farahmand, P.; Popovic, N.; Claus, P.; LePrince, P.; Nicoletti, A.; Jalife, J.; et al. Atrial fibrillation is associated with the fibrotic remodelling of adipose tissue in the subepicardium of human and sheep atria. *Eur. Heart J.* **2017**, *38*, 53–61. [CrossRef]
18. Muzurović, E.M.; Vujošević, S.; Mikhailidis, D.P. Can We Decrease Epicardial and Pericardial Fat in Patients With Diabetes? *J. Cardiovasc. Pharmacol. Ther.* **2021**, *26*, 415–436. [CrossRef]
19. Vyas, V.; Lambiase, P. Obesity and Atrial Fibrillation: Epidemiology, Pathophysiology and Novel Therapeutic Opportunities. *Arrhythm. Electrophysiol. Rev.* **2019**, *8*, 28–36. [CrossRef]
20. Nishimoto, S.; Fukuda, D.; Sata, M. Emerging roles of Toll-like receptor 9 in cardiometabolic disorders. *Inflamm. Regen.* **2020**, *40*, 18. [CrossRef]
21. Chen, M.C.; Chang, J.P.; Liu, W.H.; Yang, C.H.; Chen, Y.L.; Tsai, T.H.; Wang, Y.H.; Pan, K.L. Increased inflammatory cell infiltration in the atrial myocardium of patients with atrial fibrillation. *Am. J. Cardiol.* **2008**, *102*, 861–865. [CrossRef] [PubMed]
22. Karam, B.S.; Chavez-Moreno, A.; Koh, W.; Akar, J.G.; Akar, F.G. Oxidative stress and inflammation as central mediators of atrial fibrillation in obesity and diabetes. *Cardiovasc. Diabetol.* **2017**, *16*, 120. [CrossRef] [PubMed]
23. Hata, M.; Andriessen, E.; Hata, M.; Diaz-Marin, R.; Fournier, F.; Crespo-Garcia, S.; Blot, G.; Juneau, R.; Pilon, F.; Dejda, A.; et al. Past history of obesity triggers persistent epigenetic changes in innate immunity and exacerbates neuroinflammation. *Science* **2023**, *379*, 45–62. [CrossRef] [PubMed]
24. Heijman, J.; Muna, A.P.; Veleva, T.; Molina, C.E.; Sutanto, H.; Tekook, M.; Wang, Q.; Abu-Taha, I.H.; Gorka, M.; Künzel, S.; et al. Atrial Myocyte NLRP3/CaMKII Nexus Forms a Substrate for Postoperative Atrial Fibrillation. *Circ. Res.* **2020**, *127*, 1036–1055. [CrossRef] [PubMed]
25. Regouski, M.; Galenko, O.; Doleac, J.; Olsen, A.L.; Jacobs, V.; Liechty, D.; White, K.L.; Bunch, T.J.; Lee, P.M.; Rutigliano, H.M.; et al. Spontaneous Atrial Fibrillation in Transgenic Goats With TGF (Transforming Growth Factor)-β1 Induced Atrial Myopathy With Endurance Exercise. *Circ. Arrhythm. Electrophysiol.* **2019**, *12*, e007499. [CrossRef]
26. Ravindran, M.; Khan, M.A.; Palaniyar, N. Neutrophil Extracellular Trap Formation: Physiology, Pathology, and Pharmacology. *Biomolecules* **2019**, *9*, 365. [CrossRef]
27. Guerriero, J.L. Macrophages: Their Untold Story in T Cell Activation and Function. *Int. Rev. Cell Mol. Biol.* **2019**, *342*, 73–93. [CrossRef]
28. Bassoy, E.Y.; Walch, M.; Martinvalet, D. Reactive Oxygen Species: Do They Play a Role in Adaptive Immunity? *Front. Immunol.* **2021**, *12*, 755856. [CrossRef]
29. Moghadam, Z.M.; Henneke, P.; Kolter, J. From Flies to Men: ROS and the NADPH Oxidase in Phagocytes. *Front. Cell Dev. Biol.* **2021**, *9*, 628991. [CrossRef]
30. Fujimoto, C.; Yamasoba, T. Mitochondria-Targeted Antioxidants for Treatment of Hearing Loss: A Systematic Review. *Antioxidants* **2019**, *8*, 109. [CrossRef]
31. Gemel, J.; Su, Z.; Gileles-Hillel, A.; Khalyfa, A.; Gozal, D.; Beyer, E.C. Intermittent hypoxia causes NOX2-dependent remodeling of atrial connexins. *BMC Cell Biol.* **2017**, *18*, 7. [CrossRef] [PubMed]
32. McCauley, M.D.; Hong, L.; Sridhar, A.; Menon, A.; Perike, S.; Zhang, M.; da Silva, I.B.; Yan, J.; Bonini, M.G.; Ai, X.; et al. Ion Channel and Structural Remodeling in Obesity-Mediated Atrial Fibrillation. *Circ. Arrhythm. Electrophysiol.* **2020**, *13*, e008296. [CrossRef] [PubMed]
33. Mighiu, A.S.; Recalde, A.; Ziberna, K.; Carnicer, R.; Tomek, J.; Bub, G.; Brewer, A.C.; Verheule, S.; Shah, A.M.; Simon, J.N.; et al. Inducibility, but not stability, of atrial fibrillation is increased by NOX2 overexpression in mice. *Cardiovasc. Res.* **2021**, *117*, 2354–2364. [CrossRef] [PubMed]
34. Reilly, S.N.; Jayaram, R.; Nahar, K.; Antoniades, C.; Verheule, S.; Channon, K.M.; Alp, N.J.; Schotten, U.; Casadei, B. Atrial sources of reactive oxygen species vary with the duration and substrate of atrial fibrillation: Implications for the antiarrhythmic effect of statins. *Circulation* **2011**, *124*, 1107–1117. [CrossRef] [PubMed]
35. Li, H.M.; Huang, Q.; Tang, F.; Zhang, T.P. Altered NCF2, NOX2 mRNA Expression Levels in Peripheral Blood Mononuclear Cells of Pulmonary Tuberculosis Patients. *Int. J. Gen. Med.* **2021**, *14*, 9203–9209. [CrossRef] [PubMed]
36. Denson, L.A.; Jurickova, I.; Karns, R.; Shaw, K.A.; Cutler, D.J.; Okou, D.T.; Dodd, A.; Quinn, K.; Mondal, K.; Aronow, B.J.; et al. Clinical and Genomic Correlates of Neutrophil Reactive Oxygen Species Production in Pediatric Patients with Crohn's Disease. *Gastroenterology* **2018**, *154*, 2097–2110. [CrossRef] [PubMed]
37. Harms, J.E.; Kuczmarski, J.M.; Kim, J.S.; Thomas, G.D.; Kaufman, M.P. The role played by oxidative stress in evoking the exercise pressor reflex in health and simulated peripheral artery disease. *J. Physiol.* **2017**, *595*, 4365–4378. [CrossRef] [PubMed]
38. Cao, F.; Li, Z.; Ding, W.; Yan, L.; Zhao, Q. Angiotensin II-Treated Cardiac Myocytes Regulate M1 Macrophage Polarization via Transferring Exosomal PVT1. *J. Immunol. Res.* **2021**, *2021*, 1994328. [CrossRef]
39. Yamashita, T.; Sekiguchi, A.; Iwasaki, Y.K.; Date, T.; Sagara, K.; Tanabe, H.; Suma, H.; Sawada, H.; Aizawa, T. Recruitment of immune cells across atrial endocardium in human atrial fibrillation. *Circ. J.* **2010**, *74*, 262–270. [CrossRef]
40. McCann, K.J.; Christensen, S.M.; Colby, D.H.; McGuire, P.J.; Myles, I.A.; Zerbe, C.S.; Dalgard, C.L.; Sukumar, G.; Leonard, W.J.; McCormick, B.A.; et al. IFNγ regulates NAD+ metabolism to promote the respiratory burst in human monocytes. *Blood Adv.* **2022**, *6*, 3821–3834. [CrossRef]
41. Sanges, S.; Guerrier, T.; Launay, D.; Lefèvre, G.; Labalette, M.; Forestier, A.; Sobanski, V.; Corli, J.; Hauspie, C.; Jendoubi, M.; et al. Role of B cells in the pathogenesis of systemic sclerosis. *Rev. Med. Interne* **2017**, *38*, 113–124. [CrossRef] [PubMed]

42. Frasca, D.; Blomberg, B.B. Obesity Accelerates Age Defects in Mouse and Human B Cells. *Front. Immunol.* **2020**, *11*, 2060. [CrossRef] [PubMed]
43. McLaughlin, T.; Ackerman, S.E.; Shen, L.; Engleman, E. Role of innate and adaptive immunity in obesity-associated metabolic disease. *J. Clin. Investig.* **2017**, *127*, 5–13. [CrossRef] [PubMed]

biology

MDPI

Article

Retinol-Binding Protein-4—A Predictor of Insulin Resistance and the Severity of Coronary Artery Disease in Type 2 Diabetes Patients with Coronary Artery Disease

Sangeetha Perumalsamy [1], Wan Azman Wan Ahmad [2] and Hasniza Zaman Huri [1,3,*]

1 Department of Clinical Pharmacy & Pharmacy Practice, Faculty of Pharmacy, Universiti Malaya, Kuala Lumpur 50603, Malaysia; geetha2_20@yahoo.com
2 Cardiology Unit, Department of Medicine, Faculty of Medicine, Universiti Malaya, Kuala Lumpur 50603, Malaysia; wanazman@ummc.edu.my
3 Clinical Investigation Centre (CIC), Universiti Malaya Medical Centre, Petaling Jaya 59100, Malaysia
* Correspondence: hasnizazh@um.edu.my; Tel.: +60-3-7967-6659

Simple Summary: Cytokines are cell-signaling molecules that cause cells to migrate to inflammation, infection, or trauma sites. An imbalance of cytokines in the body can result in severe illness. Increased cytokine retinol-binding protein-4 levels cause muscle, fat, and liver cells to become unresponsive to insulin and not absorb sugar from the blood. Type 2 diabetes, the most common type of diabetes, is caused by the unresponsiveness of insulin (insulin resistance). Moreover, elevated retinol-binding protein-4 causes fat and cholesterol buildup in the arteries of the heart. This results in coronary artery disease, a type of heart disease. These two diseases are hypothesized to share a common underlying cause, but the details have not been fully elucidated. Therefore, this study was conducted to find the association between retinol-binding protein-4 with insulin resistance and the severity of coronary artery disease. We postulated that retinol-binding protein-4 is linked to insulin resistance and the severity of coronary artery disease. This study proves a definitive relationship between retinol-binding protein-4 and insulin resistance and coronary artery disease severity. Hence, retinol-binding protein-4 may serve as a valuable biological indicator to depict insulin resistance and the severity of coronary artery disease.

Abstract: (1) Background: Insulin resistance (IR) is the fundamental cause of type 2 diabetes (T2D), which leads to endothelial dysfunction and alters systemic lipid metabolism. The changes in the endothelium and lipid metabolism result in atherosclerotic coronary artery disease (CAD). In insulin-resistant and atherosclerotic CAD states, serum cytokine retinol-binding protein-4 (RBP-4) levels are elevated. The adipocyte-specific deletion of glucose transporter 4 (GLUT4) results in higher RBP-4 expression and IR and atherosclerotic CAD progression. (2) Aim: This study aimed to investigate the association of RBP-4 and clinical factors with IR and the severity of CAD. (3) Methods: Patients were recruited from diabetes and cardiology clinics and divided into three subgroups, namely (i) T2D patients with CAD, (ii) T2D-only patients, and (iii) CAD-only patients. The severity of CAD was classified as either single-vessel disease (SVD), double-vessel disease (DVD), or triple-vessel disease (TVD). An enzyme-linked immunosorbent assay was conducted to assess the concentration of serum RBP-4. Univariate (preliminary analysis) and multivariate (secondary analysis) logistic regressions were applied to assess the associations of RBP-4 and clinical factors with IR and the severity of CAD. (4) Results: Serum RBP-4 levels were associated with IR and the severity of CAD in all the three groups (all p-values are less than 0.05). Specifically, serum RBP-4 levels were associated with IR ($p = 0.030$) and the severity of CAD (SVD vs. DVD, $p = 0.044$; SVD vs. TVD, $p = 0.036$) in T2D patients with CAD. The clinical factors fasting plasma glucose (FPG) and angiotensin-converting-enzyme inhibitor (ACEI) were also associated with both IR and the severity of CAD in T2D patients with CAD. (5) Conclusion: RBP-4, FPG, and ACEI are predictors of IR and severity of CAD in T2D patients with CAD.

Citation: Perumalsamy, S.; Ahmad, W.A.W.; Huri, H.Z. Retinol-Binding Protein-4—A Predictor of Insulin Resistance and the Severity of Coronary Artery Disease in Type 2 Diabetes Patients with Coronary Artery Disease. *Biology* **2021**, *10*, 858. https://doi.org/10.3390/biology10090858

Academic Editors: Gaetano Santulli and Lei Ye

Received: 13 July 2021
Accepted: 29 August 2021
Published: 1 September 2021

Keywords: RBP-4; insulin resistance; severity of CAD; T2D; atherosclerotic CAD; endothelial dysfunction

1. Introduction

Type 2 diabetes (T2D) has become a significant public health concern in developed and developing countries over the last few decades, making it a global health priority [1]. Insulin resistance (IR) is the fundamental key feature of T2D [2]. Although the mechanism of IR leading to atherosclerosis is not fully explicated, IR is said to be involved in endothelial dysfunction and alters systemic lipid metabolism, resulting in dyslipidemia and the well-known lipid triad of high levels of plasma triglycerides (TG), low levels of high-density lipoprotein (HDL), and high levels of low-density lipoproteins (LDL). This triad and endothelial dysfunction may lead to the formation of atherosclerotic plaques [3]. The formation of atherosclerotic plaques in the coronary arteries results in atherosclerotic coronary artery disease (CAD) [4]. The CAD is defined as severe when the atherosclerotic plaques narrow down the vessels with stenosis by more than 50% [5]. The severity is also classified on the basis of the number of vessels involved as a single-vessel disease (SVD), double-vessel disease (DVD), and triple-vessel disease (TVD). The most severe type of CAD is TVD with more than 50% stenosis, followed by DVD and SVD. DVD and TVD are also referred to as multivessel diseases [6].

In T2D patients, CAD is more complex, with small diffuse, calcified, multivessel involvement that frequently necessitates coronary revascularization besides optimal medical therapy to control angina [7]. Furthermore, CAD has been shown to increase mortality in T2D patients [8]. T2D and CAD are said to share the exact pathogenesis involving inflammation, endothelial dysfunction, and the release of proinflammatory cytokines [9]. Retinol-binding protein-4 (RBP-4) is a proinflammatory cytokine that may involve in the progression of IR, T2D, atherosclerosis and CAD. It is a 21 kDa cytokine produced predominantly in the liver and adipose tissue [10,11], which acts as an adipokine and fatty acid transporter that aid retinol (vitamin A) transport in the body [12]. It has recently been proposed that RBP-4-induced inflammation causes IR and CAD [13,14].

Although the role of RBP-4 in the pathogenesis of T2D and CAD is unknown, several studies have hypothesized that RBP-4 elevation causes IR and atherosclerotic CAD via the mitogen-activated protein kinase (MAPK) pathway [15–17]. Moreover, p44/42 MAPK, c-Jun N-terminal kinase, and p38 MAPK are all part of this pathway. When the p38 MAPK pathway is activated, glucose transporter 1 (GLUT1) expression increases, while glucose transporter 4 (GLUT4) expression decreases. The downregulation of GLUT4 expression leads to an increase in RBP-4 levels in the blood [18,19]. Chadt and Al-Hasani (2020) revealed that high RBP-4 secretion by adipocytes reduced GLUT4 expression in adipose tissue, as commonly reported in T2D patients [20]. In patients with CAD, there is an increase in epicardial RBP-4 and a decrease in GLUT4 levels [21].

Numerous studies have found elevated serum RBP-4 levels in T2D and CAD patients. RBP-4 levels of more than 55 g/mL were linked to an increased risk of T2D incidence in one study [22]. In another study, higher RBP-4 levels were linked to an increased risk of CAD, and RBP-4 levels increased as the number of stenosed vessels increased [23].

Because of both disorders sharing the same hypothesized mechanisms leading to endothelial dysfunction, and, thus, inflammation, RBP-4 can be the typical cytokine of IR and the severity of CAD in patients with T2D. In this study, we examined the relationship of RBP-4 and clinical factors with both IR and the severity of CAD in Malaysian T2D patients.

2. Materials and Methods

2.1. Participants

The selection of participants and inclusion and exclusion criteria were as earlier described (PMID: 34071097) [5]. This study used the same cohort as in the PMID: 34071097.

Specifically, in this study, we investigate the cytokine (RBP-4) associations with IR and the severity of CAD, whereas PMID: 34071097 explored the genetic (rs17173608) associations with IR and the severity of CAD. The purpose of the study was explained to all participants, and they were asked to sign a written informed consent form. The patients were divided into three groups according to the presence of T2D with CAD, T2D without CAD, and CAD without T2D.

2.2. Sample Size Calculation

The dichotomous test (PS software) was used to calculate the sample size under the assumptions of a level of significance, α of 0.05; the power of study at 0.80; the probability of exposure among controls, P_0 at 0.65; and the probability of exposure among cases, P_1 at 0.95 (the probability of exposure among cases was higher than of controls by 30%). The ratio of the control group to the case group, m was 1:1. As a result of the test, the minimum sample size required was 88 cases and 88 controls. There were 300 samples collected, namely 150 cases and 150 controls (150 T2D patients with CAD (cases); 90 T2D-only patients + 60 CAD-only patients (controls).

2.3. Aim and Hypotheses

The study aimed to investigate the association of RBP-4 and clinical factors with IR and the severity of CAD. Meanwhile, this study hypothesizes that RBP-4 is associated with IR and the severity of CAD in T2D patients with CAD and that clinical factors are associated with IR and the severity of CAD in T2D patients with CAD.

2.4. Demographic and Clinical Information, and Anthropometric Measurements

The assessments of demographic and clinical information and anthropometric measurements were as earlier described (PMID: 34071097) [5]. Laboratory investigation results, such as fasting plasma glucose (FPG), fasting plasma insulin (FPI), A1C, hs-CRP and lipid profile, the types and number of comorbidities, and details on the pharmacological treatments, were also obtained.

2.5. Biochemical Parameters

Biochemical parameters' assessments were as earlier described (PMID: 34071097) [5]. Blood samples were taken after at least 8 h of fasting in the morning (7:00 a.m. to 10:00 a.m.). FPI and FPG were multiplied and divided by 22.5 to calculate the Homeostasis Model Assessment of Insulin Resistance (HOMA-IR) [23]. The concentrations of RBP-4 in the blood were determined by using an enzyme-linked immunosorbent assay (ELISA) kit and read at 450 nm with a microplate reader [24].

2.6. RBP-4 Assay Protocol

The samples were incubated for 30 min to an hour at room temperature. The blood samples were then centrifuged for 15 min at 1000× *g*. The serum was then drawn out and aliquoted. The serum was kept at −20 °C after the process. The serum samples were brought to room temperature before use. To avoid protein degradation and denaturalization, recurrent freeze–thaw cycles were avoided. For the RBP-4 ELISA assay, the quantitative sandwich enzyme immunoassay technique was used (Cusabio ELISA kit-Elabscience, Houston, TX, USA). The RBP-4 concentrations were measured in nanograms per milliliter.

2.7. Statistical Analysis

The preliminary and secondary analyses applied binary and multinomial logistic regression tests (univariate and multivariate), with age, race, BMI (body mass index), and gender adjusted as covariates. The association model of IR with RBP-4 levels and clinical factors was investigated by using binary logistic regression. Meanwhile, multinomial logistic regression analysis was used to evaluate the association model of the severity of CAD with RBP-4 levels and clinical factors. The secondary tests were based on the significant associations found in the preliminary analysis ($p \leq 0.05$).

A plotted receiver-operating characteristic (ROC) curve was used to determine the cutoff point of IR and RBP-4 levels. To determine the optimal threshold of HOMA-IR and RBP-4 levels, the point on the ROC curve with the highest Youden index (sensitivity-(1-specificity)) and the point with the shortest distance from the point (0, 1) ((1-sensitivity)2 + (1-specificity)2) were calculated (cutoff point of HOMA-IR is 7.17 and RBP-4 levels are 1.6045 ng/mL).

2.8. Operational Definitions

Operational definitions were as earlier described (PMID: 34071097) [5].

3. Results

3.1. Demographic and Clinical Factors of Study Population

The results of demographic and clinical factors were as described earlier (PMID: 34071097) [5].

3.2. The Severity of CAD

Several T2D patients with CAD had multivessel disease (DVD, 39%, and TVD, 39%). Meanwhile, 50% of CAD patients had TVD. Figure 1 describes the distribution of patients according to the severity of CAD.

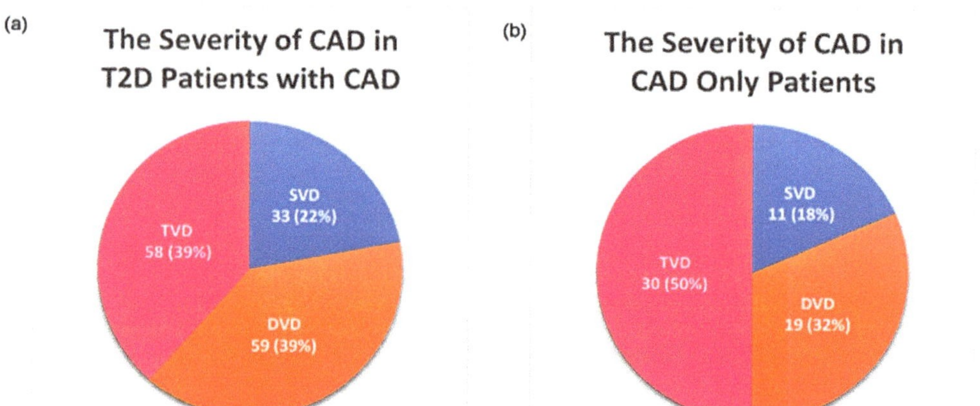

Figure 1. Distribution of patients according to the severity of CAD: (**a**) severity of CAD in T2D patients with CAD and (**b**) severity of CAD in CAD only patients.

3.3. Association of Clinical Factors with RBP-4

Glycated hemoglobin (A1C) ($p = 0.034$), high-sensitive C-reactive protein (hs-CRP) ($p < 0.001$), low-density lipoprotein cholesterol (LDL-c) ($p < 0.001$), high-density lipoprotein cholesterol (HDL-c) ($p = 0.001$), triglycerides (TG) ($p = 0.028$), biguanides + insulin therapy ($p < 0.001$), biguanides + sodium–glucose co-transporter-2 (SGLT2) + insulin therapy ($p = 0.001$), biguanides + sulphonylureas (SU) combination therapy ($p < 0.001$), biguanides monotherapy ($p < 0.001$), SU monotherapy ($p < 0.001$), nitrates ($p = 0.008$), diuretics ($p = 0.002$), and cardiac glycosides ($p < 0.001$) were all significantly associated with RBP-4 levels in patients with T2D who presented with CAD. In T2D patients with CAD, cardiac glycoside was found to be 18.444 times most likely to be associated with RBP-4 levels.

FPI and total cholesterol (TC) levels were significantly associated with RBP-4 levels in T2D-only patients. When comparing the two factors, the association of TC (OR = 1.345; $p = 0.031$) with RBP-4 levels in T2D-only patients was stronger than that of FPI (OR = 1.220; $p = 0.032$). In CAD-only patients, no significant associations were found between clinical factors and cytokines. Table 1 depicts the association of clinical factors of the study population with RBP-4 levels. Appendix A shows the significant results (p-values) of the preliminary associations.

Table 1. Association of clinical factors of study population with RBP-4 levels.

Parameter	OR (95% CI)		
	T2D + CAD (n = 150)	**T2D-Only (n = 90)**	**CAD-Only (n = 60)**
FPG (mmol/L)	1.088 (0.983–1.204)	0.937 (0.825–1.065)	0.943 (0.328–2.707)
FPI (pmol/L)	0.995 (0.987–1.002)	**1.220 (1.041–1.430)**	0.936 (0.846–1.036)
A1C (%)	**0.797 (0.654–0.970)**	1.048 (0.856–1.283)	0.869 (0.268–2.820)
hs-CRP (mg/L)	**1.317 (1.052–1.632)**	1.728 (0.776–3.845)	1.214 (0.248–5.939)
TC (mmol/L)	1.017 (0.783–1.230)	**1.345 (1.033–1.589)**	1.476 (0.715–3.047)
LDL-c (mmol/L)	**2.918 (1.428–3.385)**	0.602 (0.339–1.069)	1.261 (0.533–2.986)
HDL-c (mmol/L)	**0.490 (0.300–0.800)**	1.181 (0.265–5.264)	0.965 (0.750–2.455)
TG (mmol/L)	**1.402 (1.201–1.702)**	0.740 (0.437–1.255)	0.983 (0.435–2.218)
Hypertension	0.244 (0.040–1.251)	1.102 (0.175–6.940)	3.062 (0.248–7.879)
Dyslipidemia	2.336 (0.708–7.708)	2.800 (0.548–14.311)	0.557 (0.096–3.244)
Peripheral neuropathy	0.635 (0.325–1.240)	1.021 (0.440–2.368)	5.667 (0.239–19.655)
Chronic kidney disease (CKD)	1.615 (0.691–3.778)	1.483 (0.526–4.183)	0.160 (0.009–2.823)
Retinopathy	0.547 (0.255–1.173)	0.839 (0.338–2.083)	5.667 (0.778–10.661)
Anemia	0.783 (0.048–12.760)	0.731 (0.050–2.065)	4.076 (0.421–40.755)
Gastritis	0.786 (0.333–1.546)	4.371 (0.437–43.763)	0.327 (0.026–4.034)
Biguanides	**6.400 (2.024–20.237)**	0.613 (0.220–1.705)	–
Sulphonylureas	**17.714 (5.812–53.993)**	1.482 (0.590–3.722)	–
DPP4i	8.400 (0.056–34.855)	1.005 (0.399–2.529)	–
AGI	0.786 (0.045–1.433)	0.731 (0.088–4.033)	–
Meglitinides	0.786 (0.088–2.113)	1.031 (0.326–3.264)	–
Biguanide + SU	**10.000 (3.240–30.866)**	1.360 (0.539–3.432)	–
SU + DPP4i	9.056 (0.076–65.877)	0.907 (0.144–5.715)	–
Biguanide + insulin	**1.206 (1.093–1.458)**	–	–
Biguanide + SU + insulin	1.290 (0.310–5.365)	–	–
Biguanide + DPP4i + insulin	0.021 (0.006–0.043)	–	–
Biguanide + SGLT2 + insulin	**1.161 (1.053–1.493)**	–	–
SGLT2 + insulin	0.308 (0.034–2.821)	–	–
Antiplatelet agents	4.114 (0.469–36.102)	–	0.008 (0.002–0.015)
ACEI	0.847 (0.443–1.619)	0.724 (0.268–1.951)	0.429 (0.090–2.043)
ARB II	0.530 (0.237–1.187)	1.052 (0.429–2.579)	2.909 (0.666–12.708)
Calcium channel blockers	1.041 (0.496–2.182)	1.235 (0.534–2.859)	1.818 (0.390–8.466)
Beta blockers	1.106 (0.578–2.114)	1.029 (0.445–2.377)	0.318 (0.064–1.574)
Alpha blockers	9.664 (0.099–45.123)	0.731 0.045–3.912)	–
Nitrates	**2.657 (1.331–5.303)**	–	2.629 (0.425–16.263)
Fibrates	7.690 (0.065–33.878)	–	–
Statins	0.040 (0.012–0.077)	7.760 (0.013–27.112)	–
Diuretics	**1.297 (1.141–1.624)**	1.283 (0.522–3.153)	2.629 (0.425–16.263)
Antianginal drugs	1.983 (0.935–4.205)	–	0.500 (0.046–5.423)
Hematinic agents	3.925 (0.402–38.903)	0.676 (0.059–7.735)	0.380 (0.061–2.354)
Cardiac glycosides	18.444 (2.331–145.961)	–	–

Computed using binary logistic regression analysis. Bold font indicates significance at $p < 0.05$. OR, odds ratio; CI, confidence interval. The cutoff point of RBP-4 was used in this analysis (<1.604 ng/mL>); '–' indicates not relevant. FPG, fasting plasma glucose; FPI, fasting plasma insulin; A1C, glycated hemoglobin; hs-CRP, high-sensitive C-reactive protein; LDL-c, low-density lipoprotein cholesterol; HDL-c, high-density lipoprotein cholesterol; TG, triglycerides; TC, total cholesterol; ACEI, angiotensin-converting-enzyme inhibitor; AGI, alpha-glucosidase inhibitors; ARB II, angiotensin II receptor blockers; DPP4i, dipeptidyl peptidase-4 inhibitor; SGLT2, sodium–glucose co-transporter-2; SU, sulphonylureas.

3.4. Association of Clinical Factors with IR

In T2D patients with CAD, FPG (p = 0.011), FPI (p < 0.001), hs-CRP (p = 0.025), biguanides + dipeptidyl peptidase-4 inhibitor (DPP4i) + insulin treatment (p = 0.008), antiplatelet agents (p = 0.003), and angiotensin-converting-enzyme inhibitor (ACEI) (p = 0.026) were all linked to IR. The first-ranked factor was hs-CRP, associated with IR in T2D patients with CAD 2.378 times more likely. In the T2D-only group, FPI (p < 0.001) was significantly related to IR. FPG (p = 0.048) and FPI (p < 0.001) were linked to IR in CAD-only patients. FPG was the most powerful factor, being 2.570 times more likely to be associated with IR in CAD-only patients. Table 2 highlights the correlation of clinical factors to IR in this study population. Appendix A shows the significant results (p-values) of the preliminary associations.

Table 2. Association of clinical factors with IR in study population.

Parameter	HOMA-IR, OR (95% CI)		
	T2D + CAD (n = 150)	T2D-Only (n = 90)	CAD-Only (n = 60)
FPG (mmol/L)	**1.160 (1.031–1.306)**	1.010 (0.894–1.142)	**2.570 (1.097–5.773)**
FPI (pmol/L)	**1.233 (1.146–1.327)**	**1.376 (1.197–1.581)**	**1.368 (1.167–1.603)**
A1C (%)	1.102 (0.913–1.331)	1.098 (0.896–1.344)	2.122 (0.759–5.932)
hs-CRP (mg/L)	**2.378 (1.155–4.899)**	1.394 (0.636–3.057)	3.502 (0.909–13.493)
TC (mmol/L)	0.813 (0.622–1.063)	1.159 (0.809–1.660)	0.775 (0.443–1.356)
LDL-c (mmol/L)	0.942 (0.650–1.366)	1.119 (0.675–1.855)	0.806 (0.405–1.606)
HDL-c (mmol/L)	0.738 (0.500–1.089)	0.702 (0.159–3.100)	0.524 (0.053–5.140)
TG (mmol/L)	0.973 (0.673–1.407)	0.799 (0.482–1.324)	0.627 (0.292–1.342)
Hypertension	0.614 (0.147–2.556)	1.687 (0.268–10.617)	6.304 (0.044–19.030)
Dyslipidemia	0.515 (0.176–1.505)	1.951 (0.456–8.341)	0.684 (0.149–3.134)
Peripheral neuropathy	0.633 (0.319–1.258)	0.980 (0.426–2.253)	–
Chronic kidney disease (CKD)	0.990 (0.415–2.359)	2.645 (0.893–7.831)	2.867 (0.169–48.744)
Retinopathy	1.319 (0.629–2.768)	0.879 (0.360–2.147)	–
Anemia	1.596 (0.098–26.032)	–	–
Gastritis	–	1.098 (0.148–8.152)	–
Biguanides	0.750 (0.365–1.540)	0.689 (0.254–1.870)	–
Sulphonylureas	1.076 (0.496–2.332)	0.714 (0.292–1.747)	–
DPP4i	0.006 (0.002–0.010)	1.133 (0.455–2.822)	–
AGI	1.586 (0.066–2.887)	0.915 (0.052–4.835)	–
Biguanide + SU	1.201 (0.509–2.835)	0.413 (0.163–1.046)	–
SU + DPP4i	0.012 (0.006–0.030)	4.718 (0.506–43.984)	–
Biguanide + insulin	0.681 (0.333–1.394)	–	–
Biguanide + SU + insulin	1.630 (0.391–6.787)	–	–
Biguanide + DPP4i + insulin	**1.860 (1.043–2.027)**	–	–
Biguanide + SGLT2 + insulin	1.033 (0.445–2.395)	–	–
SGLT2 + insulin	2.455 (0.398–15.155)	–	–
Antiplatelet agents	**1.454 (1.032–1.895)**	–	6.011 (0.042–36.772)
ACEI	**1.444 (1.227–1.868)**	2.353 (0.872–6.351)	3.800 (1.006–14.351)
ARB II	1.829 (0.845–3.961)	0.524 (0.214–1.285)	0.429 (0.123–1.495)
Calcium channel blockers	0.662 (0.303–1.446)	0.644 (0.280–1.481)	1.768 (0.488–6.397)
Beta blockers	0.824 (0.426–1.594)	1.318 (0.575–3.024)	2.111 (0.509–8.751)
Alpha blockers	2.702 (0.038–16.806)	–	0.364 (0.033–1.452)
Nitrates	0.768 (0.382–1.544)	–	0.433 (0.086–2.196)
Fibrates	2.607 (0.028–6.442)	–	–
Statins	0.009 (0.004–0.014)	15.436 (0.078–66.022)	–
Diuretics	1.699 (0.853–3.384)	0.750 (0.305–1.843)	0.433 (0.086–2.195)
Antianginal drugs	0.594 (0.267–1.320)	–	9.923 (0.950–33.701)
Hematinic agents	1.607 (0.220–11.735)	0.536 (0.047–6.128)	2.308 (0.456–11.690)
Cardiac glycosides	0.263 (0.058–1.232)	–	–

Computed using binary logistic regression analysis. Bold font indicates significance at $p < 0.05$. OR, odds ratio; CI, confidence interval. HOMA-IR cutoff point: 7.17; '–' indicates not relevant. FPG, fasting plasma glucose; FPI, fasting plasma insulin; A1C, glycated hemoglobin; hs-CRP, high-sensitive C-reactive protein; LDL-c, low-density lipoprotein cholesterol; HDL-c, high-density lipoprotein cholesterol; TG, triglycerides; TC, total cholesterol; ACEI, angiotensin-converting-enzyme inhibitor; AGI, alpha-glucosidase inhibitors; ARB II, angiotensin II receptor blockers, DPP4i, dipeptidyl peptidase-4 inhibitor; SGLT2, sodium–glucose co-transporter-2; SU, sulphonylureas.

3.5. Association of Clinical Factors with the Severity of CAD

In T2D patients with CAD, FPG ([a] $p = 0.007$; [b] $p = 0.012$), FPI ([a] $p = 0.045$), DPP4i ([a] $p < 0.001$; [b] $p < 0.001$), SU + DPP4i ([a] $p = 0.011$; [b] $p = 0.016$), biguanide + DPP4i + insulin ([a] $p < 0.001$; [b] $p = 0.003$), fibrates ([b] $p < 0.001$), statins ([b] $p = 0.018$), ACEI ([a] $p = 0.032$; [b] $p = 0.029$), alpha blockers ([a] $p = 0.008$; [b] $p = 0.020$), and hematinic agents ([a] $p = 0.011$; [b] $p = 0.016$) were significantly associated with the severity of CAD. In the CAD-only group, FPG ([a] $p < 0.001$), FPI ([b] $p < 0.001$), A1C ([b] $p < 0.001$), LDL-c ([b] $p = 0.004$, HDL-c ([a] $p < 0.001$), and TG ([b] $p = 0.001$) were associated with the severity of CAD. DPP4i was found to have the strongest association with the severity of CAD in T2D patients with CAD, with an OR value of 2.149. Table 3 demonstrates the association between clinical factors and the severity of CAD in the study population. Appendix A shows the significant results (*p*-values) of the preliminary associations

Table 3. Association between clinical factors and the severity of CAD in study population.

Parameter	Severity of CAD, OR (95% CI)	
	T2D + CAD (*n* = 150)	CAD-Only (*n* = 60)
FPG (mmol/L)	[a] **1.815 (1.710–1.935)** [b] **1.875 (1.771–1.992)**	[a] **1.651 (1.201–2.110)** [b] 0.458 (0.151–1.388)
FPI (pmol/L)	[a] 1.011 (0.997–1.026) [b] **1.015 (1.001–1.030)**	[a] 0.984 (0.885–1.094) [b] **1.553 (1.054–2.105)**
A1C (%)	[a] 0.898 (0.712–1.133) [b] 0.939 (0.747–1.179)	[a] 0.626 (0.161–2.444) [b] **1.318 (1.087–1.858)**
hs-CRP (mg/L)	[a] 0.652 (0.369–1.154) [b] 0.801 (0.518–1.237)	[a] 3.229 (0.548–19.036) [b] 2.726 (0.526–14.473)
TC (mmol/L)	[a] 0.880 (0.632–1.227) [b] 0.772 (0.546–1.092)	[a] 0.685 (0.314–1.492) [b] 1.424 (0.712–2.848)
LDL-c (mmol/L)	[a] 1.169 (0.722–1.895) [b] 0.901 (0.543–1.494)	[a] 0.510 (0.197–1.321) [b] **1.722 (1.296–2.538)**
HDL-c (mmol/L)	[a] 1.007 (0.602–1.686) [b] 1.059 (0.635–1.764)	[a] **3.754 (1.185–76.172)** [b] 2.218 (0.130–3.789)
TG (mmol/L)	[a] 1.090 (0.689–1.724) [b] 0.726 (0.437–1.207)	[a] 0.736 (0.285–1.899) [b] **1.299 (1.007–1.523)**
Hypertension	[a] 2.963 (0.331–26.504) [b] 1.143 (0.100–13.105)	[a] 4.144 (0.349–20.714) [b] 5.329 (0.957–29.532)
Dyslipidemia	[a] 0.926 (0.207–4.147) [b] 1.373 (0.330–5.711)	[a] 1.875 (0.171–20.609) [b] 2.000 (0.207–19.336)
Peripheral neuropathy	[a] 0.783 (0.320–1.912) [b] 0.660 (0.271–1.609)	– –
Chronic kidney disease (CKD)	[a] 0.875 (0.272–2.818) [b] 0.763 (0.240–2.424	[a] 0.055 (0.015–0.368) [b] 0.976 (0.843–1.280)
Retinopathy	[a] 0.791 (0.298–2.097) [b] 1.006 (0.371–2.728)	– –
Anemia	[a] 1.006 (0.954–1.087) [b] 1.014 (0.076–1.632)	[a] 1.065 (0.045–2.060) [b] 0.063 (0.036–0.123)
Gastritis	– –	[a] 1.800 (0.101–31.988) [b] 2.900 (0.166–50.815)
Biguanides	[a] 1.203 (0.491–2.945) [b] 1.500 (0.602–3.740)	– –
Sulphonylureas	[a] 1.008 (0.387–2.625) [b] 1.800 (0.646–5.018)	– –
DPP4i	[a] **1.269 (1.008–1.865)** [b] **2.149 (1.320–3.326)**	– –
Biguanide + SU	[a] 1.319 (0.450–3.872) [b] 1.466 (0.490–4.387)	– –
SU + DPP4i	[a] **1.654 (1.054–2.022)** [b] **1.734 (1.109–2.503)**	– –

Table 3. *Cont.*

Parameter	Severity of CAD, OR (95% CI)	
	T2D + CAD (n = 150)	CAD-Only (n = 60)
Biguanide + insulin	[a] 0.525 (0.194–1.418) [b] 0.441 (0.164–1.184)	– –
Biguanide + SU + insulin	[a] 2.850 (0.451–17.999) [b] 1.833 (0.348–9.652)	– –
Biguanide + DPP4i + insulin	**[a] 1.545 (1.008–1.967)** **[b] 1.877 (1.210–3.116)**	– –
Biguanide + SGLT2 + insulin	[a] 1.778 (0.612–5.165) [b] 1.367 (0.487–3.837)	– –
SGLT2 + insulin	[a] 0.891 (0.078–10.210) [b] 0.875 (0.076–10.033)	– –
Antiplatelet agents	[a] 1.123 (0.098–12.872) [b] 1.745 (0.174–17.492)	[a] 4.647 (0.077–8.945) [b] 0.945 (0.768–1.490)
ACEI	**[a] 1.487 (1.085–3.532)** **[b] 1.166 (1.032–2.890)**	[a] 0.833 (0.126–5.504) [b] 0.889 (0.151–5.241)
ARB II	[a] 1.253 (0.453–3.468) [b] 1.006 (0.371–2.728)	[a] 1.607 (0.255–10.132) [b] 1.636 (0.289–9.255)
Calcium channel blockers	[a] 0.567 (0.209–1.537) [b] 1.032 (0.362–2.946)	[a] 1.200 (0.182–7.926) [b] 1.636 (0.289–9.255)
Beta blockers	[a] 0.703 (0.299–1.654) [b] 0.548 (0.231–1.299)	[a] 0.375 (0.036–3.865) [b] 0.500 (0.052–4.834)
Alpha blockers	**[a] 1.795 (1.076–4.644)** **[b] 1.900 (1.056–3.012)**	– –
Nitrates	[a] 1.153 (0.485–2.742) [b] 2.112 (0.853–5.230)	[a] 1.875 (0.171–20.609) [b] 1.111 (0.103–11.965)
Fibrates	[a] 0.029 (0.006–0.144) **[b] 1.056 (1.008–1.768)**	– –
Statins	[a] 25.265 (0.122–46.004) **[b] 1.087 (1.004–1.255)**	– –
Diuretics	[a] 0.975 (0.395–2.404) [b] 0.950 (0.385–2.346)	[a] 16.240 (3.209–57.778) [b] 13.325 (1.620–25.921)
Antianginal drugs	[a] 0.729 (0.276–1.923) [b] 1.367 (0.487–3.837)	– –
Hematinic agents	**[a] 1.540 (1.021–2.006)** **[b] 1.444 (1.058–2.244)**	[a] 0.533 (0.049–5.862) [b] 0.900 (0.084–9.692)
Cardiac glycosides	[a] 1.025 (0.277–3.797) [b] 3.862 (0.667–22.350)	– –

Computed using multinomial logistic regression analysis. Bold font indicates significance at $p < 0.05$. OR: odds ratio; CI: confidence interval. Severity of CAD: [a] DVD, [b] TVD, and SVD is the reference group; '–' indicates not relevant. FPG, fasting plasma glucose; FPI, fasting plasma insulin; A1C, glycated hemoglobin; hs-CRP, high-sensitive C-reactive protein; LDL-c, low-density lipoprotein cholesterol; HDL-c, high-density lipoprotein cholesterol; TG, triglycerides; TC, total cholesterol; ACEI, angiotensin-converting-enzyme inhibitor; AGI, alpha-glucosidase inhibitors; ARB II, angiotensin II receptor blockers; DPP4i, dipeptidyl peptidase-4 inhibitor; SGLT2, sodium–glucose co-transporter-2; SU, sulphonylureas.

3.6. *Association of RBP-4 Levels with IR and the Severity of CAD*

RBP-4 levels were associated with HOMA-IR levels (IR) in T2D patients with CAD ($p = 0.002$), T2D-only ($p = 0.042$), and CAD-only ($p = 0.031$) study population. RBP-4 levels were most associated with IR in T2D patients with CAD group (OR = 1.667). At the same time, significant associations were found between RBP-4 levels and the severity of CAD (T2D+CAD: [a] $p = 0.017$, [b] $p = 0.022$; CAD-only: [a] $p = 0.002$, [b] $p = 0.001$). RBP-4 levels formed the strongest association with severity of CAD in the CAD-only group (TVD vs. SVD); (OR = 4.111). Table 4, Figures 2 and 3 show the association of RBP-4 levels with IR and the severity of CAD. Appendix A shows the significant results (p-values) of the preliminary associations.

Table 4. Association of RBP-4 levels with IR and the severity of CAD.

Parameter	OR (95% CI)		
	T2D + CAD (n = 150)	T2D-Only (n = 90)	CAD-Only (n = 60)
Insulin resistance	**1.667 (1.341–1.303) ***	**1.594 (1.255–1.880) ***	**1.385 (1.089–1.665) ***
Severity of CAD	[a] **1.494 (1.160–2.726) ¥** [b] **1.733 (1.308–2.144) ¥**	–	[a] **1.622 (1.099–4.923) ¥** [b] **4.111 (1.381–26.379) ¥**

* Computed using binary logistic regression analysis. ¥ Computed using multinomial logistic regression analysis. HOMA-IR cutoff point: 7.17. Bold font indicates significance at $p < 0.05$. Insulin-sensitive (IS) is used as the reference group (IS vs. IR) for insulin resistance. SVD was used as the reference group ([a] SVD vs. DVD; [b] SVD vs. TVD) for the severity of CAD. Adjusted for the covariates age, race, gender, and BMI. OR, odds ratio; CI, confidence interval; '–' indicates not relevant.

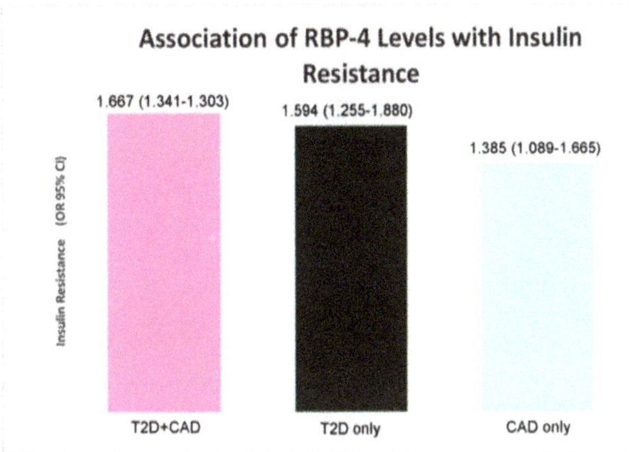

Figure 2. Association of RBP-4 levels with IR.

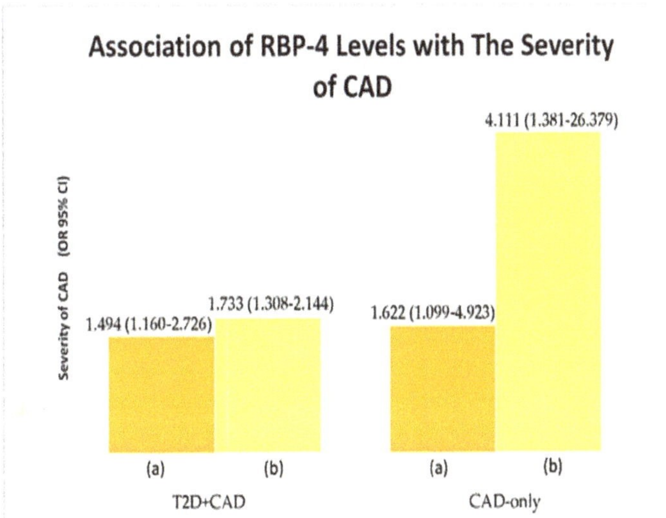

Figure 3. Association of RBP-4 levels with the severity of CAD: (**a**) DVD and (**b**) TVD; SVD is the reference group.

3.7. Association of IR and the Severity of CAD in Correlation with RBP-4 Levels and Clinical Factors (Secondary Analysis)

The significant variables from the preliminary analysis of the case group were used in the secondary analysis (T2D patients with CAD). FPG, ACEI, and RBP-4 were identified as predictors of IR in T2D patients with CAD, using a binary logistic regression model. In the multinomial analysis, the same factors were identified as predictors of the severity of CAD in T2D patients with CAD. Hence, FPG, ACEI, and RBP-4 were predictors of both IR and the severity of CAD in T2D patients with CAD. Table 5 and Figure 4 show the association of IR and the severity of CAD in correlation with RBP-4 levels. The secondary associations of IR and the severity of CAD in correlation with clinical factors were as earlier described (PMID: 34071097) [5].

Table 5. Association of IR and the severity of CAD in correlation with RBP-4 levels.

Parameter	OR (95% CI)	*p*-Value
Insulin resistance	1.166 (1.009–3.042) *	**0.030**
Severity of CAD	[a] 1.647 (1.212–1.972) ¥ [b] 1.815 (1.044–3.040) ¥	**0.044** **0.036**

* Computed using binary logistic regression analysis. ¥ Computed using multinomial logistic regression analysis. HOMA-IR cutoff point: 7.17. Bold font indicates significance at $p < 0.05$. Insulin-sensitive (IS) is used as the reference group (IS vs. IR) for insulin resistance. SVD was used as the reference group ([a] SVD vs. DVD; [b] SVD vs. TVD) for the severity of CAD, adjusted for the covariates age, race, gender, and BMI. OR, odds ratio; CI, confidence interval.

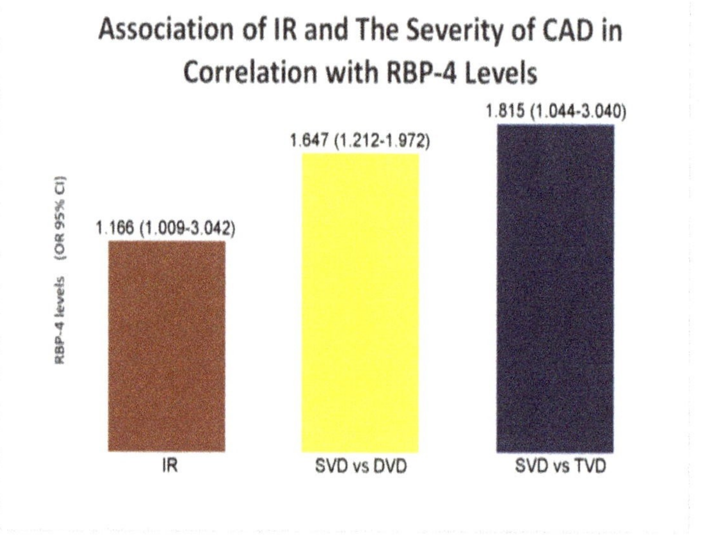

Figure 4. Association of IR and the severity of CAD in correlation with RBP-4 levels.

4. Discussion

Elevated RBP-4 levels are known to be linked to IR, T2D, atherosclerosis, and CAD. Serum RBP-4 levels were significantly associated with A1C, hs-CRP, LDL-c, HDL-c, and TG in T2D patients with CAD in the preliminary associations. A previous study found that RBP-4 levels were significantly correlated with hs-CRP, LDL-c, and A1C with T2D and CAD, which is consistent with the current study's findings [25]. At the same time, RBP-4 levels were also found to be associated with FPI and TC in T2D-only patients in this study. These findings are supported by studies conducted by Fan et al. (2019) and Wessel et al. (2019) [22,26].

Additionally, in their research, Ruijgrok et al. (2018) discovered a link between IR and FPG in their research [27]. Findings from the present study in the group of T2D patients with CAD concur with the findings of this study. This study found that FPI was associated with IR; however, no previous research on this association exists [28]. Furthermore, this study suggests FPI as a reliable and efficient test for detecting IR in people at risk of developing T2D and CAD. In a previous study, hs-CRP was correlated to IR [29]. In line with this, this study discovered an association between hs-CRP and IR. Moreover, hs-CRP is a marker that is used to determine the risk of heart disease. As inflammation appears to play a significant role in the pathogenesis of T2D and CAD, IR could have correlated with hs-CRP.

Moreover, biguanides + DPP4i + insulin treatment, antiplatelet agents, and ACEI were linked to IR in T2D patients with CAD. Thus far, no studies were conducted to investigate the relationship between IR and the mentioned OHA or pharmacological treatments in T2D patients with CAD.

In T2D-only patients, FPG was associated with IR. Khan et al. (2018) previously proposed a significant association between FPI and IR in T2D-only patients [30]. The results were comparable to the present study. Meanwhile, in the present study, FPI and FPG were significantly associated with IR in CAD-only patients. However, data are scarce on the correlation of FPI and FPG with IR in CAD-only patients. The present study's findings revealed significant associations between FPG and FPI and the severity of CAD in T2D patients with CAD. Despite being on diabetic medications, patients with DVD and TVD had higher FPG and FPI levels. However, Srinivasan et al. (2017) discovered an association between hyperinsulinemia and adverse cardiac events in T2D patients [31].

A previous study found significant correlations between laboratory parameters, such as HDL-c, hs-CRP, and TG, and the severity of CAD [32,33]. This previous study's findings differed from those of the present study. The hs-CRP levels in this study were within the normal range (< 1.0 mg/L). The successful reduction of hs-CRP levels in T2D patients with CAD was most likely due to the patients' antihypertensive (ACEIs) and lipid-lowering (statins) pharmacological treatments. Previous studies had shown that antihypertensive and lipid-lowering medications lower hs-CRP levels [34]. Aside from hs-CRP, significant improvements in TC and LDL-c levels were observed. Despite improvements in hs-CRP and TG levels in the study group, no significant associations were found between the markers and the severity of CAD. These findings agree with the previous study by Razban et al. (2016) [34].

DPP4i was described as a second- or third-line add-on treatment that provided cardiovascular benefits without increasing the risk of heart failure, hypoglycemia, or death [35]. Besides DPP4i, statins are used to reduce the frequency of cardiovascular events in T2D patients with and without CAD [36]. Meanwhile, ACEIs are the first-line treatment for hypertension in T2D and CAD patients, and they have been shown to reduce the incidence and recurrence of atherosclerotic CAD [37]. According to previous research, fenofibrate may reduce CVD risk in T2D patients [38]. Although DPP4i and fibrates were significantly associated with DVD and TVD in T2D patients with CAD in this study, the associations were with patients who did not receive the medications. By contrast, 50.8% of patients with DVD and 56.9% of patients with TVD were taking ACEIs. Meanwhile, in the T2D patients with CAD group, statins were taken by 98.3% of DVD patients and 100% of TVD patients. A majority of T2D patients with CAD were not on DPP4i and fibrates, most likely because they were on first-line diabetic pharmacological treatments and insulin. Thereafter, a majority of the patients were taking ACEIs and statins because hypertension and dyslipidemia were the two most common comorbidities of T2D and CAD [5].

In the CAD-only group, FPG, FPI, HOMA-IR, A1C, hs-CRP, LDL-c, HDL-c, and TG were significantly associated with the severity of CAD. Hyperinsulinemia was identified as an independent risk factor for the severity of coronary artery stenosis in non-diabetic CAD patients by Srinivasan et al. (2017) [31]. Fasting serum insulin level was not associated with CAD stenosis in a study by Vafaeimanesh et al. (2018); however, an association was found between the two after the prescription of glucose [39]. According to the author's knowledge, there have not been several studies on laboratory investigations associated with the severity of CAD in CAD-only patients. Hence, although the CAD-only patients are not diabetic, there are chances for them to develop T2D in the future as many of the participants of this group have higher readings of FPI, FPG, and A1C, and the CAD severity was associated with the mentioned clinical factors.

In a previous study, serum RBP-4 levels were correlated to IR in T2D patients, as well as non-diabetic populations with a strong T2D family history [40]. At the same time, elevated serum RBP-4 was associated with metabolic syndrome components [41]. Previous research found a significant positive correlation between RBP-4 levels and CAD severity [23,42–44]. All of these studies support the present study's outcomes. From the additional secondary analysis, we found that RBP-4 is associated with IR and the severity of CAD in T2D patients with CAD, together with the clinical factors FPG and ACEI [5]. Thus, increased RBP-4 levels in the patients may play an important role in the inflammatory progress and further development of IR and severe CAD.

This study shows that RBP-4 levels and clinical factors are related to IR and the severity of CAD in T2D patients with CAD in the Malaysian population. The elevated RBP-4 signifies the progression of IR and endothelial dysfunction. This could provide an initial clue to healthcare providers in optimizing the management or treatment of T2D and CAD patients. RBP-4 as an alternative to standard biomarkers, such as HOMA-IR and hs-CRP, could signal perhaps an early intervention to prevent the disease progression.

Strengths and Limitations

One of the study's major strengths is that the factors were examined by using a specific common cytokine, RBP-4, which was linked to both IR and the severity of CAD separately in previous studies. Furthermore, the study's stringent inclusion and exclusion criteria provided the most effective means of reducing the effect of confounding variables. Another notable feature of this study is the use of sandwich ELISA (Cusabio Elabscience, Texas, USA). Compared with other ELISA methods, the sandwich ELISA method has the highest specificity because it involves two antibodies that detect different epitopes on the same antigen. It also has a high degree of flexibility and sensitivity. One inherent weakness of this study is that some variable data from the electronic medical records and the National Cardiovascular Disease Database were not available. Consequently, socioeconomic factors, lifestyle factors (such as diet logs), and medication adherence were not analyzed in this study. Furthermore, no questionnaires or interviews were used in this study to collect socioeconomic and lifestyle information from the participants.

5. Conclusions

RBP-4 was found to be significantly associated with IR and the severity of CAD in both preliminary and secondary analyses together with FPG and ACEI. Consequently, this study found that RBP-4 is a significant predictor of IR and the severity of CAD in T2D patients with CAD. This study suggests that identifying RBP-4 as a common predictor of IR and the severity of CAD in T2D patients with CAD may serve as a valuable clinical indicator to predict the progression of IR and the severity of CAD. This could prevent unnecessary clinical burden to the healthcare system.

Author Contributions: S.P. was responsible for clinical sample collection, laboratory work, interpretation of data and writing the whole manuscript. H.Z.H. and W.A.W.A. were involved in the conceptual design of the study and revised the manuscript critically for important intellectual content. All authors have read and agreed to the published version of the manuscript.

Funding: This study was supported by the Universiti Malaya (Grant number: RP024B-14HTM and PG173-2016A).

Institutional Review Board Statement: The study was conducted according to the guidelines of the Declaration of Helsinki, and approved by the Institutional Review Board of UMMC (protocol code: 20158-1552, date of approval: 22/10/2015).

Informed Consent Statement: Informed consent was obtained from all subjects involved in the study.

Data Availability Statement: The data presented in this study are available on request from the corresponding author. The data are not publicly available as they contain information that could compromise the privacy of research participants.

Acknowledgments: The authors would like to thank the Universiti Malaya for the financial and technical support in completing the study (grant number: RP024B-14HTM and PG173-2016A). Additionally, the authors would like to thank all the participants of this study and the nurses of UMMC who helped with blood sample collection.

Conflicts of Interest: The authors have no other relevant affiliations or financial involvement with any organization or entity with a financial interest in or financial conflict with the subject matter of materials discussed in the manuscript apart from those disclosed. No writing assistance was utilized in the production of this manuscript.

Appendix A

Table A1. Summarization of significant outcomes from preliminary analysis of study ($p < 0.05$).

Association of RBP-4 with Clinical Factors		
Group	**Parameter**	***p*-Value**
T2D + CAD	A1C	0.034
	hs-CRP	<0.001
	LDL-c	<0.001
	HDL-c	0.001
	TG	0.028
	Bi + I	<0.001
	Bi + SGLT2 + I	0.001
	Bi + SU	<0.001
	Bi	<0.001
	SU	<0.001
	Nitrates	0.008
	Diuretics	0.002
	Cardiac glycosides	<0.001
T2D-only	FPI	0.032
	TC	0.031
CAD-only	–	–
Association of clinical factors with IR and the severity of CAD		
Group	Factor	*p*-value
T2D + CAD		
IR	FPG	0.011
	FPI	<0.001
	hs-CRP	0.025
	Bi + DPP4i + I	0.008
	Antiplatelet agents	0.003
	ACEI	0.026
Severity of CAD	FPG	[a] 0.007 [b] 0.012
	FPI	[a] 0.045
	DPP4i	[a] <0.001 [b] <0.001
	SU + DPP4i	[a] 0.011 [b] 0.010
	Bi + DPP4i + I	[a] <0.001 [b] 0.003

Table A1. *Cont.*

Association of RBP-4 with Clinical Factors		
Group	**Parameter**	***p*-Value**
	ACEI	[a] 0.032 [b] 0.029
	AB	[a] 0.008 [b] 0.020
	Fibrates	[b] <0.001
	Statins	[b] 0.018
	Hematinic agents	[a] 0.011 [b] 0.016
T2D-only		
IR	FPI	<0.001
Severity of CAD	–	–
CAD-only		
IR	FPG	0.048
	FPI	<0.001
Severity of CAD	FPG	[a] <0.001
	FPI	[b] <0.001
	A1C	[b] <0.001
	LDL-c	[b] 0.004
	HDL-c	[a] <0.001
	TG	[b] 0.001
Association of RBP-4 with IR and the severity of CAD		
Group	Factor	*p*-value
T2D + CAD		
IR	RBP-4	0.002
Severity of CAD		[a] 0.017 [b] 0.022
T2D-only		
IR	RBP-4	0.042
Severity of CAD		–
CAD-only		
IR	RBP-4	0.031
Severity of CAD		[a] 0.002 [b] 0.001

Association of clinical factors and RBP-4 with IR, association of clinical factors with RBP-4 computed by using binary logistic regression analysis. Association of clinical factors and RBP-4 with the severity of CAD computed using multinomial logistic regression analysis. SVD was used as the reference group ([a] SVD vs. DVD; [b] SVD vs. TVD); '–' indicates not relevant.

References

1. Zhang, Y.; Pan, X.F.; Chen, J.; Xia, L.; Cao, A.; Zhang, Y.; Wang, J.; Li, H.; Yang, K.; Guo, K.; et al. Combined lifestyle factors and risk of incident type 2 diabetes and prognosis among individuals with type 2 diabetes: A systematic review and meta-analysis of prospective cohort studies. *Diabetologia* **2019**, *63*, 21–33. [CrossRef]
2. Taylor, R. Insulin Resistance and Type 2 Diabetes. *Diabetes* **2012**, *61*, 778–779. [CrossRef]
3. Ormazabal, V.; Nair, S.; Elfeky, O.; Aguayo, C.; Salomon, C.; Zuñiga, F.A. Association between insulin resistance and the development of cardiovascular disease. *Cardiovasc. Diabetol.* **2018**, *17*, 122. [CrossRef] [PubMed]
4. Rafieian-Kopaei, M.; Setorki, M.; Doudi, M.; Baradaran, A.; Nasri, H. Atherosclerosis: Process, indicators, risk factors and new hopes. *Int. J. Prev. Med.* **2014**, *5*, 927–946.
5. Perumalsamy, S.; Wan Ahmad, W.A.; Zaman Huri, H. Single Nucleotide Polymorphism rs17173608 in the Chemerin Encoding Gene: Is It a Predictor of Insulin Resistance and Severity of Coronary Artery Disease in Non-Obese Type 2 Diabetes? *Healthcare* **2021**, *9*, 623. [CrossRef] [PubMed]
6. Peng, L.; Guo, X.; Gao, Y.; Guo, Q.; Zhang, J.; Fang, B.; Yan, B. Impact of right coronary dominance on triple-vessel coronary artery disease: A cross-sectional study. *Medicine* **2018**, *97*, e11685. [CrossRef] [PubMed]
7. Naito, R.; Kasai, T. Coronary artery disease in type 2 diabetes mellitus: Recent treatment strategies and future perspectives. *World J. Cardiol.* **2015**, *7*, 119–124. [CrossRef] [PubMed]

8. Einarson, T.R.; Acs, A.; Ludwig, C.; Panton, U.H. Prevalence of cardiovascular disease in type 2 diabetes: A systematic literature review of scientific evidence from across the world in 2007-2017. *Cardiovasc. Diabetol.* **2018**, *17*, 83. [CrossRef] [PubMed]
9. Tsalamandris, S.; Antonopoulos, A.S.; Oikonomou, E.; Papamikroulis, G.A.; Vogiatzi, G.; Papaioannou, S.; Deftereos, S.; Tousoulis, D. The Role of Inflammation in Diabetes: Current Concepts and Future Perspectives. *Eur. Cardiol.* **2019**, *14*, 50–59. [CrossRef] [PubMed]
10. Noy, N.; Li, L.; Abola, M.V.; Berger, N.A. Is retinol binding protein 4 a link between adiposity and cancer? *Horm. Mol. Biol. Clin. Investig.* **2015**, *23*, 39–46. [CrossRef]
11. Chang, X.; Yan, H.; Bian, H.; Xia, M.; Zhang, L.; Gao, J.; Gao, X. Serum retinol binding protein 4 is associated with visceral fat in human with nonalcoholic fatty liver disease without known diabetes: A cross-sectional study. *Lipids Health Dis.* **2015**, *14*, 28. [CrossRef]
12. Liu, Y.; Albrecht, E.; Dannenberger, D.; Hammon, H.M.; Kuehn, C.; Sauerwein, H.; Yang, R.; Zhao, Z.; Maak, S. Retinol binding protein 4 abundance in plasma and tissues is related to body fat deposition in cattle. *Sci. Rep.* **2019**, *9*, 8056. [CrossRef]
13. Zabetian-Targhi, F.; Mahmoudi, M.J.; Rezaei, N.; Mahmoudi, M. Retinol binding protein 4 in relation to diet, inflammation, immunity, and cardiovascular diseases. *Adv. Nutr.* **2015**, *6*, 748–762. [CrossRef] [PubMed]
14. Bobbert, T.; Raila, J.; Scharz, F.; Mai, K.; Henze, A.; Pfeiffer, A.F.; Schweigert, F.J.; Spranger, J. Relation between retinol, retinol-binding protein 4, transthyretin and carotid intima media thickness. *Atherosclerosis* **2010**, *213*, 549–551. [CrossRef] [PubMed]
15. Takebayashi, K.; Sohma, R.; Aso, Y.; Inukai, T. Effects of retinol binding protein-4 on vascular endothelial cells. *Biochem. Biophys. Res. Commun.* **2011**, *408*, 58–64. [CrossRef]
16. Liu, C.; Zhou, X.R.; Ye, M.Y.; Xu, X.Q.; Zhang, Y.W.; Liu, H.; Huang, X.Z. RBP4 Is Associated With Insulin Resistance in Hyperuricemia-Induced Rats and Patients With Hyperuricemia. *Front. Endocrinol.* **2021**, *12*, 653819. [CrossRef] [PubMed]
17. Wang, J.; Chen, H.; Liu, Y.; Zhou, W.; Sun, R.; Xia, M. Retinol binding protein 4 induces mitochondrial dysfunction and vascular oxidative damage. *Atherosclerosis* **2015**, *240*, 335–344. [CrossRef]
18. Bengal, E.; Aviram, S.; Hayek, T. p38 MAPK in Glucose metabolism of skeletal muscle: Beneficial or harmful? *Int. J. Mol. Sci.* **2020**, *21*, 6480. [CrossRef]
19. Inoue, E.; Yamashita, A.; Inoue, H.; Sekiguchi, M.; Shiratori, A.; Yamamoto, Y.; Tadokoro, T.; Ishimi, Y.; Yamauchi, J. Identification of glucose transporter 4 knockdown-dependent transcriptional activation element on the retinol binding protein 4 gene promoter and requirement of the 20 S proteasome subunit for transcriptional activity. *J. Biol. Chem.* **2010**, *285*, 25545–25553. [CrossRef]
20. Chadt, A.; Al-Hasani, H. Glucose transporters in adipose tissue, liver, and skeletal muscle in metabolic health and disease. *Pflügers Arch. -Eur. J. Physiol.* **2020**, *472*, 1273–1298. [CrossRef]
21. Naji, M.T.; Sami, O.M.; Shams, H.A.; Abdul-Hadi, M.H.; Al-Kuraishy, H.M.; Al-Gareeb, A.I.; Al-Harchan, N.A.A. The associations between retinol binding protein-4 and cardiometabolic profile: Intertwined-intricate relationship. *Biomed. Biotechnol. Res. J. (BBRJ)* **2020**, *4*, 95.
22. Fan, J.; Yin, S.; Lin, D.; Liu, Y.; Chen, N.; Bai, X.; Ke, Q.; Shen, J.; You, L.; Lin, X.; et al. Association of serum retinol-binding protein 4 levels and the risk of incident type 2 diabetes in subjects with prediabetes. *Diabetes Care* **2019**, *42*, 1574–1581. [CrossRef]
23. Sun, H.X.; Ji, H.H.; Chen, X.L.; Wang, L.; Wang, Y.; Shen, X.Y.; Lu, X.; Gao, W.; Wang, L.S. Serum retinol-binding protein 4 is associated with the presence and severity of coronary artery disease in patients with subclinical hypothyroidism. *Aging* **2019**, *11*, 4510–4520. [CrossRef]
24. Baingana, R.K.; Matovu, D.K.; Garrett, D. Application of Retinol-Binding Protein Enzyme Immunoassay to Dried Blood Spots to Assess Vitamin A Deficiency in a Population-Based Survey: The Uganda Demographic and Health Survey 2006. *Food Nutr. Bull.* **2008**, *29*, 297–305. [CrossRef]
25. Su, Y.; Huang, Y.; Jiang, Y.; Zhu, M. The Association between Serum retinol-binding protein 4 levels and cardiovascular events in patients with chronic kidney disease. *Lab. Med.* **2020**, *51*, 491–497. [CrossRef] [PubMed]
26. Wessel, H.; Saeed, A.; Heegsma, J.; Connelly, M.A.; Faber, K.N.; Dullaart, R.P.F. Plasma levels of retinol binding protein 4 relate to large VLDL and small LDL particles in subjects with and without type 2 diabetes. *J. Clin. Med.* **2019**, *8*, 1792. [CrossRef] [PubMed]
27. Ruijgrok, C.; Dekker, J.M.; Beulens, J.W.; Brouwer, I.A.; Coupé, V.M.H.; Heymans, M.W.; Sijtsma, F.P.C.; Mela, D.J.; Zock, P.L.; Olthof, M.R.; et al. Size and shape of the associations of glucose, HbA1c, insulin and HOMA-IR with incident type 2 diabetes: The Hoorn Study. *Diabetologia* **2017**, *61*, 93–100. [CrossRef]
28. Abdul-Ghani, M.A.; Matsuda, M.; Balas, B.; DeFronzo, R.A. Muscle and Liver Insulin Resistance Indexes Derived From the Oral Glucose Tolerance Test. *Diabetes Care* **2006**, *30*, 89–94. [CrossRef]
29. Uemura, H.; Katsuura-Kamano, S.; Yamaguchi, M.; Bahari, T.; Ishizu, M.; Fujioka, M.; Arisawa, K. Relationships of serum high-sensitivity C-reactive protein and body size with insulin resistance in a Japanese cohort. *PLoS ONE* **2017**, *12*, e0178672. [CrossRef] [PubMed]
30. Khan, A.; Khan, A.H.; Adnan, A.S.; Sulaiman, S.A.S.; Mushtaq, S.; Ahmad, N.; Khan, I. Hypertension control among euvolemic hypertensive hemodialysis patients in Malaysia: A prospective follow-up study. *J. Pharm. Policy Pract.* **2019**, *12*, 10. [CrossRef] [PubMed]
31. Srinivasan, M.; Kamath, P.; Bhat, N.; Pai, N.; Bhat, R.; Shah, T.; Manjrekar, P.; Mahabala, C. Basal hyperinsulinemia beyond a threshold predicts major adverse cardiac events at 1 year after coronary angiogram in type 2 diabetes mellitus: A retrospective cohort study. *Diabetol. Metab. Syndr.* **2017**, *9*, 38. [CrossRef]

32. Dai, W.; Zhang, Z.; Zhao, S. Baseline levels of serum high sensitivity C reactive protein and lipids in predicting the residual risk of cardiovascular events in Chinese population with stable coronary artery disease: A prospective cohort study. *Lipids Health Dis.* **2018**, *17*, 273. [CrossRef] [PubMed]
33. Toth, P.P.; Granowitz, C.; Hull, M.; Liassou, D.; Anderson, A.; Philip, S. High triglycerides are associated with increased cardiovascular events, medical costs, and resource use: A real-world administrative claims analysis of statin-treated patients with high residual cardiovascular risk. *J. Am. Heart Assoc.* **2018**, *7*, e008740. [CrossRef] [PubMed]
34. Kim-Mitsuyama, S.; Soejima, H.; Yasuda, O.; Node, K.; Jinnouchi, H.; Yamamoto, E.; Sekigami, T.; Ogawa, H.; Matsui, K. Reduction in hsCRP levels is associated with decreased incidence of cardiovascular events in Japanese hypertensive women but not in men. *Sci. Rep.* **2020**, *10*, 17040. [CrossRef] [PubMed]
35. Razban, M.M.; Eslami, M.; Bagherzadeh, A. The relationship between serum levels of Hs-CRP and coronary lesion severity. *Med. Pharm. Rep.* **2016**, *89*, 352–364. [CrossRef] [PubMed]
36. Ou, H.T.; Chang, K.C.; Li, C.Y.; Wu, J.S. Comparative cardiovascular risks of dipeptidyl peptidase 4 inhibitors with other second- and third-line antidiabetic drugs in patients with type 2 diabetes. *Br. J. Clin. Pharmacol.* **2017**, *83*, 1556–1570. [CrossRef]
37. Arnold, S.V.; Bhatt, D.L.; Barsness, G.W.; Beatty, A.L.; Deedwania, P.C.; Inzucchi, S.E.; Kosiborod, M.; Leiter, L.A.; Lipska, K.J.; Newman, J.D.; et al. Clinical management of stable coronary artery disease in patients with type 2 diabetes mellitus: A scientific statement from the American Heart Association. *Circulation* **2020**, *141*, e779–e806. [CrossRef]
38. Elam, M.B.; Ginsberg, H.N.; Lovato, L.C.; Corson, M.; Largay, J.; Leiter, L.A.; Lopez, C.; O'Connor, P.J.; Sweeney, M.E.; Weiss, D.; et al. Association of fenofibrate therapy with long-term cardiovascular risk in statin-treated patients with type 2 diabetes. *JAMA Cardiol.* **2017**, *2*, 370. [CrossRef] [PubMed]
39. Vafaeimanesh, J.; Parham, M.; Norouzi, S.; Hamednasimi, P.; Bagherzadeh, M. Insulin resistance and coronary artery disease in non-diabetic patients: Is there any correlation? *Casp. J. Intern. Med.* **2018**, *9*, 121–126. [CrossRef]
40. Majerczyk, M.; Kocełak, P.; Choręza, P.; Arabzada, H.; Owczarek, A.J.; Bożentowicz-Wikarek, M.; Brzozowska, A.; Szybalska, A.; Puzianowska-Kuźnicka, M.; Grodzicki, T.; et al. Components of metabolic syndrome in relation to plasma levels of retinol binding protein 4 (RBP4) in a cohort of people aged 65 years and older. *J. Endocrinol. Investig.* **2018**, *41*, 1211–1219. [CrossRef] [PubMed]
41. McInnes, K.J.; Smith, L.B.; Hunger, N.I.; Saunders, P.T.; Andrew, R.; Walker, B.R. Deletion of the androgen receptor in adipose tissue in male mice elevates retinol binding protein 4 and reveals independent effects on visceral fat mass and on glucose homeostasis. *Diabetes* **2012**, *61*, 1072–1081. [CrossRef]
42. Jin, Y.; Cao, J.N.; Wang, C.X.; Feng, Q.T.; Ye, X.H.; Xu, X.; Yang, C.J. High serum YKL-40 level positively correlates with coronary artery disease. *Biomark. Med.* **2017**, *11*, 133–139. [CrossRef]
43. Li, X.; Zhang, K.; Yan, J.; Wang, L.; Wang, Y.; Shen, X.; Sun, H.; Liu, L.; Zhao, C.; He, H.; et al. Serum retinol-binding protein 4 as a predictor of cardiovascular events in elderly patients with chronic heart failure. *ESC Heart Fail.* **2020**, *7*, 542–550. [CrossRef] [PubMed]
44. Liu, Y.; Wang, D.; Chen, H.; Xia, M. Circulating retinol binding protein 4 is associated with coronary lesion severity of patients with coronary artery disease. *Atherosclerosis* **2015**, *238*, 45–51. [CrossRef] [PubMed]

biology MDPI

Article

Correlation between Carbonic Anhydrase Isozymes and the Evolution of Myocardial Infarction in Diabetic Patients

Sorina Magheru [1], Calin Magheru [1], Florin Maghiar [1], Liliana Sachelarie [2,*], Felicia Marc [1], Corina Maria Moldovan [1], Laura Romila [2], Anica Hoza [1], Dorina Maria Farcas [1], Irina Gradinaru [3] and Loredana Liliana Hurjui [3]

1 Department of Medical Disciplines, Faculty of Medicine and Pharmacy, University of Oradea, 410073 Oradea, Romania
2 Departament of Preclinical Disciplines, Apollonia University, 700511 Iasi, Romania
3 Department of Medical Disciplines, Faculty of Medicine and Pharmacy, "Grigore T. Popa" University of Medicine and Pharmacy, 700115 Iasi, Romania
* Correspondence: lisachero@yahoo.com

Simple Summary: Heart disease in diabetics presents distinctive characteristics both anatomically and physiopathologically compared to non-diabetics. In people with diabetes, high blood pressure has a high incidence (approximately one-third of diabetic patients have high blood pressure) and is a risk factor for diabetic macro- and microvascular complications. The correlation of these parameters could represent early markers of the prognosis and evolution of diabetic patients with acute myocardial infarction and their routine determination could be included in the biological algorithm of acute myocardial infarction, but understanding of this aspect must be deepened in the future. The results showed that diabetic patients develop acute myocardial infarction more frequently, regardless of age. The level of the enzymes of myocardial necrosis was higher in diabetics compared to non-diabetics, and acute coronary syndrome occurs mainly in diabetics with inadequate metabolic balance. Our research may provide useful information for the medical community.

Abstract: (1) Background: Myocardial infarction was, until recently, recognized as a major coronary event, often fatal, with major implications for survivors. According to some authors, diabetes mellitus is an important atherogenic risk factor with cardiac determinations underlying the definition of the so-called "diabetic heart". The present study aims to establish a correlation between the evolution of myocardial infarction in diabetic patients, by determining whether lactic acid levels, the activity of carbonic anhydrase isoenzymes, and the magnitude of ST-segment elevation are correlated with the subsequent evolution of myocardial infarction. (2) Methods: The study analyzed 2 groups of 30 patients each: group 1 consisted of diabetic patients with acute myocardial infarction, and group 2 consisted of non-diabetic patients with acute myocardial infarction. Patients were examined clinically and paraclinical, their heart markers, lactic acid, and the activity of carbonic anhydrase I and II isozymes were determined. All patients underwent electrocardiogram and echocardiography analyses. (3) Results: The results showed that diabetics develop acute myocardial infarction more frequently, regardless of how much time has passed since the diagnosis. The value of myocardial necrosis enzymes was higher in diabetics than in non-diabetics, and acute coronary syndrome occurs mainly in diabetics with poor metabolic balance. Lethality rates in non-diabetic patients with lactic acid values above normal are lower than in diabetics. (4) Conclusions: Lactic acid correlated with the activity of isozyme I of carbonic dioxide which could be early markers of the prognosis and evolution of diabetic patients with acute myocardial infarction.

Keywords: acute myocardial infarction; diabetes; lactic acid; carbonic anhydrase

Citation: Magheru, S.; Magheru, C.; Maghiar, F.; Sachelarie, L.; Marc, F.; Moldovan, C.M.; Romila, L.; Hoza, A.; Farcas, D.M.; Gradinaru, I.; et al. Correlation between Carbonic Anhydrase Isozymes and the Evolution of Myocardial Infarction in Diabetic Patients. *Biology* **2022**, *11*, 1189. https://doi.org/10.3390/biology11081189

Academic Editor: Gaetano Santulli

Received: 25 July 2022
Accepted: 5 August 2022
Published: 8 August 2022

1. Introduction

Until recently, myocardial infarction was recognized as a major coronary event, often fatal, with major implications for survivors [1,2]. This paradigm has changed as a result of improved therapeutic strategies in patients with coronary heart disease and due to the methods of detecting or excluding myocardial necrosis. The introduction of techniques for measuring cardiac troponins has allowed the detection with high sensitivity and accuracy of even small areas of myocardial necrosis, changing the criteria for defining infarction [3].

As a consequence of applying the new definition, more patients with myocardial infarction will be identified and many more episodes of reinfarction will be diagnosed in patients with progressive coronary heart disease. According to the new definition, patients requiring myocardial revascularization (interventional or surgical) are at risk of additional myocardial destruction or myocardial infarction [4,5]. The detection of microinfarcts, possibly due to the new markers produced during the revascularization maneuvers, places these patients in the high risk group [6,7]. However, the patient's long-term prognosis can be significantly improved by interventional or surgical revascularization. For example, a patient with unstable angina and severe anterior descending artery stenosis will have a much greater benefit from coronary dilation despite a small increase in cardiac troponins. The benefit will far outweigh the negative impact of periprocedural infarction.

According to some authors, diabetes mellitus is an important atherogenic risk factor with cardiac determinations underlying the definition of the so-called "diabetic heart". Heart disease in diabetics has distinctive features from non-diabetics, both anatomically and pathophysiologically. The concept of diabetic heart muscle disease, later called diabetic cardiomyopathy, involves diffuse myocardial damage even with coronary and microangiopathic lesions being absent. The physio-pathological processes that lead to the significant increase in cardiovascular damage in patients with diabetes are represented by three important biohumoral changes: hyperglycemia, hyperinsulinemia and the increase in circulating free fatty acids [8].

Hyperglycemia is the negative factor that influences the prognosis and clinical evolution of AMI, leading to higher mortality and more severe complications. Hyperglycemia is an important factor that determines endothelial dysfunction, leading to vascular damage and microvascular obstructions, and its effect of increasing oxidative stress, inflammation and platelet aggregation is also known. Thus, hyperglycemia causes a change in protein synthesis, with increased expression of pro-inflammatory factors (IL11, IL 6, TNF alpha, fibrinogen), oxidative stress and adhesion molecules (ICAM 1, VCAM 1), all of which result in endothelial damage and the appearance of atherosclerotic lesions [9,10].

Diabetes is accompanied by an extremely high incidence of acute cardiovascular events; thus, 15–20% of patients presenting with acute myocardial infarction also have diabetes. The subsequent evolution of these patients is affected, the risk of adverse events being twice as high as in non-diabetics. The risk of cardiovascular disease is increased at blood sugar values slightly above the allowed limit and even at values above the normal limit. Hyperglycemia is associated with a poor prognosis in patients with acute myocardial infarction, regardless of the presence of diabetes [11,12].

Coronary findings in diabetes are of two types: macroangiopathy affecting the epicardial coronary vessels and microangiopathy characterized by thickening of the basement membrane of capillaries and arterioles with a diameter below 150 μm.

The clinical expression of cardiac involvement in diabetics also includes the autonomic cardiac neuropathy that appears in diabetic neuropathy and is usually associated with other types of neuropathic complications. High blood pressure has a high incidence in people with diabetes (about a third of diabetic patients have high blood pressure) and is a risk factor for the macro- and microvascular complications of the diabetic patient [13,14].

Although diabetes mellitus is frequently associated with other cardiovascular risk factors (age, abdominal obesity, hypertension, dyslipidemia), it is an independent cardiovascular risk factor that acts even in the absence of other risk factors or after other risk factors are brought under control [15,16].

Carbonic anhydrase (CA), an enzyme discovered in 1932 by Meldrum and Roughton, catalyzes one of the simplest chemical reactions, that of carbon dioxide and water [17].

The physiological relevance of CA isozymes and their possible fundamental role in the living organism derives from the discovery of some families which genetically lack cytosolic CA II [18,19]. These patients had symptoms of osteopetrosis, renal tubular acidosis and cerebral calcifications. These findings are currently included in other investigations, aimed, both at the quantitative determination of the activity of other CA isozymes in various organs and tissues of individuals lacking CA II, and in the research and study of an animal model in which one or more isozymes are completely missing [20,21].

Transport and elimination of CO_2, regulation of pH and osmotic pressure, fluid secretion and regulation of the ion exchange process [16] are therefore the most common physiological processes involving the catalytic activity of one or more isozymes of CA. Regarding the physiological role of CA isozymes, the research team led by Prof. Puscas showed that I isozymes are involved in modulating vascular processes in the body [22,23] in which substances with vasodilating effects inhibit CA I, and vasoconstrictors activate these isozymes, while at the same time, II isozymes are involved in the modulation of secretory processes in the body, all these results materializing into a new concept of signal transmission in the cell—a theory of pH [24,25].

2. Materials and Methods

2.1. Aim of the Study

This study proposed to establish whether there was a correlation between the evolution of myocardial infarction in diabetic patients and the activity of CA isoenzymes. The presence of lactic acid and the activity of CA I and CA II isozymes were determined and the values were correlated with the magnitude of the ST segment elevation and with the subsequent evolution of myocardial infarction.

2.2. Materials

We studied a total of 60 patients, who were divided into two groups. All patients were hospitalized with acute myocardial infarction at the CF Oradea Clinical Hospital, over two years.

Group 1—Included 30 diabetic patients with acute myocardial infarction
Group 2—Included 30 non-diabetic patients with acute myocardial infarction

The measurement of lactic acid and markers of cardiac necrosis (immediately after admission) and the use of the two groups of patients arose from the need to determine biological parameters that would lead us to early diagnosis. These markers could indicate a prospective unfavorable evolution of acute myocardial infarction in diabetics compared to non-diabetics, and thus allow the use of more aggressive treatment, followed by an improvement in the evolution and the quality of life of those patients. We determined the value of lactic acid, cardiac markers and the activity of CA isozymes and looked for a correlation between their levels, the size of ST-segment elevation and the subsequent evolution of patients.

We prepared a study sheet for each patient, which included data for a single hospitalization.

2.3. Methods

All patients with myocardial infarction studied were first examined clinically, undergoing a complete and complex evaluation according to the current standards, both in terms of anamnestic and objective examination, concentrating on examination of the cardiovascular system. Patients were monitored in the coronary intensive care unit for heart rate, ventricular rate, blood pressure, and oxygen saturation.

Laboratory tests were then undertaken as a matter of urgency, supplemented by subsequent data during hospitalization with the repeated measurement of certain biological parameters.

All patients with myocardial infarction underwent an electrocardiogram at the hospitalization stage, which was repeated during hospitalization and interpreted in dynamics.

After the initiation of emergency treatment and after clinical and paraclinical stabilization, the patients underwent echocardiography, and where there were indications of disease, they underwent exercise testing, 24-h Holter ECG monitoring, Doppler ultrasound of the peripheral arteries or carotids, myocardial scintigraphy, arteriography, coronary angiography, and measurement of myocardial revascularization. Lactic acid and cardiac markers were determined by spectrophotometric methods with an automatic biochemistry analyzer, Cobas model.

The measurement of carbonic anhydrase activity in red blood cells was performed using the stopped-flow method [20] with a HI-TECH SF-51MX rapid kinetics spectrophotometer provided with a RKBIN IS-1 kinetic program.

2.4. Statistical Analysis

The data were statistically analyzed using the EPIINFO application, version 6.0, a program of the Center for Disease Control and Prevention—CDC (Center of Disease Control and Prevention) in Atlanta, adapted to the processing of medical statistics. Frequency ranges, average parameter values and standard deviations were calculated. Tests of statistical significance by the $\chi 2$ method were used, and ANOVA (Brown–Forsythe) was used to compare the means. The level of statistical significance was 0.05.

3. Results

3.1. Characteristics of the Population (Age, Gender and Environment)

In our study group over 50% were women. The average age of the diabetic group (age range 37–73 years) was 56.7 ± 7.3 years, and of the non-diabetic group (age range 38–81 years) was 64.6 ± 6.9 years, and the patients came mainly from an urban environment (93.33%). This may be explained by the fact that diabetes is diagnosed faster and more easily in urban areas, while in rural areas there is still a lack of medical staff, a lack of education and a very low addressability of patients. Table 1 shows the characteristics of the diabetic and non-diabetic groups.

Table 1. Characteristics of the population.

Baseline Characteristics of the Diabetic Group and Non-Diabetic Group	Diabetic MD ± DS		Non-Diabetic MD ± DS	
Age (years)	56.7 ± 7.3		64.6 ± 6.9	
Gender Women	19	Percentage% 63.33	17	Percentage% 56.66
Men	11	36.66	13	43.33
Environment Urban	28	93.33	25	83.33
Rural	2	6.66	5	16.66

In the diabetic group, 93.3% come from urban areas and 6.66% from rural areas. In the non-diabetic group, 83.33% come from urban and 16.66% from rural areas. Their histories showed that over 55% of diabetic patients had had a diabetes diagnosis for less than 5 years, and the average duration of the disease in our study group was 5.9 years ± 1.3. Consistent with the mean low duration of their diabetes, 51.5% of patients had no complications.

The most common complications were macroangiopathic (stroke—5.4%, ischemic heart disease—18.8%, chronic obliterative arteriopathy—17.8%), followed by nephropathy (15.8%). Neuropathy and retinopathy were found in a small percentage of patients (6.4% and 4.0%, respectively), which were correlated with a shorter duration of diabetes in the patients studied. We observe a net favorable balance of macrovascular complications (stroke, obliterating arteriopathy, ischemic heart disease)—29.2%, compared to microangiopathic complications (nephropathy, neuropathy, retinopathy)—26.2%. There were patients in this study who had two or even three macrovascular complications without any microvascular complications.

3.2. Electrocardiographic Evaluation

In terms of the electrocardiographic evaluation at admission, the cases were distributed according to Table 2:

Table 2. Electrocardiographic evaluation case distribution at hospitalization.

Electrocardiographic Evaluation	Diabetic	Non-Diabetic
	%	%
Pathological Q wave	18.3	8.8
ST segment change	81.2	91.6
ST Elevation	78.7	82.8
ST Subleveling	2.5	8.8
T wave change	31.7	25.5

The electrocardiogram helped us to assess the topography and extent of the myocardial infarction. The most sensitive and specific ECG change for IMA is ST-segment elevation. Newly pathological Q waves show high sensitivity and specificity, with about 90% of patients having an evolving AMI. Pathological Q wave occurs later in the evolution of IMA than ST segment elevation. ST-segment elevation is sensitive in detecting myocardial ischemia, but cannot differentiate between AMI and unstable angina. T-wave changes can be induced by acute myocardial ischemia, in about one-third of patients developing AMI.

In our study, a significantly higher number of diabetic patients had a pathological Q wave at admission, which proves the delay in the presentation of these patients in an emergency department, and implicitly a delay in starting treatment. Non-diabetic patients who have more severe clinical manifestations than nondiabetics report to the emergency department earlier. ST-segment changes, which appear first in the evolution of an AMI, are found in a higher percentage in nondiabetic patients.

Therefore, the following locations were identified, Table 3:

Table 3. Distribution of cases according to the location of myocardial infarction.

Location of the Infarction	Diabetic	Non-Diabetic
	%	%
Lower	31.7	28.9
Previous	7.9	20.1
Anteroseptal	21.3	14.2
Previously stretched	18.3	5.9
Side	20.8	30.9

In diabetic patients, the localization of AMI is mainly in the antero-septal and previously extended territory, these infarcts being large, and implicitly their evolution is unfavorable, being accompanied by multiple complications. This is due to the late presentation in the emergency services and implicitly due to the delay in starting the treatment.

Non-diabetic patients presented with smaller infarcts, located in the lower or lateral territory. The presence of large infarcts (previously extended, antero-septal) in diabetic patients and of smaller infarcts (lateral, anterior) in nondiabetic patients was observed.

3.3. Blood Sugar Values Distribution

The distribution of cases in terms of the blood sugar values at hospitalization is presented in Figure 1:

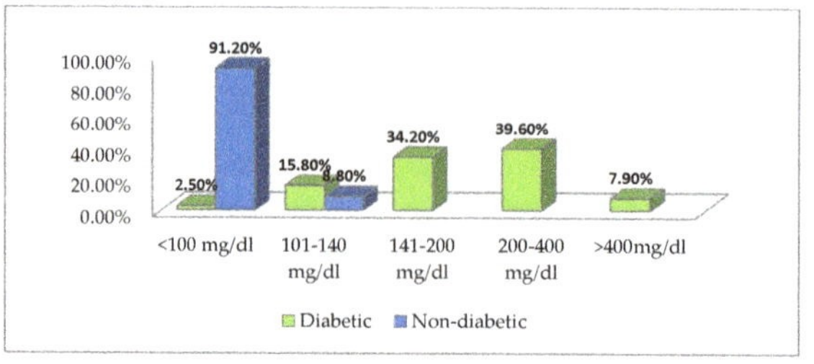

Figure 1. Case distribution according to blood sugar values at hospitalization.

We considered the blood sugar to be pathological at a value of over 100 mg/dL according to the ADA standards [11]. In diabetics, the mean blood sugar at hospitalization was 248 mg/dL (with limits of 102 mg/dL and 572 mg/dL). 2.5% of diabetics had blood sugar values below 100 mg/dL, which indicates a good control of diabetes in these patients. 15.8% had blood sugar values between 100 and 140 mg/dL, 34.2% had glycemic values between 140 and 200 mg/dL, 39.4% had values between 200–400 mg/dL and 7.9% had values above 400 mg/dL.

In non-diabetic patients the blood sugar values were below the reference value of 100 mg/dL in 22 patients (75.5%), and in 8 patients the blood sugar value was over 100 mg/dL (with an upper limit of 140 mg/dL), but these values returned below 100 mg/dL in the next two days.

The enzymes of myocardial necrosis (Table 4), that served to confirm or exclude the diagnosis of major coronary heart disease determined in the two groups of patients were:

- creatine phosphokinase (CPK): VN = 25–90 U/I;
- creatine phosphokinase-MB (CPK-MB): VN < 5% of CK;
- lactic dehydrogenase (LDH): VN = 150–240 U/I;
- cardiac-specific troponins (cTnT and cTnI): VN < 0.1 ng/mL;
- oxaloacetic glutamic transaminase (GOT): VN < 35–40 IU.

Table 4. Enzymes of myocardial necrosis.

	Normal Values		Increased Values		p Value
Parameter	**%**	**Mp ± DS**	**%**	**Mp ± DS**	
Diabetic					
CPK	23.8	68.1 ± 8.5	76.2	458.5 ± 40.7	$p < 0.001$
CPK-MB	21.3	5.2 ± 1.3	78.7	68.2 ± 7.9	$p < 0.01$
LDH	2.5	159.7 ± 21.6	97.5	625.0 ± 64.3	$p < 0.001$
cTnT- cTnI	0	-	100	2.2 ± 0.3	-
GOT	26.2	27.5 ± 3.9	73.8	116.6 ± 12.3	$p < 0.001$
Non-diabetic					
CPK	11.3	56.3 ± 6.6	88.7	399.4 ± 41.3	$p < 0.001$
CPK-MB	8.8	5.0 ± 1.2	91.2	43.4 ± 5.7	$p < 0.001$
LDH	0	-	100	575.3 ± 61.8	-
cTnT- cTnI	0	-	100	2.1 ± 0.2	-
GOT	40.2	25.7 ± 3.2	59.8	107.2 ± 11.8	$p < 0.01$

Statistical significance: $p < 0.001$—highly significant; $p < 0.01$—very significant.

CPK data for diabetic patients indicated that 76.2% of patients showed an increased value, with a mean of 458 U/I, and 23.8% had a normal value, with a mean of 68 U/I; this indicates that the latter patients had a major coronary accident more than 3–4 days ahead of other patients who had acute myocardial infarction (AMI) in the 3 days prior to the test.

CPK data for non-diabetic patients indicated that 88.7% of patients showed an increase above normal, with a mean value of 399 U/I, and the level was normal in 11.3% of patients with a mean value of 56 U/I.

CPK-MB data indicated that 78.7% of diabetic patients showed an increased value above normal, having a mean value of 68 U/I, while 21.3% of patients had a normal value with a mean of 5 U/I having a greater specificity for the AMI diagnosis.

CPK-MB data indicated that 91.2% of non-diabetic patients, having a mean value of 43 U/I and this was normal in 8.8% of patients with a mean value of 5 U/I.

LDH data presented an increased value in 97.5% of diabetic patients, with a mean value of 625 U/I, and in all non-diabetic patients with a mean value of 575 U/I.

The specific cardiac troponins determined showed elevated values above 0.1 ng/mL with an average value of 2.1 ng/mL.

GOT showed increased values in 73.8% of diabetic patients, with a mean value of 116 U/I, and in 59.8% of non-diabetic patients, with a mean value of 107 U/I.

In diabetics, severe myocardial ischemia may cause an increase in serum lactic acid (VN = 8–19.8 mg/dL) even without a decrease in blood pH. Based on these considerations, we tried to identify the place of lactic acid as a predictive marker of the evolution of diabetics with myocardial infarction. Lactic acid values were divided into 3 intervals: <25 mg/dL, 25–35 mg/dL and> 35 mg/dL (Table 5).

Table 5. Lactic acid value.

Lactic Acid	Diabetic	Non-Diabetic	*p* Value
	%	%	
<25 mg/dL	27.0	30.0	$p > 0.05$
25–35 mg/dL	29.7	33.3	$p < 0.01$
>35 mg/dL	43.2	36.7	$p < 0.001$

The main objective of analyzing biochemical markers in patients with acute chest pain according to modern standards is to stratify the risk of these patients. This means not only detecting or excluding myocardial necrosis, but also detecting patients at risk of developing a life-threatening cardiac event in the near future. In the present study, we tried to find new biochemical markers that would provide us with information on a future unfavorable evolution of IMA. Thus, we found that a higher number of diabetic patients had high lactic acid values compared to nondiabetics. We divided the high values of lactic acid into three intervals: <25 mg/dL; 25–35 mg/dL; and > 35 mg/dL.

In the diabetic patients group it was found that for 43.2% of patients the value of lactic acid increased significantly (>35 mg/dL), for 29.7% the value of lactic acid was between 25–35 mg/dL and for 27.0% the lactic acid value was below 25 mg/dL ($p < 0.001$).

In non-diabetics the value of lactic acid increased significantly for 36.7% of cases, for 33.3% the value of lactic acid was between 25–35 mg/dL and for 30% of cases the value of lactic acid was below 25 mg/dL ($p < 0.001$) (Table 5).

3.4. Correlation of Lactic Acid Values with ST-Segment Elevation

For diabetic patients with lactic acid values higher than 35 mg/dL, we found a regression below 30% of ST-segment elevation, without any obvious clinical improvement. In diabetic patients with lactic acid values between 25–35 mg/dL, we found a regression of 30–70% of ST-segment elevation, with obvious clinical improvement. There is also a linear correlation between the lactic acid value with ST elevation.

We made a correlation between the values of lactic acid and the regression of ST segment elevation. The regression (descent) of the ST segment elevation towards the isoelectric line represents a marker of favorable evolution of IMA, of repermeabilization of the hibernating areas around the myocardial necrosis. We took as a benchmark a regression of ST elevation with a percentage of 30%, on ECG performed in dynamics.

We found that diabetic patients who had high lactic acid values in the range >35 mg/dL had a small regression of ST segment elevation, by <30%, showing persistent ST segment elevation, leading to a wider necrotic area, and therefore to an unfavorable evolution of IMA.

Non-diabetic patients who had lower lactic acid values, <25 mg/dL, had a 100% regression of ST segment elevation by >30%, with a more favorable evolution and obvious clinical improvement (Table 6).

Table 6. Correlation of lactic acid values with ST-segment elevation.

Lactic Acid	Diabetic		Non-Diabetic	
ST-Segment Elevation	<30%	>30%	<30%	>30%
<25 mg/dL $p > 0.05$	20.0	80.0	-	100.0
25–35 mg/dL	36.4	63.6	20.0	80.0
>35 mg/dL $p < 0.001$	75.0	25.0	55.6	44.4

CA II activity showed higher values in diabetic patients compared to nondiabetic patients, while CA I activity was increased in both diabetic and nondiabetic patients, as shown in Table 7.

Table 7. The activity of CA isozymes in diabetic and non-diabetic patients.

Red Blood Cells (UE/mL)	Normal Values	Diabetic	Non-Diabetic	*p* Value
CA I red blood cells (UE/mL)	0.262 ± 0.011	0.582 ± 0.021 *	0.574 ± 0.018 *	$p < 0.01$
CA II red blood cells (UE/mL)	1.015 ± 0.083	1.701 ± 0.118 *	1.042 ± 0.105	$p < 0.05$

* statistically significant difference compared to normal values ($p < 0.05$).

The biological parameters determined for patients from both groups are presented in Figure 2:

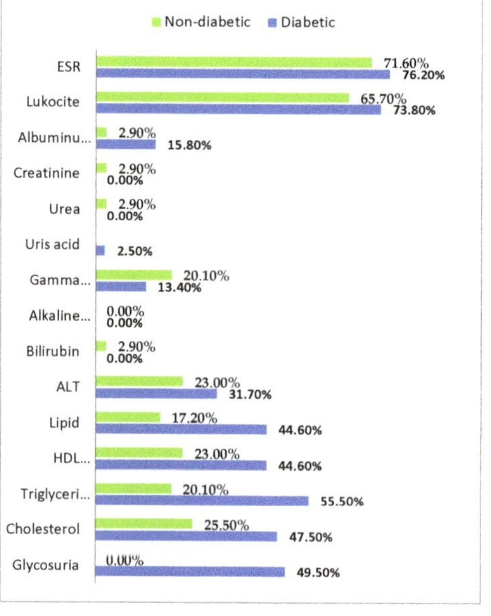

Figure 2. Distribution of cases according to biological parameters outside the limits of normality.

4. Discussion

The size of the infarction is crucial to survival. Diabetic patients have larger infarcts than non-diabetics, and since the size of myocardial infarction is closely correlated with cardiac performance and mortality, this may explain the higher mortality rate in diabetics. It has also been shown that women suffer from more severe heart attacks, which correlates well with the increase in mortality after acute myocardial infarction in diabetic women.

Diabetes is a major risk factor for the development of cardiovascular diseases. The risk of cardiovascular disease in patients with diabetes is two to three times higher than in patients without diabetes. In addition, diabetes is a strong risk factor for cardiovascular events after acute myocardial infarction [26]. Some studies indicate that diabetes is independently associated with impaired epicardial reperfusion and higher mortality [27].

Another factor associated with acute myocardial infarction that seems to be relevant in determining the increased mortality rate in diabetics is the location of the infarction. Diabetic patients most often have an anteroseptal localization of acute myocardial infarction, more commonly than non-diabetics, and mortality in such areas is higher than in any other localization [28,29].

It should be mentioned that if at hospitalization in most patients the ST segment changes and especially the ST segment elevation dominated, in both groups of diabetic patients and non-diabetic patients, at discharge the Q wave changes dominated, which overlaps with the evolution of the natural ECG of an acute myocardial infarction.

Echocardiography showed the presence of systolic dysfunction in a higher number of diabetic patients (71.3%) than non-diabetics (56.9%). The ejection fraction recorded lower values in diabetics than in non-diabetics (42.0%versus 47.5%) the difference being significant. This is consistent with some studies published in the literature [25]. The shortening fraction is higher in non-diabetic patients than in diabetics, but without significant differences between the two groups of patients ($p > 0.05$).

The analysis of paraclinical data performed on the groups of patients included in this study shows the presence of dyslipidemia (hypercholesterolemia, hypertriglyceridemia, hyperlipidemia, low HDL-cholesterol) in a higher number of diabetic patients compared to non-diabetic; and these dyslipidemias are more severe in diabetic than in non-diabetic patients. This data is consistent with the literature, given that the dyslipidemia profile of diabetic patients is frequently characterized by hypertriglyceridemia and decreased HDL-cholesterol; these dyslipidemias are independent cardiovascular risk factors in diabetic patients. In the studied groups, in the diabetic patients hypertriglyceridemia was present to a level of 55.5% and in non-diabetics to 20.1%. Low HDL-cholesterol was present in 44.6% of diabetics, compared to 23% in non-diabetics. This dyslipidemia profile of diabetic patients is more suitable for fibrate therapy (Helsinki-Heart study); however, many trials have shown that diabetics benefit equally with the non-diabetic population from statin therapy.

In the studied groups, we also observed a higher number of diabetic patients with hypercholesterolemia (47.5% versus 25.5%), so the dyslipidemia profile of the studied patients is different from the data published in the literature.

The study also focused on finding out whether there was an influence of lactic acid values on the evolution of patients with AMI. Thus, in diabetic patients we found an unfavorable evolution in those with high lactic acid values (>35 mg/dL). In patients with lactic acid above 35 mg/dL the risk of death was 1.7 times higher than in those with values between 25–35 mg/dL.

In non-diabetics, lethality rates in patients with lactic acid values above normal were lower than in diabetics (10.0% versus 18.2% at 25–35 mg/dL, and 22.2% versus 31.3% (>35 mg/dL), respectively.

It has also been observed that there is a direct correlation between CA II activity and diabetes, in the sense that diabetic patients have higher values of this isoenzyme, while CA I isoenzyme, known for its implications in vasoconstriction and vasodilation, has increased activity in all patients with myocardial infarction, regardless of their diabetic status [30–32]. The high values of CA II are also correlated with the high values of cardiac markers, meaning that the enzyme can also be an indicator of the state of damage to the heart tissue [33]. A long-term case follow-up study is needed to investigate the incidence of diabetes and cardiovascular complications of the disease [34].

5. Conclusions

Current studies indicate that there is no satisfactory biomarker that can specifically identify acute manifestations related to myocardial ischemia and its prognosis [31]. However, in this study the electrocardiogram analysis showed the presence in diabetics, especially, of much larger transmural myocardial infarctions and the fact that at the time of presentation to the doctor many diabetic patients had already developed a pathological Q wave of ECG necrosis. Lactic acid values together with the activity of CA II isoenzyme could be early markers in the prognosis and evolution of diabetic patients with acute myocardial infarction; their routine measurement may be included in the biological algorithm of acute myocardial infarction, but this should be fully researched in the future.

Author Contributions: Conceptualization, S.M. and C.M.; methodology, L.S. and C.M.; software, I.G., C.M.M. and L.S.; validation, F.M. (Florin Maghiar) and A.H.; investigation, C.M. and D.M.F.; resources, L.S. and L.R.; data curation, I.G.; writing—original draft preparation, L.S.; writing—review and editing, D.M.F., C.M., S.M., A.H., L.L.H. and L.S.; visualization, I.G.; supervision, L.S.; project administration, F.M. (Felicia Marc), C.M. and S.M. All authors have read and agreed to the published version of the manuscript. All authors contributed equally to this paper as did the first author and the corresponding author.

Funding: This research received no external funding.

Institutional Review Board Statement: The study was conducted in accordance with the Declaration of Helsinki, and approved by the Ethics Commission of CF Oradea Clinical Hospital, County, Romania, no. 12/21.10.2019.

Informed Consent Statement: Informed consent was obtained from all subjects involved in the study.

Data Availability Statement: Not applicable.

Conflicts of Interest: The authors declare no conflict of interest.

References

1. Thielmann, M.; Massoudy, P.; Neuhäuser, M.; Tsagakis, K.; Marggraf, G.; Kamler, M.; Mann, K.; Erbel, R.; Jakob, H. Prognostic value of preoperative cardiac troponin I in patients undergoing emergency coronary artery bypass surgery with non-ST-elevation or ST-elevation acute coronary syndromes. *Circulation* **2006**, *114* (Suppl. S1), I-448–I-453. [CrossRef] [PubMed]
2. Than, M.P.; Aldous, S.J.; Troughton, R.W.; Pemberton, C.J.; Richards, A.M.; Frampton, C.M.; Florkowski, C.M.; George, P.M.; Bailey, S.; Young, J.M.; et al. Detectable high-sensitivity cardiac troponin within the population reference interval conveys high 5-year cardiovascular risk: An observational study. *Clin. Chem.* **2018**, *64*, 1044–1053. [CrossRef] [PubMed]
3. Freisinger, E.; Fuerstenberg, T.; Malyar, N.M.; Wellmann, J.; Keil, U.; Breithardt, G.; Reinecke, H. German nationwide data on current trends and management of acute myocardial infarction: Discrepancies between trials and real-life. *Eur. Heart J.* **2014**, *35*, 979–988. [CrossRef]
4. De Waha, S.; Patel, M.R.; Granger, C.B.; Ohman, E.M.; Maehara, A.; Eitel, I.; Ben-Yehuda, O.; Jenkins, P.; Thiele, H.; Stone, G.W. Relationship between microvascular obstruction and adverse events following primary percutaneous coronary intervention for ST-segment elevation myocardial infarction: An individual patient data pooled analysis from seven randomized trials. *Eur. Heart J.* **2017**, *38*, 3502–3510. [CrossRef]
5. Tasar, O.; Karabay, A.K.; Oduncu, V.; Kirma, C. Predictors and outcomes of no-reflow phenomenon in patients with acute ST-segment elevation myocardial infarction undergoing primary percutaneous coronary intervention. *Coron. Artery Dis.* **2019**, *30*, 270–276. [CrossRef]
6. Califf, R.M.; Abdelimeguid, A.E.; Kuntz, R.E.; Popma, J.J.; Davidson, C.J.; Cohen, E.A.; Kleiman, N.S.; Mahaffey, K.W.; Topol, E.J.; Pepine, C.J.; et al. Myonecrosis after revascularization procedures. *J. Am. Coll. Cardiol.* **1998**, *31*, 241–251. [CrossRef]
7. Nam, K.; Jeon, Y.; Kim, W.H.; Jung, D.E.; Kwon, S.M.; Kang, P.; Cho, Y.J.; Kim, T.K. Intraoperative glucose variability, but not average glucose concentration, may be a risk factor for acute kidney injury after cardiac surgery: A retrospective study. *Can. J. Anaesth.* **2019**, *66*, 921–933. [CrossRef]

8. Stalikas, N.; Papazoglou, A.S.; Karagiannidis, E.; Panteris, E.; Moysidis, D.; Daios, S.; Anastasiou, V.; Patsiou, V.; Koletsa, T.; Sofidis, G.; et al. Association of stress induced hyperglycemia with angiographic findings and clinical outcomes in patients with ST-elevation myocardial infarction. *Cardiovasc. Diabetol.* **2022**, *21*, 140. [CrossRef]
9. Mone, P.; Gambardella, J.; Minicucci, F.; Lombardi, A.; Mauro, C.; Santulli, G. Hyperglicemia Drives Stent Restenosis in STEMI Patients. *Diabetes Care* **2021**, *44*, e192–e193. [CrossRef]
10. Xu, W.; Yang, Y.-M.; Zhu, J.; Wu, S.; Wang, J.; Zhang, H.; Shao, X.-H. Predictive value of the stress hyperglycemia ratio in patients with acute ST-segment elevation myocardial infarction: Insights from a multi-center observational study. *Cardiovasc. Diabetol.* **2022**, *21*, 48. [CrossRef]
11. Mone, P.; Pansini, A.; Rizzo, M.; Minicucci, F.; Mauro, C. ST-Elevation Myocardial Infarction Patients with Hyperglycemia: Effects of Intravenous Adenosine. *Am. J. Med. Sci.* **2022**, *363*, 122–129. [CrossRef] [PubMed]
12. Bellou, V.; Belbasis, L.; Tzoulaki, I.; Evangelou, E. Risk factors for type 2 diabetes mellitus: An exposure-wide umbrella review of meta-analyses. *PLoS ONE* **2018**, *13*, e0194127. [CrossRef] [PubMed]
13. Kodama, S.; Fujihara, K.; Horikawa, C.; Sato, T.; Iwanaga, M.; Yamada, T.; Kato, K.; Watanabe, K.; Shimano, H.; Izumi, T.; et al. Diabetes mellitus and risk of new-onset and recurrent heart failure: A systematic review and meta-analysis. *ESC Heart Fail.* **2020**, *7*, 2146–2174. [CrossRef]
14. Bouthoorn, S.; Valstar, G.B.; Gohar, A.; den Ruijter, H.M.; Reitsma, H.B.; Hoes, A.W.; Rutten, F.H. The prevalence of left ventricular diastolic dysfunction and heart failure with preserved ejection fraction in men and women with type 2 diabetes: A systematic review and meta-analysis. *Diab. Vasc. Dis. Res.* **2018**, *15*, 477–493. [CrossRef] [PubMed]
15. Alterio, V.; Monti, S.M.; de Simone, G. Thermal-stable carbonic anhydrases: A structural overview. In *Carbonic Anhydrase: Mechanism, Regulation, Links to Disease, and Industrial Applications*, 1st ed.; Frost, S.C., McKenna, R., Eds.; Springer: Dordrecht, The Netherlands, 2014; pp. 387–404.
16. Maren, T.H. CARBONI Canhydrase: Chemistry, physiology and inhibition. *Physiol. Rev.* **1967**, *47*, 595–781. [CrossRef] [PubMed]
17. Supuran, C.T. Emerging role of carbonic anhydrase inhibitors. *Clin. Sci.* **2021**, *135*, 1233–1249. [CrossRef]
18. Parkkila, S.; Kivela, A.J.; Kaunisto, K.; Parkkila, A.K.; Hakkola, J.; Rajaniemi, H.; Waheed, A.; Sly, W.S. The plasma membrane CARBONIC anhydrase in murine hepatocytes identified as isozyme XIV. *BMC Gastroenterol.* **2002**, *2*, 13–16. [CrossRef]
19. Pastoreková, S.; Parkkila, S.; Parkkila, A.K.; Opavský, R.; Zelník, V.; Saarnio, J.; Pastorek, J. Carbonic anhydrase IX, MN/CA IX: Analysis of stomach complementary DNA sequence and expression in human and rat alimentary tracts. *Gastroenterology* **1997**, *112*, 398–408. [CrossRef]
20. Parkkila, S. An overview of the distribution and function of CARBONIC anhydrase in mammals. In *The Carbonic Anhydrases*; Chegwidden, W.R., Carter, N.D., Edwards, Y.H., Eds.; Birkhäuser: Basel, Switzerland, 2000; pp. 79–93.
21. Puscas, I.; Coltau, M.; Baican, M.; Pasca, R.; Domuta, G.; Hecht, A. Vasoconstrictive drugs increase CARBONICic anhydrase I in vascular smooth muscle while vasodilating drugs reduce the activity of this isozyme by a direct mechanism of action. *Drugs Exp. Clin. Res.* **2001**, *27*, 231–238.
22. Benkovic, S.; Hammes-Schiffer, S. A perspective on enzyme catalysis. *Science* **2003**, *301*, 1196–1202. [CrossRef]
23. Aamand, R.; Dalsgaard, T.; Jensen, F.B.; Simonsen, U.; Roepstorff, A.; Fago, A. Generation of nitric oxide from nitrite by carbonic anhydrase: A possible link between metabolic activity and vasodilation. *Am. J. Physiol. Heart Circ. Physiol.* **2009**, *297*, H2068–H2074. [CrossRef] [PubMed]
24. Demandt, J.A.F.; Dubois, L.J.; van Kuijk, K.; Zat'ovičová, M.; Jin, H.; Parkkila, S.; van der Laan, S.W.; Jelenska, L.; Mees, B.M.E.; Reutelingsperger, C.P.M.; et al. The hypoxia-sensor carbonic anhydrase IX affects macrophage metabolism, but is not a suitable biomarker for human cardiovascular disease. *Sci. Rep.* **2021**, *11*, 425. [CrossRef] [PubMed]
25. De Luca, G.; Małek, L.A.; Maciejewski, P.; Wąsek, W.; Niewada, M.; Kamiński, B.; Wiecki, J.D.; Kośmider, M.; Kubica, J.; Rużyłło, W.; et al. Impact of diabetes on survival in patients with ST-segment elevation myocardial infarction treated by primary angioplasty: Insights from the POLISH STEMI registry. *Atherosclerosis* **2010**, *210*, 516–520. [CrossRef]
26. Liu, Z.; Zhang, Y.; Qiu, C.; Zhu, H.; Pan, S.; Jia, H.; Kang, H.; Guan, G.; Hui, R.; Zhu, L.; et al. Diabetes mellitus exacerbates post-myocardial infarction heart failure by reducing sarcolipin promoter methylation. *ESC Heart Fail.* **2020**, *7*, 1935–1948. [CrossRef] [PubMed]
27. Khalifah, R.G. The CARBONIC dioxide hidration activity of CARBONIC anhydrase: Stop-flow kinetic studies on the native human isozymes B and C. *J. Biol. Chem.* **1971**, *246*, 2561–2573. [CrossRef]
28. Supuran, C.T. Carbonic anhydrases: Novel therapeutic applications for inhibitors and activators. *Nat. Rev. Drug Discov.* **2008**, *7*, 168–181. [CrossRef]
29. Supuran, C.T. Carbonic anhydrases—An overview. *Curr. Pharm. Des.* **2008**, *14*, 603–614. [CrossRef]
30. Kosiborod, M.; Gomes, M.B.; Nicolucci, A.; Pocock, S.; Rathmann, W.; Shestakova, M.V.; Watada, H.; Shimomura, I.; Chen, H.; Cid-Ruzafa, J.; et al. Vascular Complications in Patients with Type 2 Diabetes: Prevalence and Associated Factors in 38 Countries (the DISCOVER Study Program). *Cardiovasc. Diabetol.* **2018**, *17*, 150. [CrossRef]
31. El-Lebedy, D.; Raslan, H.M.; Mohammed, A.M. Apolipoprotein E Gene Polymorphism and Risk of Type 2 Diabetes and Cardiovascular Disease. *Cardiovasc. Diabetol.* **2016**, *15*, 12. [CrossRef]
32. Yu, B.; Xu, C.; Tang, X.; Liu, Z.; Lin, X.; Meng, H.; Shi, C.; Ma, K.; Xiao, B.; Li, L. Endoplasmic reticulum stress -related secretory proteins as biomarkers of early myocardial ischemia-induced sudden cardiac deaths. *Int. J. Legal Med.* **2022**, *136*, 159–168. [CrossRef]

33. Xu, C.; Yu, B.; Zhao, X.; Lin, X.; Tang, X.; Liu, Z.; Gao, P.; Ge, J.; Wang, S.; Li, L. Valosin Containing Protein as a Specific Biomarker for Predicting the Development of Acute Coronary Syndrome and Its Complication. *Front. Cardiovasc. Med.* **2022**, *9*, 803532. [CrossRef] [PubMed]
34. Song, L.; Zhang, X.; Chen, R.; Li, J.; Zhou, J.; Liu, C.; Zhou, P.; Wang, Y.; Chen, Y.; Zhao, H.; et al. Association of PCSK9 with inflammation and platelet activation markers and recurrent cardiovascular risks in STEMI patients undergoing primary PCI with or without diabetes. *Cardiovasc. Diabetol.* **2022**, *21*, 80. [CrossRef] [PubMed]

biology

MDPI

Article

The Effect of a 13-Valent Conjugate Pneumococcal Vaccine on Circulating Antibodies Against Oxidized LDL and Phosphorylcholine in Man, A Randomized Placebo-Controlled Clinical Trial

Hendrika W. Grievink [1,2], Pim Gal [1,3], Maria Ozsvar Kozma [4], Erica S. Klaassen [1], Johan Kuiper [2], Jacobus Burggraaf [1,5,6], Christoph J. Binder [4] and Matthijs Moerland [1,3,*]

1 Centre for Human Drug Research, Zernikedreef 8, 2333 CL Leiden, The Netherlands; wgrievink@chdr.nl (H.W.G.); pgal@chdr.nl (P.G.); eklaassen@chdr.nl (E.S.K.); kb@chdr.nl (J.B.)
2 Division of BioTherapeutics, Leiden Academic Center for Drug Research, Leiden University, Einsteinweg 55, 2333 CC Leiden, The Netherlands; j.kuiper@lacdr.leidenuniv.nl
3 Clincal Pharmacy and Toxicology, Leiden University Medical Center, Albinusdreef 2, 2333 ZA Leiden, The Netherlands
4 Department of Laboratory Medicine, Medical University of Vienna, 1090 Vienna, Austria; maria.ozsvarkozma@meduniwien.ac.at (M.O.K.); christoph.binder@meduniwien.ac.at (C.J.B.)
5 Department of Surgery, Leiden University Medical Center, Albinusdreef 2, 2333 ZA Leiden, The Netherlands
6 Division of Pharmacology, Leiden Academic Center for Drug Research, Leiden Universtiy, Einsteinweg 55, 2333 CC Leiden, The Netherlands
* Correspondence: mmoerland@chdr.nl

Received: 9 September 2020; Accepted: 17 October 2020; Published: 22 October 2020

Simple Summary: Atherosclerosis is the main underlying mechanism for cardiovascular disease. The main cause for atherosclerosis development is oxidized low density lipoprotein (oxLDL) accumulation in the vessel wall and a subsequent immune response. It has been established that immunoglobulin M antibodies against oxLDL help protect against atherosclerosis. It has been found in mice that vaccination with *Streptococcus pneumoniae* results in an increase of these protective antibodies and thereby decreases the development of atherosclerosis. In this study, we investigated if this increase of antibodies can be found in human as well. Twenty-four healthy male volunteers were vaccinated with Prevenar-13, a pneumococcal vaccine, using different dosing regimens. An increase in anti-Prevenar antibodies was found, showing that the vaccination worked. However, no increase in protective anti-phosphorylcholine or anti-oxLDL antibodies was observed. This work shows that vaccination against pneumococcal does not seem to be a suitable treatment option to help prevent atherosclerosis development, although further research would be required to test alternative pneumococcal-based vaccines, vaccination regimens or study populations.

Abstract: In mice vaccination with *Streptococcus pneumoniae* results in an increase in anti-oxLDL IgM antibodies due to mimicry of anti-phosphorylcholine (present in the cell wall of *S. pneumoniae*) and anti-oxLDL IgM. In this study we investigated the human translation of this molecular mimicry by vaccination against *S. pneumoniae* using the Prevenar-13 vaccine. Twenty-four healthy male volunteers were vaccinated with Prevenar-13, either three times, twice or once in a double-blind, placebo-controlled, randomized single center clinical study. Anti-pneumococcal wall, oxLDL and phosphorycholine antibody levels were measured at a fixed serum dilution, as well as circulating lipid levels over the course of 68 weeks. A significant increase in anti-oxLDL IgG and IgM was seen in the group receiving two doses six months apart compared to the placebo. However, these differences were not observed in the groups receiving a single dose, two doses one month apart, or three doses. This study shows that vaccination with Prevenar-13 does not result in robust anti-oxLDL IgM levels in humans. Further research would be required to test alternative pneumococcal-based vaccines, vaccination regimens or study populations, such as cardiovascular disease patients.

Keywords: cardiovascular disease; vaccine; clinical trials; translational medicine; atherosclerosis; oxLDL; phosphorylcholine

---

1. Introduction

Oxidized low density lipoprotein (oxLDL) particles play a key role in the etiology of atherosclerosis [1]. In the vessel wall, oxLDL is recognized and phagocytosed by macrophages primarily via scavenger receptors leading to foam cell formation [2]. Macrophage foam cells are hallmark cells of atherosclerotic lesions and participate in the inflammatory responses that mediate smooth muscle cell migration and proliferation, and extracellular matrix production, and thereby stimulate atherosclerotic plaque progression.

Several mouse studies showed that IgM antibodies against oxLDL are atheroprotective [3,4]. IgM antibodies against oxidized particles facilitate the clearance of apoptotic cells, thereby promoting the resolution of inflammation [5,6]. Additionally, these antibodies neutralize the proinflammatory effects of oxidized phospholipids [7,8]. Inhibition of scavenger receptor-mediated oxLDL uptake by macrophages prevents the formation of foam cells and subsequent progression of atherosclerotic plaque formation [3,9]. In clinical studies, oxLDL-specific IgM has been reported to be a protective factor for atherosclerosis development, correlating with cardiovascular disease incidence and clinical outcome [10–13].

In contrast to IgM, the role of oxLDL–specific IgG in atherosclerosis is thought to be atherogenic. OxLDL–IgG complexes have been shown to induce survival of plaque-resident monocytes [14] and secretion of proatherogenic cytokines by mast cells [15]. Clinical studies showed a correlation between oxLDL–IgG antibodies and acute coronary syndrome, suggesting an untoward role of this antibody in plaque destabilization [16]. In human, oxLDL–specific IgG antibody titers correlated inversely to the oxLDL serum concentration [17] and, in mouse, serum cholesterol levels [18], which suggests that oxLDL–specific IgG facilitates phagocytosis of oxLDL by macrophages.

Mouse experiments showed that certain IgM clones binding oxLDL bind phosphorylcholine (PC) of oxidized—but not unoxidized—phospholipids [3,19]. Importantly, Binder et al. showed in Ldlr knockout mice that vaccination against *S. pneumoniae* using pneumoccocal extracts induced high titers of oxLDL–specific IgM, subsequently leading to a decrease in atherosclerotic lesions [19]. This effect was explained by the fact that PC is present as part of the capsular polysaccharide of *S. pneumoniae*. Moreover, immunization with PC conjugated to carrier proteins also induced oxLDL–IgM and decreased the extent of atherosclerosis in ApoE knockout mice [20,21].

Autoantibodies against PC are also found in humans, where low levels of PC–IgM autoantibodies correlate with a higher incidence of cardiovascular disease [22–25]. Moreover, pneumococcal-specific IgG and oxLDL-specific antibody titers correlated significantly in subjects who had received pneumococcal vaccination [26], although there are also reports of an absent effect of pneumococcal vaccination on oxLDL–specific IgM levels [27].

The present proof-of-concept study investigated the human translation of the observed effects of pneumococcal immunization in mice. Healthy human volunteers were vaccinated with a 13-valent conjugated pneumococcal vaccine (Prevenar-13®), and the induction of PC- and oxLDL-specific antibodies was measured.

2. Materials and Methods

This investigation was a double-blind, randomized, placebo-controlled, parallel, single-center study with twenty-four healthy males between 18 and 45 years of age. The study was performed at the Centre for Human Drug Research in Leiden, The Netherlands. Participants were recruited via advertisements and social media. Participants were assessed to be generally healthy based on a complete medical screening and had no previous exposure to the 13-valent pneumococcal vaccine. All participants gave written informed consent prior to any study-related activity. The study was approved by the Ethics Committee of the Leiden University Medical Centre (LUMC) and Declaration of Helsinki principles were followed. The study is registered in the Dutch Trial Registry (Nederlands Trial Register, NTR) under study number NTR5643 and took place for all participants simultaneously between March 2016 and October 2017. This study was funded by the European Union, call FP7-HEALTH-2013-INNOVATION, project ID 603131.

2.1. Vaccination Schedule

The 13-valent conjugated pneumococcal vaccine (Prevenar-13®) used in this study was from a single batch (batch no. MU7958). The presence of residual PC in the vaccine preparation was confirmed by ELISA using the PC-specific mAb IgM E06. Placebo consisted of 0.9% NaCl solution. Since there are clear visual differences between these vaccinations, three physicians were unblinded for administration of the vaccine. These physicians were not otherwise involved in the study.

Vaccinations took place at three time points: at baseline, at four weeks and at 28 weeks. Subjects were randomized in a consecutive order based on eligibility. The randomization code was generated using SAS v9.4 for Windows (SAS Institute Inc., Cary, NC, USA) by an independent statistician. The randomization code was only made available for data analysis after study completion. There were five different treatment arms, as displayed in Figure 1. In the mouse study three immunizations were enough for oxLDL–specific IgM induction [19]. In the mouse study proteinase-treated *S. pneumoniae* extracts were used. In this design, the power to detect differences between placebo and active treatment arms was optimized between baseline and at the 28 week time point (PP vs. AA vs. AP, n = 8 per group).

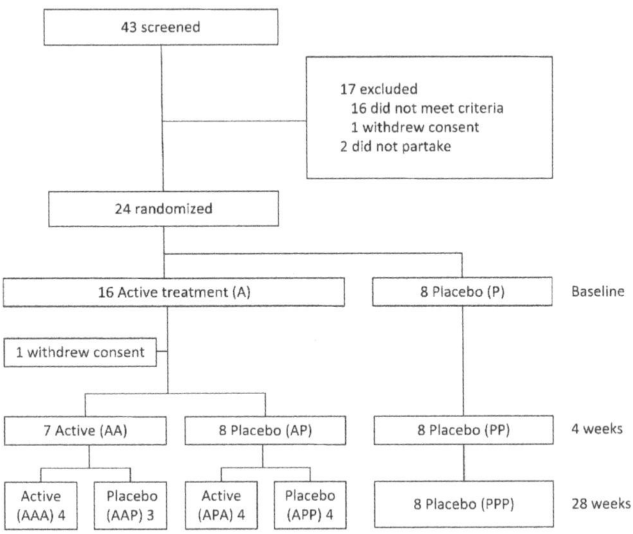

Figure 1. Study flowchart.

2.2. Antibody Measurements

K2EDTA plasma antibody levels to Prevenar, PC-BSA, and CuSO4-oxidized LDL (oxLDL) were measured by chemiluminescent ELISA as reported previously [28]. In brief, Prevenar (Pfizer) was coated at 1:5000, PC–BSA (Biosearch Technologies, Novato, CA, USA) and oxLDL at 5 ug/ml in PBS/EDTA. IgM antibodies were measured at a dilution of 1:500 and IgG antbodies at 1:1000. Binding of

IgG subclasses to Prevenar was measured at a dilution of 1:100 and to PC–BSA at 1:100 for IgG3 and IgG4, and 1:500 for IgG2.

Serum levels of total cholesterol, low density lipoprotein (LDL), high density lipoprotein (HDL) and triglycerides were measured by the chemistry lab of the Leiden University Medical Center on the Cobas P800 analyzer (Hoffmann–La Roche, Basel, Switzerland).

2.3. Power Calculation

In humans, the median anti-oxLDL IgG levels in the healthy, unvaccinated population is around 50 U/l, with an interquartile range of around 25–75 U/l [26]. Anticipating an immune response minimally resulting in a five-fold rise in IgG and IgM antibody levels, and based on an inter-subject variability of 50% in basal IgG and IgM levels [19,24,26], a sample size of 4–8 subjects per group (dependent on the contrast) will be sufficient to meet the study objectives. This was a conservative approach considering in the magnitude of the oxLDL-specific IgM response observed in the murine model [19].

2.4. Statistical Analysis

Data are presented as mean ± standard deviation (SD). In case of non-normal distribution, parameters were log-transformed. Repeatedly measured variables were analyzed with a mixed model analysis of variance with fixed factors treatment group, time and the interaction of treatment group and time as fixed factor and subject as random factor. Primary endpoints (Prevenar-specific Ig levels, oxLDL-specific Ig levels and PC-specific Ig levels) were compared between treatment groups for the 0–4 week window, the 4–28 week window, and the 28–68 week window. As a secondary endpoint, lipid levels in circulation were measured. Estimated differences were calculated between the groups. A positive value indicates a higher estimated value for the active group, a negative value indicates a lower value for the active group. The analysis was performed in SAS v9.4 (SAS Institute, Cary, NC, USA).

3. Results

Twenty-four healthy volunteers were included in the study; their baseline characteristics can be found in Table 1. One subject withdrew consent after two weeks for non-study related reasons (Figure 1). This subject was randomized to the active-active-placebo treatment arm and was not replaced.

Table 1. Baseline characteristics.

Parameter	n = 24
Age (years)	28.5 ± 8.5
Gender male (%)	100
Ethnicity Caucasian (%)	100
Height (cm)	180.5 ± 5.3
Weight (kg)	75.0 ± 11.0
BMI (kg/m^2)	23.0 ± 3.2
Heart rate (min^{-1})	58.5 ± 9.0
Systolic blood pressure (mmHg)	123 ± 9.3
Diastolic blood pressure (mmHg)	75.1 ± 6.7

3.1. Anti-Prevenar Antibodies

Prevenar-specific IgG was significantly increased in all subjects who received any active treatment compared to placebo-treated subjects, see Table 2 and Figure 2A. Prevenar-specific IgM was significantly increased in subjects who received any active treatment up to 28 weeks, however after 68 weeks only subjects receiving three active doses had a significantly increased IgM level compared to placebo.

Table 2. Estimated differences for Prevenar-specific IgG and IgM levels.

	IgG			IgM		
	ED	%95 CI	p Value	ED	%95 CI	p Value
0–4 weeks (A vs. P)	19,187.8	14,098.7–24,276.9	<0.0001	13,128.5	5356.9–20,900.1	0.0020
4–28 weeks (AP vs. PP)	16,656.9	11,626.7–21,687.1	<0.0001	5480.2	1143.1–9817.3	0.0158
4–28 weeks (AA vs. PP)	17,155.5	12,449.3–21,861.7	<0.0001	9272.0	4568.5–13,975.5	0.0005
28–68 weeks (AAA vs. PPP)	22,733.6	17,512.7–27,954.5	<0.0001	8189.1	1407.0–14,971.1	0.0209
28–68 weeks (AAP vs. PPP)	18,439.7	12,869.1–24,010.2	<0.0001	10,320.1	−23.1–20,663.3	0.0505
28–68 weeks (APA vs. PPP)	25,805.3	19,200.6–32,409.9	<0.0001	987.0	−6717.0–8691.0	0.7903
28–68 weeks (APP vs. PPP)	17,116.7	11,720.9–22,512.5	<0.0001	2233.9	−4392.3–8860.1	0.4862

ED = estimated difference, A = active, P = placebo, CI = confidence interval.

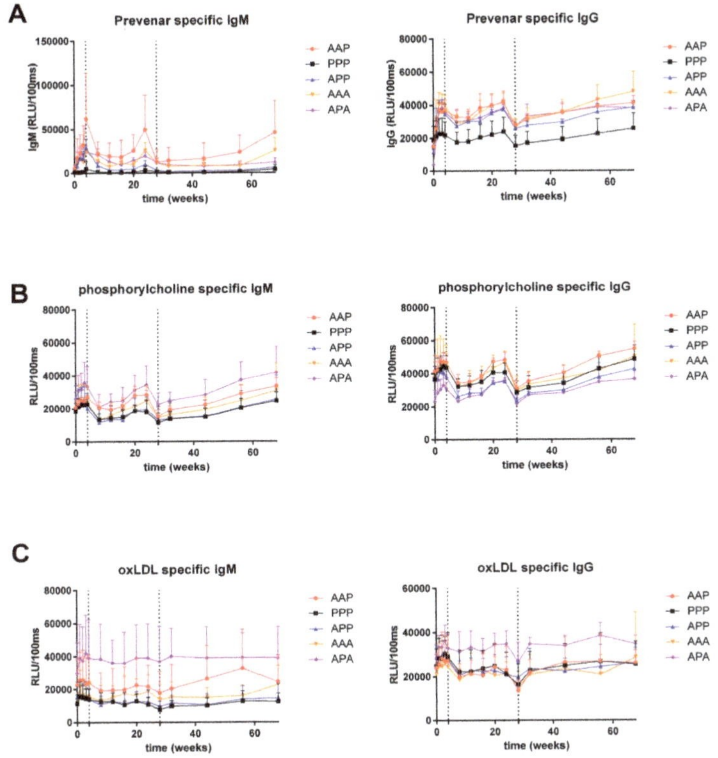

Figure 2. Anti-Prevenar (quantified by antibodies against pneumococcal wall saccharide) (**A**); anti-phosphorylcholine (**B**); and anti–oxLDL responses (**C**). Mean + SD; n = 4 per group, n = 3 for AAP (active-active-placebo) group and n = 8 for PPP (placebo-placebo-placebo) group. RLU/100ms: relative light units/100 ms 'A': active treatment, 'P': placebo treatment. Dotted lines indicate vaccination times (baseline, 4 weeks, 28 weeks). Statistical analysis using a mixed model analysis of variance with fixed factors: treatment group and time, and the interaction of treatment group and subject as random factor. For p-values for Prevenar responses, see Table 2. No significant changes were found for oxLDL and PC responses, except for PC-specific IgM in the APA (active-placebo-active) group compared to PPP (p = 0.005), and oxLDL-specific IgM and IgG responses at 68 weeks in the APA group compared to PPP (p = 0.007 for IgG, p = 0.005 for IgM).

3.2. Anti-OxLDL and Anti-PC Antibodies

No difference was observed in PC-specific IgG levels compared to placebo, for any of the active treatment groups. Similarly, no difference was observed for PC-specific IgM levels between active treatment groups and placebo, with the exception of IgM levels being higher in the APA group compared to PPP during the study period (ED: 9409.7, 95% CI: 3227.5–15,591.9, $p = 0.005$) (Figure 2B).

There were no differences in oxLDL-specific IgG and IgM antibodies between active and placebo treated subjects up to 28 weeks. However, at 68 weeks, subjects who received an active treatment at baseline and after 28 weeks (APA) had an increased oxLDL-specific IgG level compared to subjects receiving three placebo injections (PPP) with an estimated difference (ED) of 9913 (95% CI: 3141–16,686; $p = 0.007$). As shown in Figure 2C difference between these groups were also observed for oxLDL-specific IgM levels (ED: 12235, 95% CI: 4179–20,290; $p = 0.005$).

3.3. Lipids

The levels of total cholesterol (**A**), LDL (**B**), HDL (**C**) and triglycerides (**D**) of all groups during the study are depicted in Figure 3. No significant differences were found between treatment groups, with the exception of subjects receiving a single active treatment at the beginning of the study (APP) who had a significant higher triglyceride level compared to placebo (PPP) (ED: 7.9%, 95 %CI 18%–171%; $p = 0.009$).

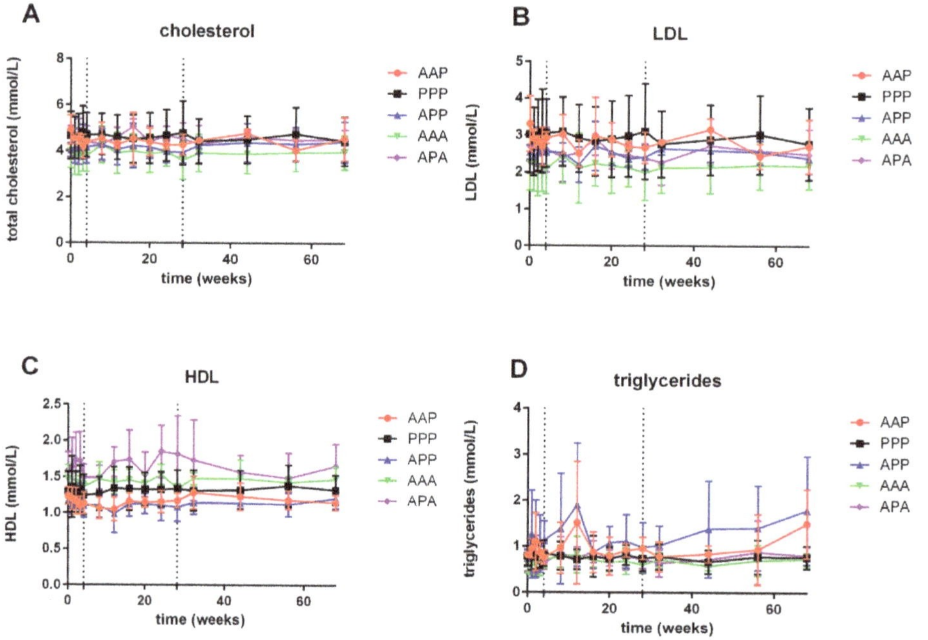

Figure 3. Cholesterol (**A**), LDL (low density lipoproteins) (**B**), HDL (high density lipoproteins) (**C**) and triglycerides plasma levels (**D**) (mean + SD). $n = 4$ per group, $n = 3$ for AAP group and $n = 8$ for PPP group. 'A': active treatment, 'P': placebo treatment. Dotted lines indicate vaccinations (baseline, 4 weeks, 28 weeks). Statistical analysis using a mixed model analysis of variance with fixed factors: treatment group and time, and the interaction of treatment group and subject as random factor. No significant differences were found, with the exception of APP versus PPP for triglycerides ($p = 0.009$).

4. Discussion

The present study evaluated the effect of a 13-valent conjugate pneumococcal vaccine on the induction of anti-oxLDL and anti-PC antibodies and cholesterol levels in humans. Several vaccination regimens were tested, where subjects received either one, two or three doses of Prevenar-13 over a period of 28 weeks compared to placebo. Despite the induction of an adequate anti-Prevenar 13 antibody response, there was no evident induction of either PC-specific or oxLDL-specific antibodies. Prevenar-13 immunization induced a significant IgG2 response when subjects were immunized at least twice, while levels of IgG3 and IgG4 were not altered (Figure S1). In one active treatment group a statistically significant difference in PC-specific and oxLDL-specific antibody levels was observed compared to placebo. This occurred in the group receiving two doses of the vaccine at the start of the study and after 28 weeks (APA). A significant increase in PC-specific IgM and oxLDL-specific IgM and IgG was observed at 68 weeks. Interestingly, we observed a significant increase in PC-specific IgG3 for the group receiving two vaccinations four weeks apart (AAP) compared to placebo (PPP) (Figure S1 panel B). Elevated oxLDL-specific IgM is believed to be atheroprotective [10–13], but the role of oxLDL-specific IgG levels is not fully elucidated. Laczik et al. [16] showed that increased oxLDL-specific IgG levels correlate with acute coronary syndrome, while immunization with oxLDL, resulting in an increase in oxLDL-specific IgG, resulted in decreased plaque development in several mouse models [18]. Furthermore, oxLDL levels are inversely correlated with oxLDL-specific IgG serum levels [17]. Binder et al. [19] showed *S. pneumoniae* immunization in mice induced a much stronger oxLDL-specific IgM response than an oxLDL-specific IgG response (100,000 vs. 10,000 RLU/100ms). The current clinical data are at odds with this observation, though the observed anti-oxLDL responses in the APA group may indicate that the timing of vaccination could be important.

Previous studies have investigated the 23-valent polysaccharide pneumococcal vaccine as a means to elicit oxLDL-specific antibodies in humans, with conflicting results [26,27,29]. One study reported that, after vaccination with the 23-valent vaccine, an increased oxLDL-specific IgG antibody titer was observed compared to healthy, unvaccinated individuals (248 U/l vs. 55 U/l). An effect of vaccination on IgM was not reported. In the other two studies, no association between vaccination status and oxLDL-specific antibodies was observed. There are, however, key differences between these studies and the present study. First, patients in all three referenced studies only got a single vaccination, whereas in the present study, up to three vaccinations were given. Second, there is a major difference between the 23-valent polysaccharide vaccine that was used in these studies and the 13-valent polysaccharide conjugate vaccine that was used in the present study. The latter is constituted of cell wall polysaccharides that have been conjugated to a protein. The 13-valent vaccine is thereby considered to be more powerful in eliciting an antibody response against *Streptococcus* pneumonia [30,31]. These data were the basis for the selection of the 13-valent vaccine for the current clinical study.

Although the sample size per group was relatively small in the current study ($n = 4$ for treatment groups, $n = 8$ for placebo), the study was sufficiently powered to detect Prevenar-13-induced rises in IgM titers, had these occurred as in the murine experiments [19].

The induction of anti-pneumococcal wall saccharide antibodies demonstrates that the 13-valent vaccine was effective for its intended use. However, the vaccine did not elicit a robust oxLDL-specific IgM response, as observed in mouse experiments. One explanation for the poor induction of oxLDL-specific antibody responses by Prevenar 13 may be that the murine immune response is poorly translatable to humans. Human and mice have numerous discrepancies in their innate and adaptive immune systems, such as cytokine receptor and costimulatory molecule expression and function [32,33]. Moreover, murine studies are commonly performed in inbred strains, with limited genetic variability between mice. This, and the fact that mice are kept in a more-sterile environment, results in a smaller immune diversity compared to humans [34]. On the other hand, the mild but significant induction of oxLDL-specific antibody responses in one active treatment group (APA) does support further clinical investigation of mimicry between pneumococcal vaccination and oxLDL. Furthermore, it could be hypothesized that the PC content in the Prevenar-13 vaccine used was not high enough to induce robust PC- and oxLDL-specific responses.

5. Conclusions

Vaccination of humans with Prevenar-13 did not significantly increase PC-specific antibodies and oxLDL-specific antibodies nor resulted in significant changes in plasma lipids. Nevertheless, subgroup analyses suggested an induction of PC-specific and oxLDL-specific IgM and IgG in individuals receiving two doses six months apart. Future research should investigate alternative pneumococcal vaccines (driving more significant anti-PC antibody responses), vaccination regimens, or study populations, to confirm or refute the hypothesis that molecular mimicry underlying pneumococcal-driven anti-oxLDL responses as observed in mice occurs in humans.

Supplementary Materials: The following are available online at http://www.mdpi.com/2079-7737/9/11/345/s1, Figure S1: Prevenar-specific and phosphorylcholine-specific IgG subclass responses.

Author Contributions: Conceptualization: P.G., M.M., J.K., J.B. and C.J.B.; methodology: J.K., J.B., C.J.B., P.G., M.M.; software: E.S.K.; formal analysis: H.W.G., P.G., M.O.K., E.S.K.; investigation: P.G., C.J.B., M.O.K.; resources: J.B., C.J.B., J.K.; writing-original draft preparation: H.W.G., P.G., writing-review and editing: all authors, visualization: H.W.G., E.S.K.; supervision: M.M., J.B., J.K., C.J.B.;.; project administration: P.G., M.M., J.K., C.J.B.; funding acquisition: J.B., J.K., B.C. All authors have read and agreed to the published version of the manuscript.

Funding: This research was funded by the European union, Call: FP7-HEALTH-2013-INNOVATION, project ID 603131.

Conflicts of Interest: The authors declare no conflict of interest.

References

1. Lusis, A.J. Atherosclerosis. *Nat. Cell Biol.* **2000**, *407*, 233–241. [CrossRef] [PubMed]
2. Ketelhuth, D.F.; Hansson, G.K. Cellular immunity, low-density lipoprotein and atherosclerosis: Break of tolerance in the artery wall. *Thromb. Haemost.* **2011**, *106*, 779–786. [CrossRef] [PubMed]
3. Shaw, P.X.; Hörkkö, S.; Chang, M.-K.; Curtiss, L.K.; Palinski, W.; Silverman, G.J.; Witztum, J.L. Natural antibodies with the T15 idiotype may act in atherosclerosis, apoptotic clearance, and protective immunity. *J. Clin. Investig.* **2000**, *105*, 1731–1740. [CrossRef] [PubMed]
4. Van Leeuwen, M.; Kemna, M.J.; De Winther, M.P.J.; Boon, L.; Duijvestijn, A.M.; Henatsch, D.; Bos, N.A.; Gijbels, M.J.J.; Tervaert, J.W.C. Passive Immunization with Hypochlorite-oxLDL Specific Antibodies Reduces Plaque Volume in LDL Receptor-Deficient Mice. *PLoS ONE* **2013**, *8*, e68039. [CrossRef]
5. Egrönwall, C.; Evas, J.; Silverman, G.J. Protective Roles of Natural IgM Antibodies. *Front. Immunol.* **2012**, *3*, 66. [CrossRef]
6. Litvack, M.L.; Post, M.; Palaniyar, N. IgM Promotes the Clearance of Small Particles and Apoptotic Microparticles by Macrophages. *PLoS ONE* **2011**, *6*, e17223. [CrossRef]
7. Chang, M.-K.; Bergmark, C.; Laurila, A.; Hörkkö, S.; Han, K.-H.; Friedman, P.; Dennis, E.A.; Witztum, J.L. Monoclonal antibodies against oxidized low-density lipoprotein bind to apoptotic cells and inhibit their phagocytosis by elicited macrophages: Evidence that oxidation-specific epitopes mediate macrophage recognition. *Proc. Natl. Acad. Sci. USA* **1999**, *96*, 6353–6358. [CrossRef]
8. Hörkkö, S.; Bird, D.A.; Miller, E.; Itabe, H.; Leitinger, N.; Subbanagounder, G.; Berliner, J.A.; Friedman, P.; Dennis, E.A.; Curtiss, L.K.; et al. Monoclonal autoantibodies specific for oxidized phospholipids or oxidized phospholipid–protein adducts inhibit macrophage uptake of oxidized low-density lipoproteins. *J. Clin. Investig.* **1999**, *103*, 117–128. [CrossRef]
9. Iseme, R.A.; McEvoy, M.; Kelly, B.J.; Agnew, L.L.; Walker, F.R.; Handley, T.; Oldmeadow, C.; Attia, J.; Boyle, M. A role for autoantibodies in atherogenesis. *Cardiovasc. Res.* **2017**, *113*, 1102–1112. [CrossRef]
10. Karvonen, J.; Päivänsalo, M.; Kesäniemi, Y.A.; Hörkkö, S. Immunoglobulin M Type of Autoantibodies to Oxidized Low-Density Lipoprotein Has an Inverse Relation to Carotid Artery Atherosclerosis. *Circulation* **2003**, *108*, 2107–2112. [CrossRef]

11. Su, J.; Georgiades, A.; Wu, R.; Thulin, T.; De Faire, U.; Frostegård, J. Antibodies of IgM subclass to phosphorylcholine and oxidized LDL are protective factors for atherosclerosis in patients with hypertension. *Atherosclerosis* **2006**, *188*, 160–166. [CrossRef] [PubMed]
12. Tsimikas, S.; Brilakis, E.S.; Lennon, R.J.; Miller, E.R.; Witztum, J.L.; McConnell, J.P.; Kornman, K.S.; Berger, P.B. Relationship of IgG and IgM autoantibodies to oxidized low density lipoprotein with coronary artery disease and cardiovascular events. *J. Lipid Res.* **2006**, *48*, 425–433. [CrossRef]
13. Soto, Y.; Condé, H.; Aroche, R.; Brito, V.; Luaces, P.; Nasiff, A.; Obregón, A.; Vazquez, A. Abstract: P821 Autoantibodies to oxidized low density lipoprotein in relation with coronary artery disease. *Atherosclerosis Suppl.* **2009**, *10*, e1210. [CrossRef]
14. Oksjoki, R.; Kovanen, P.T.; Lindstedt, K.A.; Jansson, B.; Pentikäinen, M.O. OxLDL–IgG Immune Complexes Induce Survival of Human Monocytes. *Arter. Thromb. Vasc. Biol.* **2006**, *26*, 576–583. [CrossRef] [PubMed]
15. Lappalainen, J.; Lindstedt, K.A.; Oksjoki, R.; Kovanen, P.T. OxLDL–IgG immune complexes induce expression and secretion of proatherogenic cytokines by cultured human mast cells. *Atherosclerosis* **2011**, *214*, 357–363. [CrossRef] [PubMed]
16. Laczik, R.; Szodoray, P.; Veres, K.; Szomják, E.; Csípő, I.; Sipka, S.; R, Y.S.; Szekanecz, Z.; Soltész, P. Assessment of IgG antibodies to oxidized LDL in patients with acute coronary syndrome. *Lupus* **2011**, *20*, 730–735. [CrossRef] [PubMed]
17. Shoji, T.; Nishizawa, Y.; Fukumoto, M.; Shimamura, K.; Kimura, J.; Kanda, H.; Emoto, M.; Kawagishi, T.; Morii, H. Inverse relationship between circulating oxidized low density lipoprotein (oxLDL) and anti-oxLDL antibody levels in healthy subjects. *Atherosclerosis* **2000**, *148*, 171–177. [CrossRef]
18. Zhou, X.; Caligiuri, G.; Hamsten, A.; Lefvert, A.K.; Hansson, G.K. LDL Immunization Induces T-Cell–Dependent Antibody Formation and Protection Against Atherosclerosis. *Arter. Thromb. Vasc. Biol.* **2001**, *21*, 108–114. [CrossRef]
19. Binder, C.J.; Hörkkö, S.; Dewan, A.; Chang, M.-K.; Kieu, E.P.; Goodyear, C.S.; Shaw, P.X.; Palinski, W.; Witztum, J.L.; Silverman, G.J. Pneumococcal vaccination decreases atherosclerotic lesion formation: Molecular mimicry between Streptococcus pneumoniae and oxidized LDL. *Nat. Med.* **2003**, *9*, 736–743. [CrossRef]
20. Caligiuri, G.; Khallou-Laschet, J.; Vandaele, M.; Gaston, A.-T.; Delignat, S.; Mandet, C.; Kohler, H.V.; Kaveri, S.V.; Nicoletti, A. Phosphorylcholine-Targeting Immunization Reduces Atherosclerosis. *J. Am. Coll. Cardiol.* **2007**, *50*, 540–546. [CrossRef]
21. Faria-Neto, J.R.; Chyu, K.-Y.; Li, X.; Dimayuga, P.C.; Ferreira, C.; Yano, J.; Cercek, B.; Shah, P.K. Passive immunization with monoclonal IgM antibodies against phosphorylcholine reduces accelerated vein graft atherosclerosis in apolipoprotein E-null mice. *Atherosclerosis* **2006**, *189*, 83–90. [CrossRef] [PubMed]
22. Sjöberg, B.G.; Su, J.; Dahlbom, I.; Gronlund, H.; Wikström, M.; Hedblad, B.; Berglund, G.; De Faire, U.; Frostegård, J. Low levels of IgM antibodies against phosphorylcholine—A potential risk marker for ischemic stroke in men. *Atherosclerosis* **2009**, *203*, 528–532. [CrossRef] [PubMed]
23. Grønlund, H.; Hallmans, G.; Jansson, J.H.; Boman, K.K.; Wikström, M.; De Faire, U.; Frostegård, J. Low levels of IgM antibodies against phosphorylcholine predict development of acute myocardial infarction in a population-based cohort from northern Sweden. *Eur. J. Cardiovasc. Prev. Rehabil.* **2009**, *16*, 382–386. [CrossRef]
24. Frostegård, J.; Tao, W.; Georgiades, A.; Råstam, L.; Lindblad, U.; Lindeberg, S. Atheroprotective natural anti-phosphorylcholine antibodies of IgM subclass are decreased in Swedish controls as compared to non-westernized individuals from New Guinea. *Nutr. Metab.* **2007**, *4*, 7. [CrossRef] [PubMed]
25. Rahman, M.; Sing, S.; Golabkesh, Z.; Fiskesund, R.; Gustafsson, T.; Jogestrand, T.; Frostegård, A.G.; Hafström, I.; Liu, A.; Frostegård, J. IgM antibodies against malondialdehyde and phosphorylcholine are together strong protection markers for atherosclerosis in systemic lupus erythematosus: Regulation and underlying mechanisms. *Clin. Immunol.* **2016**, *166*, 27–37. [CrossRef] [PubMed]
26. Suthers, B.; Hansbro, P.; Thambar, S.; McEvoy, M.; Peel, R.; Attia, J. Pneumococcal vaccination may induce anti-oxidized low-density lipoprotein antibodies that have potentially protective effects against cardiovascular disease. *Vaccine* **2012**, *30*, 3983–3985. [CrossRef] [PubMed]
27. Damoiseaux, J.; Rijkers, G.; Tervaert, J.W.C. Pneumococcal vaccination does not increase circulating levels of IgM antibodies to oxidized LDL in humans and therefore precludes an anti-atherogenic effect. *Atherosclerosis* **2007**, *190*, 10–11. [CrossRef] [PubMed]

28. Eichinger, S.; Kyrle, P.; Kammer, M.; Eischer, L.; Kozma, M.O.; Binder, C.J. Natural antibodies to oxidation-specific epitopes: Innate immune response and venous thromboembolic disease. *J. Thromb. Haemost.* **2017**, *16*, 31–35. [CrossRef]
29. Nguyen, J.T.; Myers, N.; Palaia, J.; Georgopoulos, A.; Rubins, J.B.; Janoff, E.N. Humoral responses to oxidized low-density lipoprotein and related bacterial antigens after pneumococcal vaccine. *Transl. Res.* **2007**, *150*, 172–179. [CrossRef]
30. Nuorti, J.P.; Whitney, C.G. Updated recommendations for prevention of invasive pneumococcal disease among adults using the 23-valent pneumococcal polysaccharide vaccine (PPSV23). *MMWR. Morb. Mortal. Wkly. Rep.* **2010**, *59*, 1102–1106.
31. Musher, D.M.; Sampath, R.; Rodriguez-Barradas, M.C. The Potential Role for Protein-Conjugate Pneumococcal Vaccine in Adults: What Is the Supporting Evidence? *Clin. Infect. Dis.* **2011**, *52*, 633–640. [CrossRef] [PubMed]
32. Mestas, J.; Hughes, C.C.W. Of Mice and Not Men: Differences between Mouse and Human Immunology. *J. Immunol.* **2004**, *172*, 2731–2738. [CrossRef] [PubMed]
33. Khanna, R.; Burrows, S.R. Human immunology: A case for the ascent of non-furry immunology. *Immunol. Cell Biol.* **2011**, *89*, 330–331. [CrossRef] [PubMed]
34. Sellers, R.S. Translating Mouse Models. *Toxicol. Pathol.* **2016**, *45*, 134–145. [CrossRef] [PubMed]

MDPI
St. Alban-Anlage 66
4052 Basel
Switzerland
Tel. +41 61 683 77 34
Fax +41 61 302 89 18
www.mdpi.com

Biology Editorial Office
E-mail: biology@mdpi.com
www.mdpi.com/journal/biology

www.ingramcontent.com/pod-product-compliance
Lightning Source LLC
Chambersburg PA
CBHW040827250926
51850CB00007BA/1281
* 9 7 8 3 0 3 6 5 7 5 7 5 9 *